AF598290

Methods in Cell Biology

Correlative Light and Electron Microscopy III

Volume 140

Series Editors

Leslie Wilson

Department of Molecular, Cellular and Developmental Biology
University of California
Santa Barbara, California

Phong Tran

University of Pennsylvania
Philadelphia, USA &
Institut Curie, Paris, France

Methods in Cell Biology

Correlative Light and Electron Microscopy III

Volume 140

Edited by

Thomas Müller-Reichert

Experimental Center,
Medical Faculty Carl Gustav Carus,
Technische Universität Dresden,
Dresden, Germany

Paul Verkade

Wolfson Bioimaging Facility,
School of Biochemistry,
Biomedical Sciences Building, University Walk,
University of Bristol, Bristol, UK

Academic Press is an imprint of Elsevier
50 Hampshire Street, 5th Floor, Cambridge, MA 02139, United States
525 B Street, Suite 1800, San Diego, CA 92101-4495, United States
125 London Wall, London EC2Y 5AS, United Kingdom
The Boulevard, Langford Lane, Kidlington, Oxford OX5 1GB, United Kingdom

First edition 2017

ISBN: 978-0-12-809975-9
ISSN: 0091-679X

For information on all Academic Press publications visit our website at https://www.elsevier.com/books-and-journals

Publisher: Zoe Kruze
Acquisition Editor: Zoe Kruze
Editorial Project Manager: Katie Chan
Production Project Manager: Surya Narayanan Jayachandran
Senior Cover Designer: Greg Harris

Contents

Contributors

Nicholas R. Ader
MRC Laboratory of Molecular Biology, Cambridge, United Kingdom; National Institutes of Health, Bethesda, MD, United States

Nicholas Ariotti
The University of Queensland, Brisbane, QLD, Australia

Christopher P. Arthur
Thermo Fisher Scientific, Hillsboro, OR, United States; Genentech, San Francisco, CA, United States

Michael C. Ashby
University of Bristol, Bristol, United Kingdom

Nicolas Barois
Univ. Lille, CNRS UMR 8204, Inserm U1019, CHU Lille, Institut Pasteur de Lille – CIIL – Center for Infection and Immunity of Lille, Lille, France

Vivien Bauer
University of Würzburg, Würzburg, Germany

Anne Greet Bittermann
ETH Zurich, Zurich, Switzerland

Antonino Bongiovanni
Univ. Lille, CNRS UMR 8204, Inserm U1019, CHU Lille, Institut Pasteur de Lille – CIIL – Center for Infection and Immunity of Lille, Lille, France

Cedric Bouchet-Marquis
Thermo Fisher Scientific, Hillsboro, OR, United States

Filip Braet
The University of Sydney, Sydney, NSW, Australia

Sebastian Britz
University of Würzburg, Würzburg, Germany

Delfine Cheng
The University of Sydney, Sydney, NSW, Australia

Lucy M. Collinson
The Francis Crick Institute, London, United Kingdom

Alex de Marco
Monash University, Clayton, VIC, Australia

Markus Engstler
University of Würzburg, Würzburg, Germany

Jacky G. Goetz
MN3T, Inserm U1109, Strasbourg, France; Université de Strasbourg, Strasbourg, France; LabEx Medalis, Université de Strasbourg, Strasbourg, France; Fédération de Médecine Translationnelle de Strasbourg (FMTS), Université de Strasbourg, Strasbourg, France

Joe W. Gray
Oregon Health and Sciences University, Portland, OR, United States

Maja Günthert
ETH Zurich, Zurich, Switzerland

Thomas E. Hall
The University of Queensland, Brisbane, QLD, Australia

Xavier Heiligenstein
Institut Curie, PSL Research University, CNRS UMR 144 & Cell and Tissue Imaging Facility, Paris, France

Gregor Heiss
Thermo Fisher Scientific, Hillsboro, OR, United States

Frederik Helmprobst
University of Würzburg, Würzburg, Germany

Minh Huynh
The University of Sydney, Sydney, NSW, Australia

Sébastien Janel
Univ. Lille, CNRS UMR 8204, Inserm U1019, CHU Lille, Institut Pasteur de Lille – CIIL – Center for Infection and Immunity of Lille, Lille, France

Errin Johnson
University of Oxford, Oxford, United Kingdom

Nicola G. Jones
University of Würzburg, Würzburg, Germany

Matthia A. Karreman
European Molecular Biology Laboratory, Heidelberg, Germany

Rainer Kaufmann
University of Oxford, Oxford, United Kingdom

Takaaki Kinoshita
Soka University, Hachioji-shi, Japan

Androniki Kolovou
European Molecular Biology Laboratory, Heidelberg, Germany

Wanda Kukulski
MRC Laboratory of Molecular Biology, Cambridge, United Kingdom

Sunjong Kwon
Oregon Health and Sciences University, Portland, OR, United States

Frank Lafont
Univ. Lille, CNRS UMR 8204, Inserm U1019, CHU Lille, Institut Pasteur de Lille – CIIL – Center for Infection and Immunity of Lille, Lille, France

Robert M. Lees
University of Bristol, Bristol, United Kingdom

Claudia S. López
Oregon Health and Sciences University, Portland, OR, United States

Miriam S. Lucas
ETH Zurich, Zurich, Switzerland

Pedro Machado
European Molecular Biology Laboratory, Heidelberg, Germany

Sebastian M. Markert
University of Würzburg, Würzburg, Germany

Nassirhadjy Memtily
National Institute of Advanced Industrial Science and Technology (AIST), Tsukuba, Japan; University of Tsukuba, Tsukuba, Japan; Traditional Uyghur Medicine Institute of Xinjiang Medical University, Urumqi, China

Luc Mercier
MN3T, Inserm U1109, Strasbourg, France; Université de Strasbourg, Strasbourg, France; LabEx Medalis, Université de Strasbourg, Strasbourg, France; Fédération de Médecine Translationnelle de Strasbourg (FMTS), Université de Strasbourg, Strasbourg, France

Marco Morsch
Macquarie University, Sydney, NSW, Australia

Thomas S. Muenz
University of Würzburg, Würzburg, Germany

Shoko Nishihara
Soka University, Hachioji-shi, Japan

Robert G. Parton
The University of Queensland, Brisbane, QLD, Australia

Perrine Paul-Gilloteaux
Structure Fédérative de Recherche François Bonamy, INSERM, CNRS, Université de Nantes, Nantes, France

Christopher J. Peddie
The Francis Crick Institute, London, United Kingdom

Gaia Pigino
Max Planck Institute of Molecular Cell Biology and Genetics, Dresden, Germany

Lee Pullan
Thermo Fisher Scientific, Hillsboro, OR, United States

Graça Raposo
Institut Curie, PSL Research University, CNRS UMR 144 & Cell and Tissue Imaging Facility, Paris, France

Jessica L. Riesterer
Thermo Fisher Scientific, Hillsboro, OR, United States

Wolfgang Rössler
University of Würzburg, Würzburg, Germany

Bernhard Ruthensteiner
Zoologische Staatssammlung München, Munich, Germany

Carsten Sachse
European Molecular Biology Laboratory, Heidelberg, Germany

Jean Salamero
Institut Curie, PSL Research University, CNRS UMR 144 & Cell and Tissue Imaging Facility, Paris, France

Rachel Santarella-Mellwig
European Molecular Biology Laboratory, Heidelberg, Germany

Chikara Sato
National Institute of Advanced Industrial Science and Technology (AIST), Tsukuba, Japan; University of Tsukuba, Tsukuba, Japan

Mari Sato
National Institute of Advanced Industrial Science and Technology (AIST), Tsukuba, Japan

Markus Sauer
University of Würzburg, Würzburg, Germany

Nicole L. Schieber
European Molecular Biology Laboratory, Heidelberg, Germany

Martin Schorb
European Molecular Biology Laboratory, Heidelberg, Germany

Yannick Schwab
European Molecular Biology Laboratory, Heidelberg, Germany

Gerald Shami
The University of Sydney, Sydney, NSW, Australia

Frank Sieckmann
Leica Microsystems GmbH, Mannheim, Germany

Gergely Solecki
University Hospital Heidelberg, Heidelberg, Germany; German Cancer Research Center (DKFZ), Heidelberg, Germany

Ludek Stepanek
Max Planck Institute of Molecular Cell Biology and Genetics, Dresden, Germany

Anna M. Steyer
European Molecular Biology Laboratory, Heidelberg, Germany

Christian Stigloher
University of Würzburg, Würzburg, Germany

Shinya Sugimoto
The Jikei University School of Medicine, Minato-ku, Japan

Abul Tarafder
European Molecular Biology Laboratory, Heidelberg, Germany

Guillaume Thibault
Oregon Health and Sciences University, Portland, OR, United States

Patrick Trimby
The University of Sydney, Sydney, NSW, Australia

Paul Verkade
University of Bristol, Bristol, United Kingdom

Roger Wepf
The University of Queensland, Brisbane, QL, Australia

Elisabeth Werkmeister
Univ. Lille, CNRS UMR 8204, Inserm U1019, CHU Lille, Institut Pasteur de Lille – CIIL – Center for Infection and Immunity of Lille, Lille, France

Frank Winkler
University Hospital Heidelberg, Heidelberg, Germany; German Cancer Research Center (DKFZ), Heidelberg, Germany

Toshiko Yamazawa
The Jikei University School of Medicine, Minato-ku, Japan

Preface

The Key Lies in the Right Combination

Following the publication of two MCB volumes on correlative light and electron microscopy (CLEM), we are happy to present the third volume on this topic. This volume and the previous ones emphasize the need and value of CLEM for modern cell biology and illustrate the variety of microscopic approaches that are currently in use to study diverse biological problems.

As for this volume, a diversity of imaging techniques is covered, such as serial block face imaging by scanning electron microscopy, superresolution light microscopy, atomic force microscopy, atmospheric EM, and transmission electron microscopy. These imaging modalities are combined in different ways and include different techniques to tackle critical steps in a CLEM workflow. These steps include how to prepare the specimens, to increase the time-resolution of the CLEM experiments, or to enhance the precision for the relocation positions within specimens. The precise superimposition/correlation of light and electron microscopic images is also covered and, last but not least, strategies to automatically collect data and label proteins of interest for both light and electron microscopic studies are also discussed. All these approaches have in common that the microscopy of one given specimen by two or more imaging modalities can not only increase the throughput of experiments but also enhance the understanding of the biological processes purely by the combination of imaging modalities.

In parallel to this third MCB volume, it is our intention to continue to teach CLEM approaches during practical courses at various occasions, such as the EMBO-sponsored practical course. We certainly hope that our publications and courses will further stimulate the application of CLEM approaches for the years to come.

Thomas Müller-Reichert and Paul Verkade
Dresden and Bristol,
March 17, 2017

CHAPTER

1

Millisecond time resolution correlative light and electron microscopy for dynamic cellular processes

Ludek Stepanek, Gaia Pigino[1]

Max Planck Institute of Molecular Cell Biology and Genetics, Dresden, Germany

[1]*Corresponding author: E-mail: pigino@mpi-cbg.de*

CHAPTER OUTLINE

Methods in Cell Biology, Volume 140, ISSN 0091-679X, http://dx.doi.org/10.1016/bs.mcb.2017.03.003

Abstract

Molecular motors propel cellular components at velocities up to microns per second with nanometer precision. Imaging techniques combining high temporal and spatial resolution are therefore indispensable to understand the cellular mechanics at the molecular level. For example, intraflagellar transport (IFT) trains constantly shuttle ciliary components between the base and tip of the eukaryotic cilium. 3-D electron microscopy has revealed IFT train morphology and position, but was unable to correlate these features with the direction of train movement. Here, we present the methodology required to combine live-cell imaging at millisecond frame rates with electron tomography. Using this approach, we were able to correlate the direction of movement of every IFT train in a flagellum with its morphology and microtubule track. The method is ready to be further adapted for other experimental systems, including studies of single molecule dynamics.

INTRODUCTION

While light microscopy (LM) allows prolonged observation of living specimen, electron microscopy (EM) is inevitably destructive, providing a single high-resolution snapshot of the sample at the time of fixation. As all cellular structures are dynamic, different strategies have been developed to add temporal information to the EM images. For example, the working cycle of isolated macromolecular structures can be reconstructed from EM images of different conformational states, provided that these snapshots are taken at defined time points and sorted accordingly. In an in vitro system, a typical workflow would include mixing of reagents required to start the biochemical reaction and rapid freezing of the reaction mixture at increasing time points. Using this approach, a time resolution of 9.4 ms has been achieved in a cryo-EM study of ribosomal assembly (Shaikh et al., 2014). Time courses of processes triggered in intact cells can be studied in a similar way, as demonstrated by the combination of optogenetic stimulation and high pressure freezing of neuronal cells in *Caenorhabditis elegans* (Watanabe, 2016). However, most of cellular dynamics is not controllable by the observer, and correlation with LM imaging is required to determine the state of the sample before the point of fixation.

The ultrastructure of dynamic cellular events can be studied when the time-lapse LM is performed before fixation and embedding for EM (Polishchuk et al., 2000).

Such process is known as time-resolved correlative light and electron microscopy (CLEM), or video-CLEM. The image acquired by EM is still static, but the shape and position of structures can be traced back in time through the correlation with live-cell imaging. Since the time of the pioneering work of Polishchuk et al. (2000), several other studies followed (Beznoussenko & Mironov, 2015; Guizetti, Mäntler, Müller-Reichert, & Gerlich, 2010; Kukulski, Schorb, Kaksonen, & Briggs, 2012; Mironov et al., 2003; van Rijnsoever, Oorschot, & Klumperman, 2008). Verkade proposed a solution for rapid (~4 s) sample transfer between the light microscope and the high-pressure-freezing machine (Verkade, 2008). All these mentioned works studied membrane trafficking or cell division processes, in which time resolution on the order of seconds (Polishchuk et al., 2000) to minutes (Mironov et al., 2003) was satisfactory. Compared to these systems, intraflagellar transport (IFT) presents a greater technical challenge: it takes place in the confined space of a cilium (0.3 × 10 μm) and at any given time point, multiple anterograde and retrograde trains are passing each other at speeds ranging between 2.5 and 4 μm/s. The time-resolved CLEM methods mentioned above would not be fast enough to investigate IFT dynamics. We therefore worked to improve the time resolution of CLEM to the millisecond scale. In the next sections, we discuss step by step our workflow for correlating 3-D EM with time-lapse fluorescence microscopy at the limit of diffraction with theoretically unlimited time resolution.

1. METHODS

The reported CLEM procedure is carried out in five main steps:

1. Imaging chamber preparation (Fig. 1A).
2. LM imaging and sample fixation (Fig. 1B).
3. Sample preparation for electron tomography (Fig. 1C–D).
4. Electron tomography imaging.
5. Correlation of light and electron microscopy images.

In our experience, cell fixation (Step 2) and thin sectioning (Step 3) are the most critical steps of the procedure. Careful handling of the sample is required during the addition of the fixative (Fig. 1B): touching the sample with the pipette or a too strong fixative flow might shift the sample out of focus, causing ambiguity of the determination of IFT train directionality. Very careful alignment of the diamond knife is also necessary to successfully cut thick sections containing whole flagella (Fig. 1D–E).

1.1 *CHLAMYDOMONAS REINHARDTII* CULTURE

The green alga *Chlamydomonas reinhardtii* is a prominent cell biology model organism with particular significance for the field of cilia and flagella (Rosenbaum, Moulder, & Ringo, 1969). IFT motility was observed for the first time in *Chlamydomonas* by DIC microscopy (Kozminski, Johnson, Forscher, & Rosenbaum, 1993).

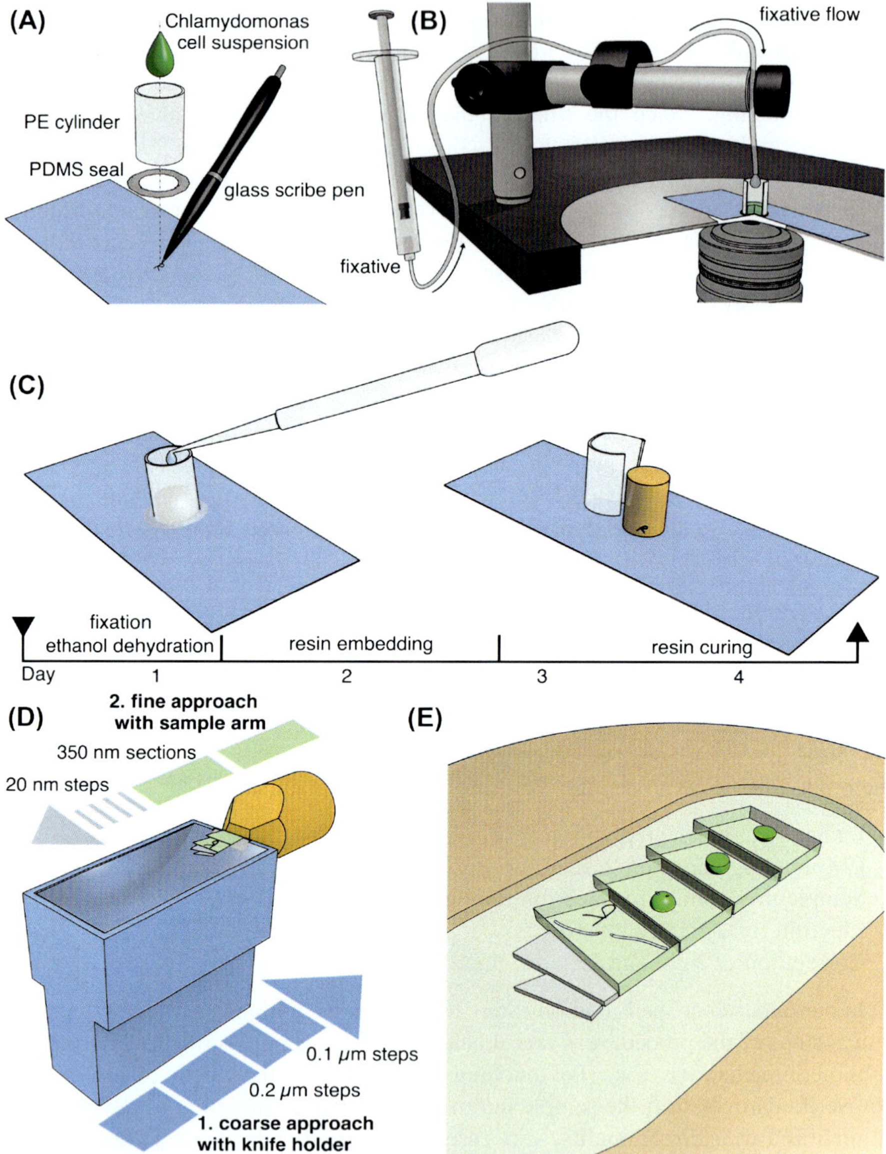

FIGURE 1 Overview of sample processing for correlative light and electron microscopy.

(A) Imaging chamber assembly and marking of the coverslip with a diamond pen. (B) Primary fixation by adding glutaraldehyde solution during the course of live-cell imaging. (C) Timeline of secondary fixation and resin embedding. The final result is cells embedded in a block of solid resin attached to the coverslip. The block and the glass slide are separated by immersion in liquid nitrogen before the next step. (D) Very accurate alignment is obtained by observing the reflection of the diamond knife on the block face. Additionally, the ultramicrotome is set up to advance 20 nm each cutting cycle to cover the last few nanometers. As the first 20-nm section appears, a full 350-nm section is cut starting exactly from the sample surface. (E) Serial *en face* sections are collected on a Formvar-coated single slot electron microscopy grid. (The dimensions of the sections and flagella are not to scale.)

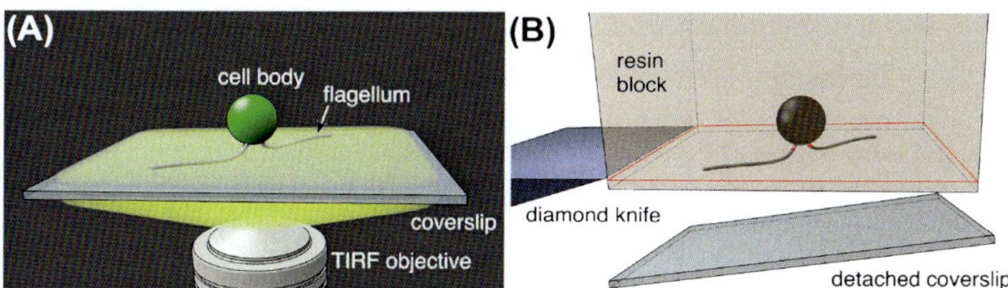

FIGURE 2 Gliding *Chlamydomonas* flagella are suitably positioned for total internal reflection fluorescence (TIRF) microscopy and *en face* thin sectioning.

(A) *Chlamydomonas* flagella, but not the autofluorescent cell body, are illuminated by the evanescent light during TIRF microscopy. (B) Full-length flagella are contained within the first 350-nm layer of resin block after embedding and removal of the coverslip. The *red lines* mark the plane of the first ultramicrotome cut.

An important milestone that enhanced the value of *Chlamydomonas* for IFT research was the introduction of *Chlamydomonas*-optimized green fluorescent protein (GFP) (Fuhrmann, Oertel, & Hegemann, 1999) and subsequent tagging of several IFT genes (Lechtreck et al., 2009; Mueller, Perrone, Bower, Cole, & Porter, 2005; Qin, Wang, Diener, & Rosenbaum, 2007).

Fluorescently tagged *Chlamydomonas* flagella are particularly suitable for total internal reflection fluorescence (TIRF) microscopy imaging (Engel et al., 2009), since they tightly adhere to the coverslip surface and are no thicker than 300 nm (Fig. 2A). For the same reasons, the flagella are good sample for flat embedding and transmission EM (Rogowski, Scholz, & Geimer, 2013), where the sample thickness is restricted to 400 nm when using a 300 kV electron source for electron tomography (Fig. 2B).

For the experiment described here, *Chlamydomonas* strain IFT27-GFP mt+ (Qin et al., 2007) is used to visualize IFT traffic with TIRF microscopy. The cells are cultured in TAP medium (Gorman & Levine, 1965), under simulated daylight illumination (14 h light/10 h dark). We recommend harvesting the cells within the first few hours of the light cycle, as the freshly hatched cells adhere to surfaces more readily.

1.2 IMAGING AND EMBEDDING CHAMBER PREPARATION

Imaging at the light microscope and subsequent embedding of the sample for EM are performed in a single, small chamber to simplify the relocalization of the cell of interest and to minimize the time of preparation. The chamber is prepared from a BEEM polyethylene capsule, which is resistant to the chemicals used throughout the process and gives the final resin block a suitable shape that fits into the microtome sample holder.

The relative position of other *Chlamydomonas* cell bodies on the glass slide and then in the resin block provides sufficient spatial information to identify the cell of interest prior to the sectioning using the ultramicrotome. However, to simplify cell relocalization in the resin block, an asymmetric symbol (a letter "R") is gently scratched into the coverslip. The imprint of the symbol is still visible in the resin when it is detached from the coverslip.

Procedure:

1. Scratch a glass coverslip with the tungsten carbide pen to introduce an orientation mark (e.g., the asymmetric letter "R"), of 1–2 mm in size.
2. Cut off the lid and the bottom part of a BEEM capsule to obtain a hollow cylinder.
3. Use PDMS elastomer to glue the cylinder to the coverslip, centering the R mark in the middle of the chamber.
4. Polymerize the PDMS for c. 10 s on a hotplate at 100°C.

1.3 LIGHT MICROSCOPY

Time-lapse movies are acquired using TIRF microscopy to track IFT trains before and during specimen fixation. The analysis of the movies allows the identification of anterograde, retrograde, and still IFT trains and, in the last frame of the movie, the exact position of each train in the fixed cilium, which is then later correlated with the EM data. As the anterograde and retrograde IFT trains might overlap in the cilium, it is essential to follow each IFT train until all intraflagellar movement stops completely to make the assignment unambiguous. The fixation buffer must therefore be introduced gently to avoid any sample movement and subsequent loss of focus. It must also act rapidly, otherwise the *Chlamydomonas* cells might sense the change of medium composition and detach from the coverslip in an escape reaction. Our approach is to use an open imaging chamber and drop the fixative from above; being heavier than water, the glutaraldehyde fixative sinks quickly to the coverslip and stops all cellular motion within 700 ± 300 ms after addition.

Procedure:

1. Fill the chamber with 200 μL of *Chlamydomonas* cell suspension in TAP medium from a fresh culture.
2. Mount the chamber on an inverted TIRF microscope.
3. Prepare a 2.5% glutaraldehyde solution in TAP and load 300 μL of the solution into a PTFE tubing with a syringe.
4. Mount the tubing system with the syringe on the microscope stage as shown in Fig. 1B. The tip of the tubing points to the sample, approximately 3 mm above the cell suspension level.
5. Start live time-lapse acquisition with exposure time of 20–35 ms.
6. While the acquisition is running, look for a cell of interest close to the R mark on the coverslip.
7. As soon as the cell of interest is in the field of view and in focus, press the syringe to drop the fixative on the sample (Fig. 1B). Within a second, all IFT movement stops.

8. Acquire the fluorescence signal for a few more seconds to get good signal-to-noise ratio in the final fixed image. Also, acquire brightfield images at 100 ×, 20 ×, and 10 × magnification to record the position of the cell of interest, the surrounding cells, and the R mark—this will help to relocalize the imaged cell later (see Fig. 3).

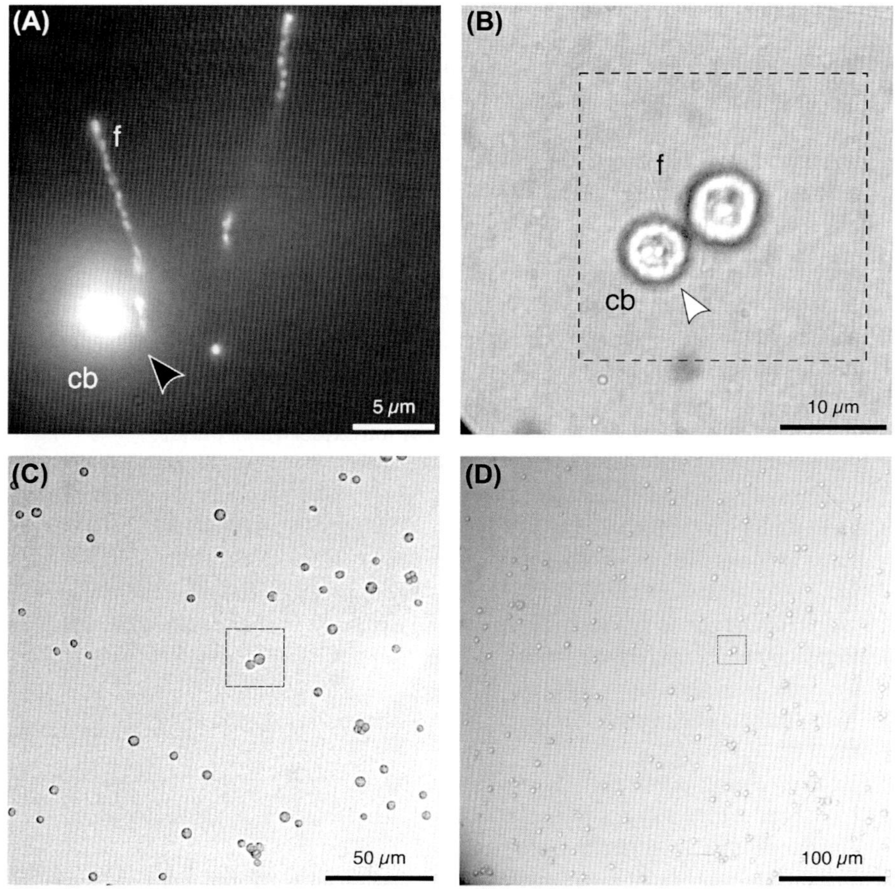

FIGURE 3 Localization of the cell of interest.

(A) The final frame of a total internal reflection fluorescence (TIRF) movie. The cell of interest is indicated by the *arrowhead*. The flagellum [f] with fixed intraflagellar transport trains, and the autofluorescent cell body [cb] are visible. (B–D) Brightfield image of the region of interest at decreasing magnifications. The original TIRF field of view is indicated by the *dashed rectangle*. The horizontally inverted mark "R" is partly visible in the panel (D). The field if view and magnification in (D) is comparable to what is later seen through the optics of an ultramicrotome.

1.4 SAMPLE PREPARATION FOR ELECTRON MICROSCOPY

The preparation of the samples for room temperature electron tomography is performed in three steps: (1) sample postfixation and embedding, (2) sectioning at the ultramicrotome, and (3) heavy metal staining of the sections and application of fiducial markers.

1.4.1 Chemical postfixation and embedding

The sample is postfixed, dehydrated, and resin-embedded in the original imaging chamber (Fig. 1C). The cured resin block needs to be detached from the glass slide before trimming and sectioning. We use liquid nitrogen immersion, which disrupts the resin−glass bond due to the difference in thermal expansion. After detachment, the resin block can be directly mounted in the ultramicrotome holder (procedure adapted from Pigino et al., 2009).

Procedure:

1. Incubate the sample fixed during LM for 30 min at room temperature.
2. Substitute the fixation solution with 300 μL fresh 2.5% glutaraldehyde/TAP.
3. Incubate 30 min at room temperature.
4. Remove the fixative solution.
5. Wash several times with 300 μL ddH_2O.
6. Incubate 30 min with 300 μL 1% osmium tetroxide/water on ice.
7. Remove the osmium solution.
8. Wash with ice-cold water.
9. Dehydrate with 30%, 50%, 70%, 90%, 96%, 100%, 100% ethanol washing steps (always 300 μL) with 1 min incubations. 30%−50% steps are done on ice, 70%−100% in the freezer (−20°C). Try to avoid moisture condensation at high ethanol concentration by minimizing exposure of the sample to warm air.
10. Replace the 100% ethanol with 1:1 mixture of 100% ethanol:LX112 resin. Incubate overnight at −20°C.
11. Bring the sample to room temperature, exchange for 1:2 ethanol:resin. Incubate for 2 h.
12. Exchange with pure resin, incubate for 1 h.
13. Exchange with fresh resin, incubate for 1 h in the vacuum desiccator.
14. Exchange with fresh resin, cure in the 60°C oven for 48 h.
15. Take the sample, still attached to the coverslip out from the oven, and cool it down to room temperature.
16. Use cyanoacrylate to glue the coverslip to a 1-mm-thick microscope slide. This reinforces the 0.17-mm-thick coverslip and prevents it from breaking into pieces, which can be difficult to remove from the resin block.
17. Drop the slide into liquid nitrogen to detach the resin from the coverslip.
18. Inspect the face of the resin block under a stereomicroscope. The flagella cannot be seen due to the lack of contrast, but the cell bodies and replica of the "R" inscribed into the coverslip are still visible in the resin.

1.4.2 Thin sectioning

While the flagella are not visible in the embedded sample anymore, the cell bodies are and so is the replica of the mark scratched into the coverslip. These cues are used to locate and trim the region of interest. There is no need to determine the z-position, as the gliding flagella are always located in the first 350 nm surface layer of the resin block. On the other hand, an almost perfect alignment of the diamond knife to the block surface is required to obtain the very first section of exactly 350 nm in thickness (Fig. 1D).

Procedure:

1. Trim the resin block to c. 0.3 × 0.3 mm area around the cell of interest.
2. Make a drawing or take a photograph showing the exact position of the cell on the trimmed block.
3. Adjust the diamond knife blade carefully to have it parallel to the sample surface. Approach as close as possible to the sample without touching the sample surface.
4. Start automated cutting with a 20 nm thickness setting. After several cutting cycles, the last few nm between the knife and the sample are covered. As soon as the knife hits the sample, stop cutting and change to the target slice thickness—350 nm. The knife blade is now exactly in plane with the sample, and it can cut the 350 nm section (containing the flagellum) at full thickness.
5. Let the microtome cut c. five slices. The first slice should contain the flagellum; the following sections contain parts of the cell body. It is important to see the positions of the cell bodies in EM to compare them with the LM images and correctly localize the cell of interest.
6. Collect the sections on a Formvar-coated single slot grid. The flagellum is present only in the first section, so that one should be as close as possible to the center on the grid (Fig. 1E).
7. Let the grid dry.

1.4.3 Staining of the sections and application of fiducial markers

The sections are stained with uranyl acetate and lead citrate according as described previously (Pigino et al., 2009). If tomography imaging is intended, then gold particles are added as fiducial markers.

Procedure:

1. Prepare fresh 2% uranyl acetate in 70% methanol (1 h on the rocking plate to dissolve).
2. Load the grids on the staining chamber, wet with 70% methanol.
3. Stain the grids for 10 min with the uranyl acetate solution.
4. Wash 3 × in the chamber with 70% methanol.
5. Wash vigorously in a sequence of clean beakers each containing the following decreasing methanol concentrations: 70%, 70%, 50%, 30%, 0% in water.
6. Stain with the Reynold's lead citrate for 5 min.

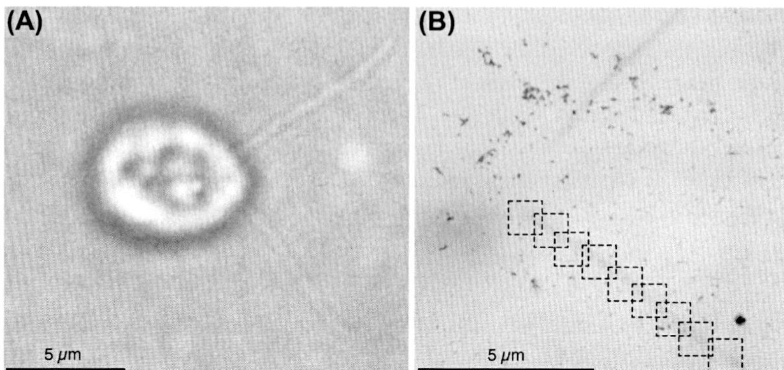

FIGURE 4 Light and electron microscopy images of a *Chlamydomonas* cell.

(A) Brightfield image of a *Chlamydomonas* cell with two flagella after glutaraldehyde fixation. (B) Low magnification EM image of the same cell. *Dashed rectangles* indicate where the tomograms were taken at higher magnification.

7. Wash 3 × in chamber and 3 × in beakers with water.
8. Let the sections dry.
9. Immerse for 70 s into undiluted 10-nm gold beads solution, blot excess solution with a filter paper, let dry.

1.5 ELECTRON TOMOGRAPHY

Tomographic tilt series are acquired on a Tecnai F30 (FEI) transmission electron microscope, operated at 300 kV, equipped with 2048 × 2048 Gatan CCD camera and SerialEM software (Mastronarde, 2005). The use of a 300 kV TEM allows tomographic acquisition of thicker section ranging between 350 and 400 nm. The series are recorded in single tilt axis geometry, with a pixel size of 7Å, a tilt range of 120–130 degrees and tilt steps of 1 degree. To minimize the missing wedge artifact, the flagellum should be carefully aligned along the tilt axis of the microscope. The magnification used and camera chip size yields an effective field of view of 1.4 × 1.4 μm, therefore it takes a mosaic of 7–13 overlapping tomograms to cover the volume of a complete flagellum (Fig. 4). We use the IMOD software package (Kremer, Mastronarde, & McIntosh, 1996) for reconstruction, joining, segmentation, and visualization of the tomograms.

1.6 REGISTRATION OF LIGHT AND ELECTRON MICROSCOPY IMAGES

Time-laps TIRF microscopy movies are represented as space–time plots, also called kymographs (Fig. 6A). Fourier transform–based analysis of the kymographs is used to separate anterograde, retrograde, and standing fluorescence signal of the trains

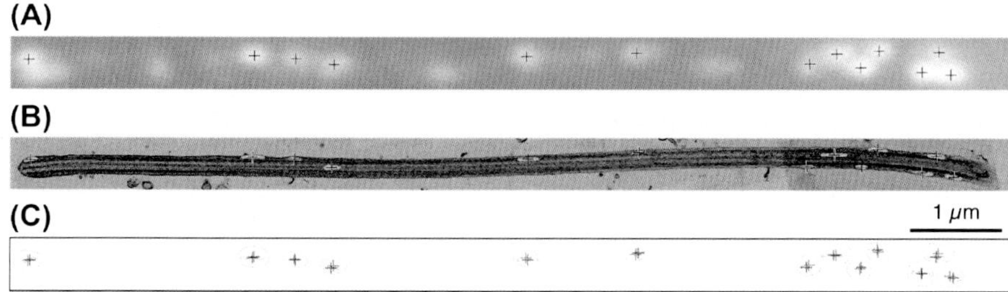

FIGURE 5 Registration of light microscopy and electron microscopy (EM) images.

(A) The centers of mass of the anterograde trains in total internal reflection fluorescence microscopy are identified. (B) The centers of mass of the corresponding compact trains (highlighted in green) in the EM are localized. (C) Similarity transform between the pairs of points is calculated. Green crosses indicate the fluorescence microscopy data, magenta crosses the EM data.

(Chenouard, Buisson, Bloch, Bastin, & Olivo-Marin, 2010) (Fig. 6A–H). The result of this approach is an image with color-coded train directionalities (Fig. 6H) per each TIRF microscopy movie.

For initial registration of EM and fluorescence images, we aligned the flagellar tip and base, features directly visible in both LM and EM images. The short compact electron-dense trains were identified first. Their relative coordinates matched the coordinates of the anterograde fluorescence signals and these trains were therefore identified as anterograde. Using FIJI software (Schindelin et al., 2012), the centers of mass of the fluorescence of anterograde trains and of the corresponding compact trains in EM are identified. The positions of the centers of mass are then used as transform point pairs to calculate precise image registration in MATLAB. The registered fluorescence image and the 3-D reconstruction in IMOD are then overlaid and used to locate the less obvious retrograde trains and the previously unknown standing trains (Fig. 5).

1.7 DATA ANALYSIS

The image with color-coded train directionalities (Fig. 6H) is overlaid with the 3-D EM image of whole flagellum (Fig. 6I), using the registration transforms obtained in the previous section. Highlighted areas of the 3-D volume are then manually searched for IFT train structures. One of such structures was found in each of the searched areas, while none was found outside. This confirms efficiency of our correlative approach and also validity of the IFT27 protein as a marker for all IFT trains. The resulting dataset allowed us to describe the ultrastructure of anterograde, retrograde, and standing trains (Fig. 6J–L) and reveal how the anterograde and retrograde trains avoid collisions by traveling on B- and A-tubules, respectively (Stepanek & Pigino, 2016).

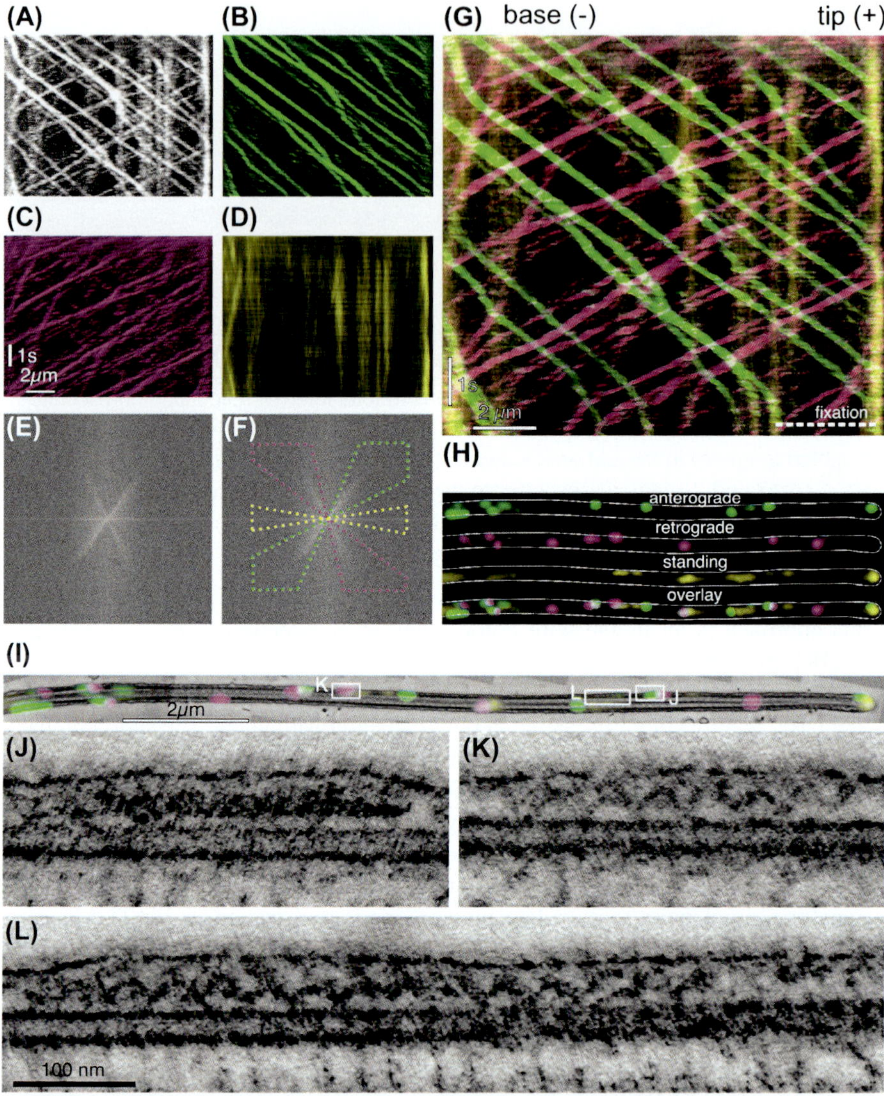

FIGURE 6 Anterograde, retrograde, and standing train ultrastructure.

(A) The space–time plot (kymograph) of the original movie. (B–D) Anterograde, retrograde, and standing tracks, obtained by inverse fast Fourier transform of regions isolated according to (F). (E) FFT of a kymograph in (A). (F) Regions corresponding to anterograde (green), retrograde (magenta), and standing (yellow) signal. (G) False-color overlay of the isolated tracks. The time point of fixation is indicated by the *dashed line*. (H) Manual tracking of the intraflagellar transport (IFT) trains. Position of each train in the last time point of the kymograph is plotted. Schematic flagellum outline is drawn in white. (I) Longitudinal section through the cilium 3-D reconstruction, assembled from 12 tomograms, overlaid with results of the IFT train tracking. The areas marked by rectangles are magnified to show representative anterograde train (J), retrograde train (K), and standing train (L).

Adapted from Stepanek, L., & Pigino, G. (2016). Microtubule doublets are double-track railways for intraflagellar transport trains. Science (New York, N.Y.), 352*(6286), 721–724. https://dx.doi.org/10.1126/science.aaf4594.*

1.8 PREPARATION OF FLAGELLAR CROSS SECTIONS FOR CORRELATIVE LIGHT AND ELECTRON MICROSCOPY ANALYSIS

Sometimes it is advantageous to obtain sections through the sample in other than *en face* orientation. For example, fine structural details, such as the links between IFT trains and microtubules are difficult to resolve in the transversal views of tomograms due to the missing wedge artifact. Cutting the flagellum transversally and imaging the flagellar cross section directly provides higher resolution and isotropic image quality. A 10 μm long flagellum can be reconstructed from c. 30 such tomograms. It is, however, challenging to locate and section the 10 × 0.3 μm flagellum of interest in the block of resin. We have modified our CLEM protocol by attaching a PDMS grid with 50 μm (width) × 60 μm (height) slots to the coverslip. This allows the imaging of individual *Chlamydomonas* cells in a single slot on the coverslip (Fig. 7A–B). After resin embedding, any excess resin and the PDMS grid are removed to expose an array of resin pillars, including the one containing the cell of interest. The pillars that do not contain the cell of interest are scraped off and the remaining pillar is reembedded in such an orientation that the flagellum is as close as possible to the tip of the resulting resin block. The cured block is immediately ready for cutting. The sections coming from the tip of the block are very narrow, which allowed at least 15 to fit on a single grid. The task of locating the cross sections of a flagellum is simplified, as they are close to the narrow edge of each section. Detailed description of the procedure follows below.

Procedure:

1. Deposit a 60 μm layer of the SU-8 photoresist on a glass slide or silicon wafer, and soft bake it for 10 min at 95°C on a hotplate.
2. Using the copper grid as photomask, illuminate the photoresist with UV light. We use 6 s exposure time with a 100 W mercury burner lamp, a 360–370 nm DAPI filter and a 10 ×/0.25 A-Plan Olympus objective as the illumination setup. Different exposure times might be required for other illumination settings.
3. Use Propylene glycol monomethyl ether acetate (PGMEA) to wash away the uncrosslinked photoresist. The array of hexagonal pillars should remain on the surface after a successful procedure.
4. Place a tiny (~3 μL) droplet of PDMS next to the pattern, and guide it gently with pipette tip to connect with the first pillars. The PDMS will start to flow between the pillars due to capillary forces.
5. Watch the PDMS flow. As soon as all the space between the pillars is filled, place the slide on a heating plate prewarmed to 100°C. Fast curing of PDMS is important, otherwise it might overflow the pillars.
6. Carefully lift the resulting PDMS structure (Fig. 7A–C) with fine forceps, and place it on a 24 × 60 mm coverslip.

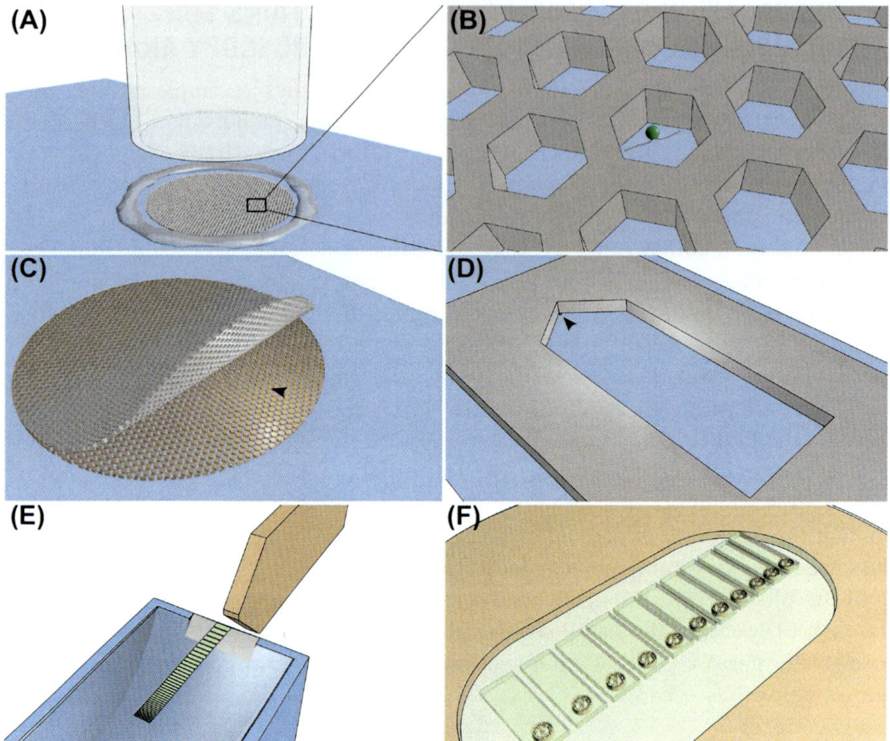

FIGURE 7 Preparation of flagellar cross sections for correlative light and electron microscopy analysis.

(A) The imaging chamber is prepared with a PDMS grid on the glass surface instead of a scratched mark. (B) The grid openings are 50 μm wide, just enough to accept a single gliding *Chlamydomonas* cell. (C) After fixation and embedding according to Fig. 1B–C, the PDMS grid is lifted to expose array of resin pillars. (D) All the pillars except the one containing the cell of interest are removed, and the cell is reembedded using a PDMS mold. (E) As the cell is positioned in the tip of the new block, ultrathin cutting can be started directly. (F) 300-nm serial cross sections of a flagellum; dimensions not to scale.

7. Assemble the rest of the imaging chamber and proceed with imaging, fixation, and sample preparation up to the resin infiltration as described above. Record the coordinates of the hole containing the cell of interest, e.g., 5th from left, 10th from top.
8. Wipe away most of the resin before curing, leaving only the holes filled.
9. Cure the resin and lift off the PDMS, exposing resin pillars (Fig. 7C).
10. Scrape away most of the pillars, leaving only the one containing the cell of interest.
11. Place a PDMS or rubber mold with pointed end over the slide in a way that the flagellum of interest points toward the sharp end of the mold (Fig. 7D).

12. Fill the mold with fresh resin and let it cure.
13. Remove the resin from coverslip as described above.
14. No trimming is needed this time. Start cutting 300 nm sections right from the tip of the resin block. Flagella-containing cross sections will come off the block after the first few micrometers of empty resin are cut away.
15. Collect the sections and prepare them for tomography as described above.

2. INSTRUMENTATION AND MATERIALS

2.1 *CHLAMYDOMONAS REINHARDTII* CULTURE

C. reinhardtii strain IFT27-GFP mt+ (Qin et al., 2007).
TAP medium (Gorman & Levine, 1965).

2.2 IMAGING AND EMBEDDING CHAMBER PREPARATION

1. Tungsten carbide tip (Glascribe) pen.
2. BEEM embedding capsule size #3.
3. 24 × 60 mm coverslips with tightly controlled (±0.005 mm) thickness (Menzel-Gläser).
4. Sylgard 184 PDMS elastomer (Dow Corning).
5. Hotplate.

2.3 LIGHT MICROSCOPY

1. Olympus BX71 inverted microscope with custom-made TIRF condenser, 150 ×/1.45 TIRF objective, 491-nm solid state laser excitation, 525/30 emission filter, Andor iXon Ultra3 CCD camera, and iQ3 acquisition software.
2. IFT27-GFP *Chlamydomonas* cells cultured in TAP medium.
3. 2.5% glutaraldehyde (EMS #16220)—prepare fresh by diluting 25% aliquot (kept at −20°C) in appropriate buffer (TAP for *Chlamydomonas* experiments).
4. PTFE tubing 0.8-mm inner diameter, 0.4-mm wall thickness (Bola #S1810-10).
5. 1-mL syringe with a needle that fits in the PTFE tubing.

2.4 SAMPLE PREPARATION FOR ELECTRON MICROSCOPY

2.4.1 Chemical postfixation and embedding

1. 2.5% or glutaraldehyde (EMS #16220)—prepare fresh by dilution of 25% aliquot (kept at −20°C) in appropriate buffer (TAP medium for *Chlamydomonas* experiments).
2. 1% osmium tetroxide (EMS #191910)—prepare fresh by dilution of 4% aliquot (kept at −20°C) in water.

3. 30%–96% ethanol (VWR Chemicals #20821)—prepare stock solutions by dilution of 96% ethanol. Keep 30% and 50% solutions at 4°C, 70%–95% at −20°C.
4. 100% ethanol—prepare stock solution by overlaying molecular sieve beads with 96% ethanol. Keep tightly sealed at −20°C.
5. Cyanoacrylate glue (Permabond 105).
6. Liquid nitrogen.
7. Glass slide.
8. Vacuum desiccator.
9. Polymerization oven with vapor exhaust.
10. LX112 resin:
    a. Component A: Add 61.8 g of DDSA (EMS #13710) and 48.0 g of LX112 (LADD research). Mix thoroughly for 5 min.
    b. Component B: Add 51.6 g of NMA (EMS #19000) and 60.0 g of LX112. Mix thoroughly for 5 min.
    c. Combine 100 g of component A and 100 g of component B. Mix thoroughly for 5 min.
    d. Add 2.8 g of DMP-30 (EMS #13600). Mix thoroughly for 5 min.
    e. Prepare stock aliquots, store at −20°C.

2.4.2 Thin sectioning

1. Cyanoacrylate glue (Permabond 105).
2. Liquid nitrogen.
3. Ultramicrotome (Leica UCT) with stereomicroscope (Leica MZ6).
4. Diamond knife (Diatome Ultra 35 degrees).
5. Single slot copper grids (Science Services #G2010-Cu) coated with Formvar (EMS #15830) membrane.

2.4.3 Staining of the sections and application of fiducial markers

1. Staining chamber (Pelco #22510).
2. Uranyl acetate (Polysciences #21447).
3. Lead citrate (EMS #512265) solution prepared after Reynolds (Reynolds, 1963).
4. Methanol (VWR Chemicals #20846.307).
5. 10-nm gold particles (Sigma-Aldrich #752584).
6. Rocking plate.

2.5 ELECTRON TOMOGRAPHY

Tecnai F30 (FEI) transmission electron microscope, operated at 300 kV, equipped with 2048 × 2048 Gatan CCD camera and SerialEM software (Mastronarde, 2005), software package (Kremer et al., 1996) for reconstruction, joining, segmentation, and visualization of the tomograms.

2.6 REGISTRATION OF LIGHT AND ELECTRON MICROSCOPY IMAGES

FIJI software (Schindelin et al., 2012).
MATLAB 2014b (MathWorks, Inc).

2.7 DATA ANALYSIS

FIJI software (Schindelin et al., 2012).
IMOD software package (Kremer et al., 1996).
MacBook Pro computer, 2.3 GHz Intel Core i7, 16 GB RAM, 256 GB SSD hard disk.

2.8 PREPARATION OF FLAGELLAR CROSS SECTIONS FOR CORRELATIVE LIGHT AND ELECTRON MICROSCOPY ANALYSIS

1. Sylgard 184 PDMS elastomer (Dow Corning).
2. SU-8 2025 photoresist (Microchem).
3. PGMEA (Sigma-Aldrich #484431).
4. EM copper grid 400 mesh hexagonal (Science Services).

3. DISCUSSION AND OUTLOOK

The method presented here combines LM at the limits of diffraction and camera acquisition speed with electron tomography. With this method, we were able to record motion patterns of IFT trains and to study the morphology and position of the trains with nanometer resolution in 3-D EM (Stepanek & Pigino, 2016). We showed that anterograde trains move along the B-tubule of the microtubule doublets and the anterograde trains move along the A-tubule, therefore revealing how bidirectional transport in the cilium is regulated.

Our method can be used to study other dynamic cellular processes in small organisms, cells, or cell-free systems.

As the portfolio of available imaging techniques becomes more and more diversified, the space of their possible combinations grows as well, allowing addressing increasingly complex biological questions (Nixon-Abell et al., 2016). The method presented here, for instance, could be easily adapted to bridge techniques other than light and electron microscopy, perhaps more than two of them in a single experiment. As an example, superresolution fluorescence observation could be performed between the time-resolved LM and EM.

The current pace of microscopic technology development raises the question if the combinatorial approach of time-resolved CLEM can be replaced by a single imaging system. As of now, the fundamental limitations of contemporary techniques do not seem to be easy to overcome.

The rapidly growing field of superresolution optical microscopy [reviewed in (Fujita, 2016)] recently focuses on improving the temporal as well as spatial resolution (Wang et al., 2016) (Liu & Wu, 2016). In some cases it might offer enough resolution to eliminate the need for EM imaging. However, the mutual exclusivity of high temporal resolution with high spatial resolution and low phototoxicity remains a common trait of light-based techniques. Moreover, most of the superresolution solutions refine the precise localization of molecular structures rather than their structure and cellular context.

An alternative approach, and the ultimate tool for time-resolved electron microcopy would be true live-cell electron imaging. It has long been considered technically infeasible, since living cells cannot withstand the high vacuum inside the transmission electron microscope. This has changed with the advent of the liquid cell EM, which uses microfluidic cells integrated into TEM holders to support living cells in their native environment (de Jonge & Ross, 2011) (Peckys & de Jonge, 2014), reviewed in (Ross, 2015). However, the resolution of this technique is far behind conventional thin-slice TEM, and it does not offer the biochemical specificity of fluorescent labeling at present time.

ACKNOWLEDGMENTS

We thank J. Rosenbaum and D. Diener for providing the *Chlamydomonas* IFT27-GFP strain; P. Kiesel, the MPI-CBG EM facility, J. Meissner, the MPI-CBG LM facility for technical support; and J. Howard, S. Diez for fruitful discussion and comments; D. Diener for helpful comments on the manuscript. This work was supported by the Max Planck Society and a fellowship of the Dresden International Graduate School for Biomedicine and Bioengineering (GS97), granted by the German Research Foundation to L.S.

REFERENCES

Beznoussenko, G. V., & Mironov, A. A. (2015). Correlative video-light-electron microscopy of mobile organelles. *Methods in Molecular Biology (Clifton, N.J.), 1270*, 321–346. http://dx.doi.org/10.1007/978-1-4939-2309-0_23 (Chapter 23).

Chenouard, N., Buisson, J., Bloch, I., Bastin, P., & Olivo-Marin, J. C. (2010). Curvelet analysis of kymograph for tracking bi-directional particles in fluorescence microscopy images. In *2010 17th IEEE International Conference on Image Processing (ICIP 2010)* (pp. 3657–3660). http://dx.doi.org/10.1109/ICIP.2010.5652479.

Engel, B. D., Lechtreck, K.-F., Sakai, T., Ikebe, M., Witman, G. B., & Marshall, W. F. (2009). Total internal reflection fluorescence (TIRF) microscopy of *Chlamydomonas* flagella. *Methods in Cell Biology, 93*, 157–177. http://dx.doi.org/10.1016/S0091-679X(08)93009-0.

Fuhrmann, M., Oertel, W., & Hegemann, P. (1999). A synthetic gene coding for the green fluorescent protein (GFP) is a versatile reporter in *Chlamydomonas reinhardtii*. *The Plant Journal : for Cell and Molecular Biology, 19*(3), 353–361.

Fujita, K. (2016). Follow-up review: Recent progress in the development of super-resolution optical microscopy. *Microscopy (Oxford, England), 65*(4), 275–281. http://dx.doi.org/10.1093/jmicro/dfw022.

Gorman, D. S., & Levine, R. P. (1965). Cytochrome f and plastocyanin: Their sequence in the photosynthetic electron transport chain of *Chlamydomonas reinhardi. Proceedings of the National Academy of Sciences of the United States of America, 54*(6), 1665–1669.

Guizetti, J., Mäntler, J., Müller-Reichert, T., & Gerlich, D. W. (2010). Correlative time-lapse imaging and electron microscopy to study abscission in HeLa cells. *Methods in Cell Biology, 96*, 591–601. http://dx.doi.org/10.1016/S0091-679X(10)96024-X.

de Jonge, N., & Ross, F. M. (2011). Electron microscopy of specimens in liquid. *Nature Nanotechnology, 6*(11), 695–704. http://dx.doi.org/10.1038/nnano.2011.161.

Kozminski, K. G., Johnson, K. A., Forscher, P., & Rosenbaum, J. L. (1993). A motility in the eukaryotic flagellum unrelated to flagellar beating. *Proceedings of the National Academy of Sciences of the United States of America, 90*(12), 5519–5523.

Kremer, J. R., Mastronarde, D. N., & McIntosh, J. R. (1996). Computer visualization of three-dimensional image data using IMOD. *Journal of Structural Biology, 116*(1), 71–76. http://dx.doi.org/10.1006/jsbi.1996.0013.

Kukulski, W., Schorb, M., Kaksonen, M., & Briggs, J. A. G. (2012). Plasma membrane reshaping during endocytosis is revealed by time-resolved electron tomography. *Cell, 150*(3), 508–520. http://dx.doi.org/10.1016/j.cell.2012.05.046.

Lechtreck, K.-F., Johnson, E. C., Sakai, T., Cochran, D., Ballif, B. A., Rush, J., … Witman, G. B. (2009). The *Chlamydomonas reinhardtii* BBSome is an IFT cargo required for export of specific signaling proteins from flagella. *The Journal of Cell Biology, 187*(7), 1117–1132. http://dx.doi.org/10.1083/jcb.200909183.

Liu, Y., & Wu, J.-Q. (2016). Cytokinesis: Going super-resolution in live cells. *Current Biology: CB, 26*(21), R1150–R1152. http://dx.doi.org/10.1016/j.cub.2016.09.026.

Mastronarde, D. N. (2005). Automated electron microscope tomography using robust prediction of specimen movements. *Journal of Structural Biology, 152*(1), 36–51. http://dx.doi.org/10.1016/j.jsb.2005.07.007.

Mironov, A. A., Beznoussenko, G. V., Trucco, A., Lupetti, P., Smith, J. D., Geerts, W. J. C., … Luini, A. (2003). ER-to-Golgi carriers arise through direct en bloc protrusion and multistage maturation of specialized ER exit domains. *Developmental Cell, 5*(4), 583–594.

Mueller, J., Perrone, C. A., Bower, R., Cole, D. G., & Porter, M. E. (2005). The FLA3 KAP subunit is required for localization of kinesin-2 to the site of flagellar assembly and processive anterograde intraflagellar transport. *Molecular Biology of the Cell, 16*(3), 1341–1354. http://dx.doi.org/10.1091/mbc.E04-10-0931.

Nixon-Abell, J., Obara, C. J., Weigel, A. V., Li, D., Legant, W. R., Xu, C. S., … Schwartz, J. (2016). Increased spatiotemporal resolution reveals highly dynamic dense tubular matrices in the peripheral ER. *Science (New York, N.Y.), 354*(6311), aaf3928. http://dx.doi.org/10.1126/science.aaf3928.

Peckys, D. B., & de Jonge, N. (2014). Liquid scanning transmission electron microscopy: Imaging protein complexes in their native environment in whole eukaryotic cells. *Microscopy and Microanalysis, 20*, 346–365.

Pigino, G., Geimer, S., Lanzavecchia, S., Paccagnini, E., Cantele, F., Diener, D. R., … Lupetti, P. (2009). Electron-tomographic analysis of intraflagellar transport particle trains in situ. *The Journal of Cell Biology, 187*(1), 135–148. http://dx.doi.org/10.1083/jcb.200905103.

Polishchuk, R. S., Polishchuk, E. V., Marra, P., Alberti, S., Buccione, R., Luini, A., & Mironov, A. A. (2000). Correlative light-electron microscopy reveals the tubular-saccular ultrastructure of carriers operating between Golgi apparatus and plasma membrane. *The Journal of Cell Biology, 148*(1), 45–58.

Qin, H., Wang, Z., Diener, D., & Rosenbaum, J. (2007). Intraflagellar transport protein 27 is a small G protein involved in cell-cycle control. *Current Biology: CB, 17*(3), 193–202. http://dx.doi.org/10.1016/j.cub.2006.12.040.

Reynolds, E. S. (1963). The use of lead citrate at high pH as an electron-opaque stain in electron microscopy. *The Journal of Cell Biology, 17*, 208–212.

van Rijnsoever, C., Oorschot, V., & Klumperman, J. (2008). Correlative light-electron microscopy (CLEM) combining live-cell imaging and immunolabeling of ultrathin cryosections. *Nature Methods, 5*(11), 973–980. http://dx.doi.org/10.1038/nmeth.1263.

Rogowski, M., Scholz, D., & Geimer, S. (2013). Electron microscopy of flagella, primary cilia, and intraflagellar transport in flat-embedded cells. *Methods in Enzymology, 524*, 243–263. http://dx.doi.org/10.1016/B978-0-12-397945-2.00014-7.

Rosenbaum, J. L., Moulder, J. E., & Ringo, D. L. (1969). Flagellar elongation and shortening in *Chlamydomonas*. The use of cycloheximide and colchicine to study the synthesis and assembly of flagellar proteins. *The Journal of Cell Biology, 41*(2), 600–619.

Ross, F. M. (2015). Opportunities and challenges in liquid cell electron microscopy. *Science (New York, N.Y.), 350*(6267), aaa9886. http://dx.doi.org/10.1126/science.aaa9886.

Schindelin, J., Arganda-Carreras, I., Frise, E., Kaynig, V., Longair, M., Pietzsch, T., … Cardona, A. (2012). Fiji: An open-source platform for biological-image analysis. *Nature Methods, 9*(7), 676–682. http://dx.doi.org/10.1038/nmeth.2019.

Shaikh, T. R., Yassin, A. S., Lu, Z., Barnard, D., Meng, X., Lu, T.-M., … Agrawal, R. K. (2014). Initial bridges between two ribosomal subunits are formed within 9.4 milliseconds, as studied by time-resolved cryo-EM. *Proceedings of the National Academy of Sciences of the United States of America, 111*(27), 9822–9827. http://dx.doi.org/10.1073/pnas.1406744111.

Stepanek, L., & Pigino, G. (2016). Microtubule doublets are double-track railways for intraflagellar transport trains. *Science (New York, N.Y.), 352*(6286), 721–724. http://dx.doi.org/10.1126/science.aaf4594.

Verkade, P. (2008). Moving EM: The rapid transfer system as a new tool for correlative light and electron microscopy and high throughput for high-pressure freezing. *Journal of Microscopy, 230*(Pt 2), 317–328. http://dx.doi.org/10.1111/j.1365-2818.2008.01989.x.

Wang, W., Shen, H., Shuang, B., Hoener, B. S., Tauzin, L. J., Moringo, N. A., … Landes, C. F. (2016). Super temporal-resolved microscopy (STReM). *The Journal of Physical Chemistry Letters*, 4524–4529. http://dx.doi.org/10.1021/acs.jpclett.6b02098.

Watanabe, S. (2016). Flash-and-Freeze: Coordinating optogenetic stimulation with rapid freezing to visualize membrane dynamics at synapses with millisecond resolution. *Frontiers in Synaptic Neuroscience, 8*(24). http://dx.doi.org/10.3389/fnsyn.2016.00024.

CHAPTER

3D subcellular localization with superresolution array tomography on ultrathin sections of various species

2

Sebastian M. Markert[1], Vivien Bauer, Thomas S. Muenz, Nicola G. Jones, Frederik Helmprobst, Sebastian Britz, Markus Sauer, Wolfgang Rössler, Markus Engstler, Christian Stigloher[1]

University of Würzburg, Würzburg, Germany

[1]*Corresponding authors: E-mail: sebastian.markert@uni-wuerzburg.de; christian.stigloher@uni-wuerzburg.de*

CHAPTER OUTLINE

Methods in Cell Biology, Volume 140, ISSN 0091-679X, http://dx.doi.org/10.1016/bs.mcb.2017.03.004

Abstract

Array Tomography (AT) is a relatively easy-to-use and yet powerful method to put molecular identity in its full ultrastructural context. Ultrathin sections are stained with fluorophores and then imaged by light and afterward by electron microscopy to obtain a correlated view of a region of interest: its ultrastructure and specific staining. By combining AT with high-pressure freezing for superior structural preservation and superresolution light microscopy, even small subcellular structures can be mapped in 3D. We established protocols for the application of superresolution AT on ultrathin plastic sections of *Caenorhabditis elegans*, *Trypanosoma brucei*, and brain tissue of *Cataglyphis fortis* and *Apis mellifera*. All steps are described in detail from sample preparation to 3D reconstruction, including species-specific modifications. We thus showcase the versatility of our protocol and give some examples for biological questions that can be answered with this technique. We offer a step-by-step recipe for superresolution AT that can be easily applied for *C. elegans*, *T. brucei*, *C. fortis*, and *A. mellifera* and adapted for other model systems.

INTRODUCTION AND RATIONALE

A variety of correlative light and electron microscopy (CLEM) techniques have been established in recent years. They offer valuable new insights into biological questions of all kinds. However, many of these protocols are technically challenging or require sophisticated equipment specifically designed for certain applications.

For obtaining subcellular resolution for localization of molecules in their ultrastructural context, several postembedding immunofluorescence studies have been published for various model systems (Albrecht, Seulberger, Schwarz, & Risau, 1990; Camilli, Cameron, & Greengard, 1983; Fialka et al., 1996; Haraguchi & Yokota, 2002; Herken, Fussek, Barth, & Götz, 1988; Kurth, Schwarz, Schneider, & Hausen, 1996; Ojeda, Ros, & Icardo, 1989; Schwarz & Humbel, 2014). Array Tomography (AT) (Micheva & Smith, 2007) is a method that builds on these techniques to obtain large 3D volumes of multiple signals. In AT, ultrathin sections are stained with fluorophores and then imaged by light and afterward by electron microscopy (EM) to obtain a correlated view of a region of interest: its ultrastructure and specific staining. The possibility to image multiple epitopes by immunofluorescence staining and consecutively the full ultrastructural context by scanning electron microscopy (SEM) on the very same section allows one to answer many biological localization questions by achieving high accuracy and precision. An important technical aspect of AT is that immunofluorescence imaging is performed on semi- and ultrathin resin sections of typically 100 nm thickness, or even below. Thus the z-resolution is determined by the section thickness, which is well below the diffraction limit.

The resolution limit in the lateral x and y dimensions for the immunofluorescence step can now be readily overcome due to recent technical advances in superresolution light microscopy (Nanguneri, Flottmann, Horstmann, Heilemann, & Kuner, 2012; Perkovic et al., 2014). In combination with new and readily applicable superresolution light microscopy techniques, such as structured illumination microscopy (SIM) (Gustafsson, 2000), AT can be used advantageously to bridge the resolution gap between light and EM (Markert et al., 2016). Here, we describe in detail our general workflow, as well as model-specific modifications, for superresolution AT (for an overview, see Fig. 1). In particular, we focus on the precise overlay of the two images. This correlation aspect is of crucial importance especially with increasing resolutions of the respective principal imaging modalities, i.e., SIM and SEM. Instead of using fiducial markers, we focus on the use of cell and tissue intrinsic landmarks that are clearly detectable in both the immunofluorescence and the consecutive electron microscopic analysis.

1. METHODS—CORE PROTOCOL

1.1 HIGH-PRESSURE FREEZING

Before freezing, the inside surface of the platelets (Leica Microsystems) is coated with lecithin (dissolve about 3 mg of lecithin in 1 mL chloroform; pipet onto platelets and let dry) to facilitate removal of the sample after freeze substitution. Fresh samples are placed in the platelet, and the recess is overfilled with freeze protectant so that no air bubbles remain in the sandwich after the lid is placed on top. Again, depending on the sample, we use bacteria paste (see below), BSA solution, PVP, or hexadecene as a filler and cryoprotectant. The samples are then frozen with a cooling rate of >20,000 K/s and a pressure of >2100 bar and stored in liquid nitrogen.

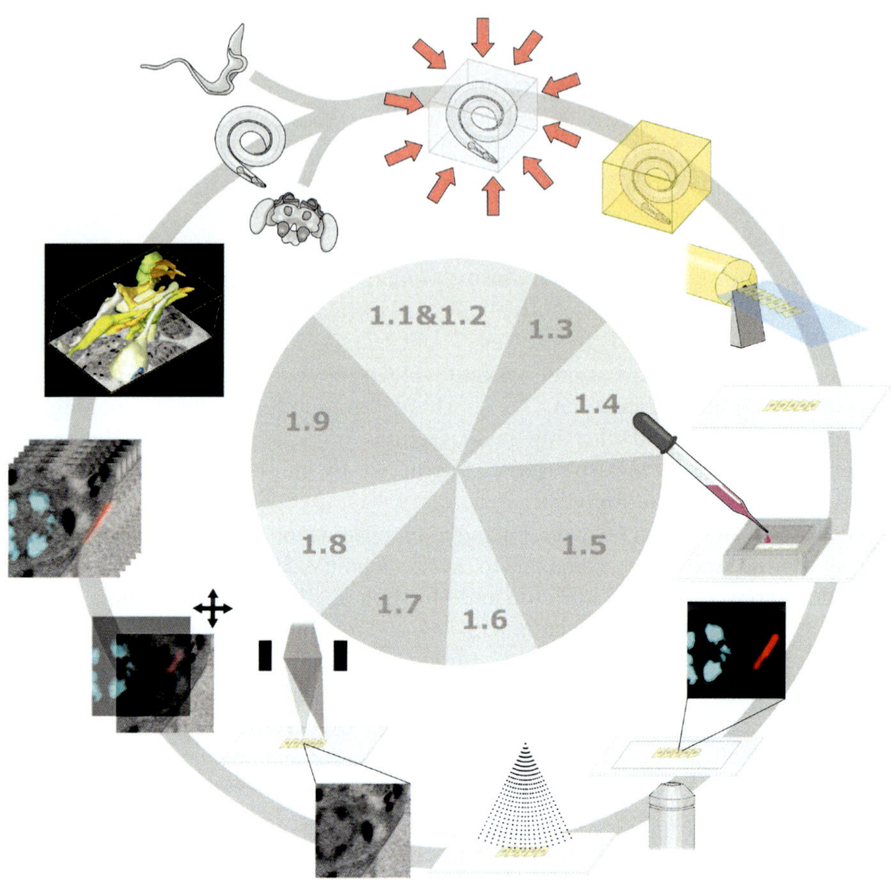

FIGURE 1

Schematic overview of the general workflow of superresolution Array Tomography. Chapter sections corresponding to the specific steps in the workflow are indicated according to Methods—Core Protocol subsections.

Although high-pressure freezing allows for vitrification of thicker samples compared to other methods such as plunge freezing, it is still advisable to make your sample as thin as possible to suppress ice crystal formation. Depending on the model organisms or tissue used here, we achieve acceptable results for samples up to 200 μm in thickness with an EM HPM100 machine (Leica Microsystems).

1.2 FREEZE SUBSTITUTION

For freeze substitution we use the EM AFS2 freeze substitution system (Leica Microsystems). The following protocol is an adaption of a previously published method (Weimer, 2006). Metal washing containers with bottom plates and flow-through plastic capsules (Leica Microsystems) are filled with a solution of 0.1% $KMnO_4$ in anhydrous acetone and cooled down to −90°C. Freezing platelets with the samples are transferred with precooled forceps from liquid nitrogen to the cups. Individual cups can be identified by the number of notches that were carved into the rim of the cups. Care needs to be taken that the samples never get much warmer (within a few degrees) than −90°C, or ice crystals will form and damage them. After all samples are dropped in the freeze substitution mix, we cover the cups loosely with Teflon discs to prevent evaporation. After incubation at −90°C for about 16 h the solution is changed once. All solutions need to be precooled prior to contact with the samples. We place an additional empty metal container inside the freeze substitution system to receive the solutions.

After 80 h in the freeze substitution mix, the temperature is linearly ramped up to −45°C over the course of 11 h. Then the samples are washed with acetone, until the solution becomes clear. We do four washing steps over the course of 3 h. Since acetone can inhibit polymerization of LR White resin, the acetone is exchanged with ethanol. The samples are first washed with one-third ethanol in acetone, then with two-thirds, and finally two times with pure ethanol (96% ethanol is sufficient). Then, over the course of 16 h, the temperature is increased to 4°C, and the samples are transferred into 50% LR White resin in ethanol. It is important that the samples stay at 4°C at all times. Then the samples are removed from the freezing platelets, if they did not get detached during the washing steps. For this we use glass pipets. To get larger openings, the tips of conventional Pasteur pipets are snapped off and the edges are melted smooth with a Bunsen burner. If pipetting up and down does not suffice to remove the samples from the platelet, we carefully use a mounted needle. It is common for the samples to break into smaller pieces during the whole process, although whole worms, cells, and insect nervous tissue tend to stay intact. All fragments are then collected in small glass vials with a lid and incubated for 16 h at 4°C. Then the samples are washed three times with LR White after 1, 4, and 16 h to allow complete resin infiltration. In case of UV polymerization (see below) fresh LR White is mixed just before embedding with the accelerator as provided by the manufacturer. We use one drop of accelerator per 10 mL resin and mix well before use. The samples are washed with this accelerated LR White once more and then immediately embedded. Otherwise, the samples are transferred to room temperature and embedded in fresh LR White without the additional accelerator.

1.3 EMBEDDING

After infiltration the samples are transferred to gelatin capsules with a glass pipet. The resin LR White does not polymerize in the presence of oxygen, so the embedding capsules have to be tightly locked. We fill the capsule to the rim with LR White, add a small paper strip with a label to mark the block, and then put on the lid tightly. It is not necessary to remove air bubbles from the capsule. As long as an influx of new oxygen is prevented, the resin will polymerize. The capsules are then cured upright either under UV light or thermally. For UV polymerization, we cure them at 4°C for 48–72 h and then ramp up to room temperature and cure for an additional 24 h. For polymerization by heat we cure the samples at 48–52°C for at least 48 h.

In our hands, thermal curing leads to more extraction of the tissue, which might be desirable depending on the questions. For example, in *Caenorhabditis elegans* nervous tissue synaptic vesicles tend to be lost, but in turn microtubules are much more clearly visible (Markert et al., 2016). However, we did not observe any obvious difference in antigenicity between these two curing methods.

1.4 ULTRAMICROTOMY

If the sample has the desired orientation, the capsule can be sectioned without further remounting. We remove the gelatin at least from the tip of the capsule with a razor blade and then start cutting. This results in circular sections. Since the sample is usually located very close to the tip of the capsule, the sections will contain tissue very soon. We check for that by staining sections with methylene blue. If the region of interest cannot be identified this way, it can be helpful to check some sections by transmission electron microscopy. This has the additional benefit that the structure preservation can be judged beforehand. Once the region of interest is located, the block is trimmed so that a trapezoid block face is achieved. Since LR White is less hydrophobic than epoxy resins such as Epon, the sections usually do not form ribbons well. To alleviate this, we add some glue to at least one edge of the block face. We use ordinary contact adhesive glue (Pattex Gel Compact) and thin it out with xylene in a ratio of about 1:1 so that the glue is very smooth and barely stringy. The glue and the xylene can easily be mixed with a toothpick in a small glass bowl. It can be very helpful to add a dash of pigment to the glue mixture to aid in localizing the sections during imaging (see Fig. 2). We use Spinel Black 47400 (deepest black) pigment to get a dark color even when cutting ultrathin sections. Then, the glue is applied to one edge of the block with a very thin needle. Care must be taken to add glue only to the cutting edge and maybe also the opposite one, but not to the other edges of the block. If that happens, the sections tend to crush during cutting. It is generally not avoidable that some glue spills onto the block face. This glue is removed with the first section, which is then lost for analysis, however.

For AT, sections are collected on a solid support, such as a glass slide, and not on grids. This way, the sections are very robust and can be handled for staining without loss. Before sectioning starts, the glass piece that collects the sections should already be submerged in the knife boat. When using a microscopic slide or a big coverslip, a large boat is required. We use the histo Jumbo diamond knife from DiATOME (Hatfield, USA), which was designed to collect sections/ribbons on microscope slides. If small coverslips are used, they can be mounted on a micromanipulator or glued to a slide and submerged that way. Then, sectioning is started. The length

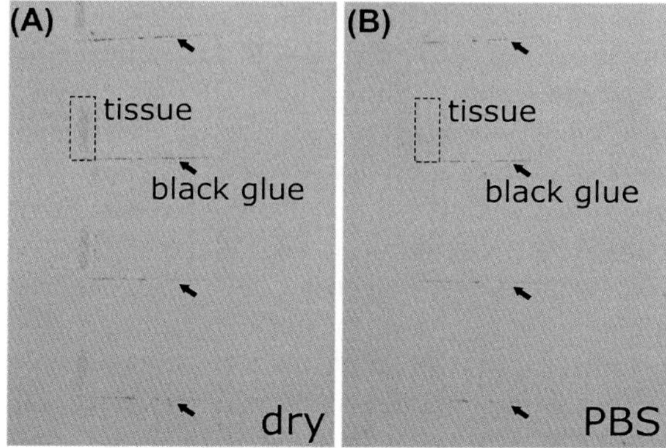

FIGURE 2

Ribbon of ultrathin sections in a bright field light microscope. (A) Dry sections mounted on a glass slide. (B) Same sections mounted on a coverslip in phosphate-buffered saline. The *dashed box* marks the position of tissue within a section and the *arrows* mark the black-stained glue between the serial sections. Once the sections are hydrated, everything but the glue becomes practically invisible.

of the ribbon is only limited by the dimensions of the support (i.e., the used piece of glass). However, it is also possible to put several ribbons in parallel on one slide. Thus, if the block face is small, hundreds of sections can be collected. We typically apply only 5–50 sections per slide, however. This way the imaging time per slide is more manageable. The thinner the sections, the better the resolution in z-dimension; but thin sections are always at risk of being lost because they are more easily crushed during cutting. In addition, very thin sections negatively impact the contrast of the electron micrographs. We usually use 100 nm sections as this has turned out as a good compromise (Micheva & Smith, 2007).

After obtaining a ribbon, it is carefully detached from the knife's edge with a mounted eyelash and pushed toward the collecting glass slide. The first section is attached to the slide by pushing it to the border of the water film so that the front edge of the section touches the glass directly. Then, the entire ribbon can be attached to the slide by slowly removing the water from the boat with a syringe or by carefully lifting up the slide from the boat. The slide is subsequently dried for at least 30 min at 50°C to allow sections to irreversibly adhere to the glass surface. Sections on dried slides can be stored for many weeks, but they begin to show reduced quality of immunolabeling after a few days, so it is best to stain right after sectioning if the staining is critical.

1.5 LIGHT MICROSCOPY

Slides and coverslips with serial sections are stained by using standard immunolabeling procedures. We use a modified version of the protocol established by Micheva and

Smith (Micheva & Smith, 2007). To place solutions on top of the sections, they are circled with a hydrophobic pen (PAP pen), or a PDMS polymer chamber is applied to form a well around the sections. The samples are then placed in a dark humid chamber. Then, the staining procedure is started. First, the sections are rehydrated and blocked by applying a blocking solution (0.1% BSA and 0.05% Tween 20 in 50 mM Tris buffer, pH = 7.6) for 10 min. If the sample was fixed with aldehydes, it is advisable to start with a glycine treatment (50 mM glycine in Tris buffer for 5 min), before the blocking solution is applied. The primary antibody is then diluted in the blocking solution, centrifuged at maximum speed in a table top centrifuge (13,000–16,000 × *g*) for 2 min to pellet debris and conglomerates, and applied to the sections for 1 h. Once the procedure starts, it is important that the sections do not dry out at any point. We use two pipets, one to drain the sample and the other to simultaneously add new solution. After incubation with the primary antibody, the sections are washed five times in 5 min intervals with Tris buffer. In the meantime, the secondary antibody is also diluted in blocking solution, centrifuged, and then applied for 30 min at RT in the dark. Afterward, the sections are washed as before with Tris buffer and finally once with ddH_2O to remove salt. The water is then removed almost completely and the sections are mounted in a medium of choice, such as Mowiol, glycerol, or Vectashield. Samples are stored at 4°C until further use. For best results the SIM imaging of the stained section should be performed within 3 days after sample preparation. It can be difficult to find the sections under the microscope because they are completely transparent. A most useful feature is the glue between the serial sections, especially when black pigment has been added to the glue (see above and Fig. 2).

1.6 CONTRASTING AND CARBON COATING

After all light microscopic image acquisition has been completed, the coverslips are detached from the slides, and the mounting medium is carefully removed from the sections by rinsing with either water or ethanol, depending on the mounting medium. In the case of Mowiol we use water to wash it off. Then, the sections are dried and can be stored for at least a few months. For contrasting we use a standard protocol with 2.5% uranyl acetate in ethanol for 15 min and 50% Reynolds' lead citrate (Reynolds, 1963) in water for 10 min. After contrasting the sections, they are rinsed with water and dried again. As microscope slides and large coverslips are too big for our SEM, we cut them with a diamond pen into smaller pieces. This can also be done before contrasting. The glass pieces are then mounted to SEM specimen holders and surrounded with a contact adhesive, such as silver paint, to reduce charging. Then they are carbon coated to further reduce charging. This is essential for good SEM imaging results. We apply a relatively thick carbon layer, so that a white indicator paper added alongside the sections shows a dark gray color. If unsure about carbon layer thickness, a thinner layer can be tried first; and if there is any charging, the sections can be coated repeatedly until the carbon layer is sufficient to stop charging of the specimen.

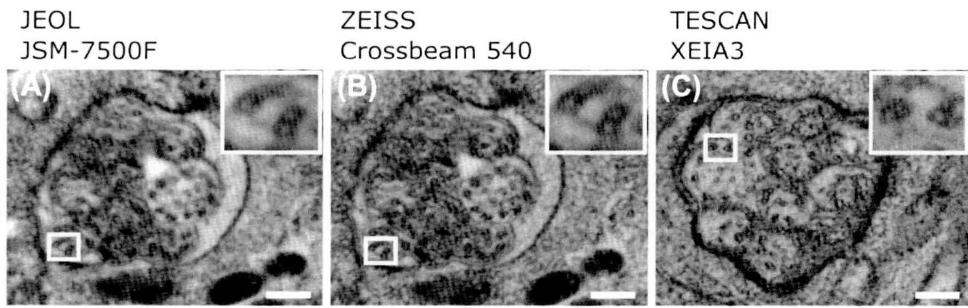

FIGURE 3

Comparison of imaging results using different scanning electron microscopy setups on ultrathin sections of *Caenorhabditis elegans* amphid channel cilia. (A) Section as visualized with a LABE detector of a JEOL JSM-7500F at 5 kV and a probe current of 300 pA at a working distance of 6.0 mm. (B) Same section as in (A) imaged with the retractable BSE detector of a ZEISS Crossbeam 540 at 5 kV and a probe current of 500 pA at a working distance of 6.5 mm. (C) Section of the same structure as in (A) and (B), but a few hundred nanometers posterior in the worm. Imaged with the Mid-angle BSE detector of a TESCAN XEIA 3 at 5 kV and a probe current of 500 pA at a working distance of 5.0 mm. Insets show microtubules. The lumen of microtubules could be resolved in all three setups. Scale bars: 200 nm.

1.7 SCANNING ELECTRON MICROSCOPY

We use an SEM to image the ultrastructure of the sectioned samples. For this, a detector for backscattered electrons at low angles is required. We use a field emission scanning electron microscope JSM-7500F (JEOL, Japan) with a LABE detector (for backscattered electron imaging at extremely low acceleration voltages). By far the best results with our machine are achieved with an acceleration voltage of 5 kV, a probe current of 0.3 nA, and a working distance of 6–8 mm. This may vary for different microscopes, but we were able to achieve comparable results with two other SEM configurations (Fig. 3). For SEM imaging, we always look first at the light microscopic images to get an idea of what area of interest should be imaged and at which magnification. It is advisable to take a few test SEM images first for an initial correlation with the light microscopic images before imaging the whole array of serial sections by EM (see also below).

1.8 IMAGE PROCESSING AND CORRELATION OF STRUCTURED ILLUMINATION MICROSCOPY AND SCANNING ELECTRON MICROSCOPY IMAGES

SIM imaging typically produces z-stacks. Although a section is only 100 nm thick in our case and would thus fit in one z-layer, stacks are required for proper image processing to generate the superresolved images. For correlation, we just choose the brightest z-layer of each channel and export it with ImageJ in portable network graphics (PNG) format for further analysis. It is also conceivable to do a maximum intensity projection or use more complex image processing algorithms to project the information onto one layer.

For unbiased manual correlation we use the free and open source vector graphics editor Inkscape (version 0.91; http://www.inkscape.org). For one section, the SEM image(s) and one SIM image per color channel are dropped onto the Inkscape canvas. For an unbiased correlation, all the channels of interest are hidden underneath the channel(s) containing the intrinsic landmarks, usually a DNA staining (see below). All channel images are perfectly aligned to each other and the landmark channel is placed on top. All these images are grouped into one object. Now the opacity can be reduced so that the SEM image is easily visible, when the channel stack is placed on top of it. Due to the grouping only the topmost (i.e., the landmark channel) will show up. The channel stack can now be rotated and resized (caution: lock the aspect ratio first) to fit the structures both seen in SIM and SEM. Once the result is optimal, the grouping is released and the channels of interest can be brought to the front to reveal where their signals illuminate the SEM image. To avoid bias, any further manipulations of image positions are not allowed anymore at this point. Now, the correlated images can be exported and overlaid in any desired configuration with appropriate software, e.g., the free

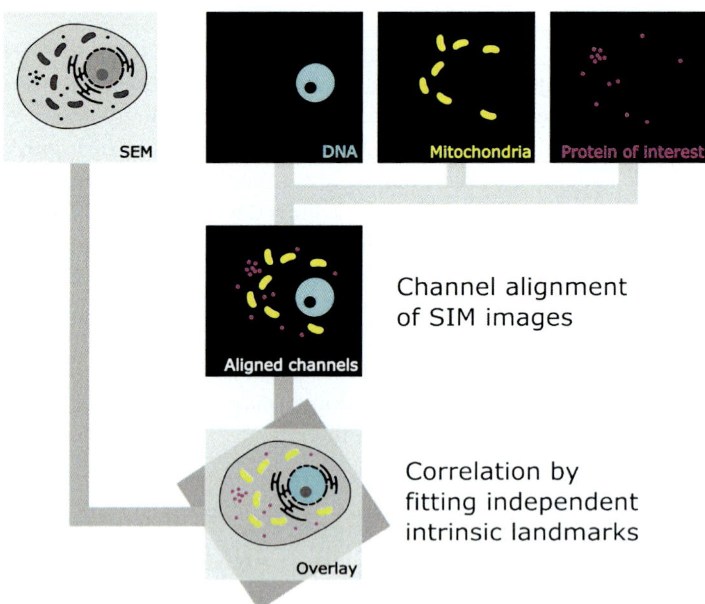

FIGURE 4

Schematic overview of the correlation strategy used in this chapter. Besides the protein of interest, one or more channels with independent intrinsic landmarks are imaged. After channel alignment the fluorescence image is correlated to the scanning electron microscopy image by using exclusively the intrinsic landmarks for guidance.

and open source image editor GIMP (http://www.gimp.org). Fig. 4 schematically depicts our correlation strategy.

Of course, correlation can also be performed without Inkscape, using only GIMP or ImageJ or similar image processing software. There are also several programs for semiautomatic correlation, for example, the ec-CLEM plugin for the software Icy (http://icy.bioimageanalysis.org/plugin/ec-CLEM#documentation).

Again, before image processing and correlation of the whole dataset is performed, it is advisable to try it first on a few sections. This will reveal the ideal settings for image processing and SEM image magnification. Also, oftentimes the importance of certain structures only becomes apparent after analyzing a few correlations, e.g., because they show unexpected labeling. Such structures can then be included in the SEM imaging right away, instead of having to go back and image the sections again later.

1.9 ALIGNMENT AND 3D RECONSTRUCTION

We perform image stack alignment and 3D modeling using the software package IMOD (Kremer, Mastronarde, & McIntosh, 1996) or Fiji (Schindelin et al., 2012) and AMIRA (FEI, Visualization Sciences Group) in combination.

1.9.1 IMOD

The first step toward a good reconstruction is the alignment of the image stack. The eTomo software (included in the IMOD package; version 4.7; http://bio3d.colorado.edu/imod/) can be used to align serial sections. Just start the software, select the corresponding function, and load the image stack. If nothing is specified, default settings should be used. In the "Align" tab, tick the box "Search for" and choose an option. If the serial images were imaged with the same magnification, the option "Rotation/translation" should be sufficient to align the stack. Otherwise choose "Rotation/translation/magnification." The "full linear transformation" option will transform your images to smoothen the alignment. This option should be used with discretion because it causes distortions of the images. It can be very useful to obtain smoother 3D models, especially when segmenting small structures. Start the alignment by clicking on "Initial Auto Alignment." Then, click on "Midas" and manually adjust coarse mistakes, if necessary. Save your adjustments and close the stack. Click on "Refine with Auto Alignment" to finish the alignment. To create the aligned stack, switch to the "Make Stack" tab and choose the option "Global alignments (remove all trends)." This will make sure that your stack does not show any drift. Just click on "Make Aligned Stack" and then view the result with "Open Aligned Stack". If satisfactory, the aligned stack can be directly used for segmentation with the 3dmod software (included in the IMOD package). Otherwise, repeat the alignment and try to improve the manual adjustments with "Midas." Sometimes it might be necessary to exclude certain images, if they disrupt the alignment too much, e.g., due to folds. Fig. 5 depicts the steps described here in a flow chart.

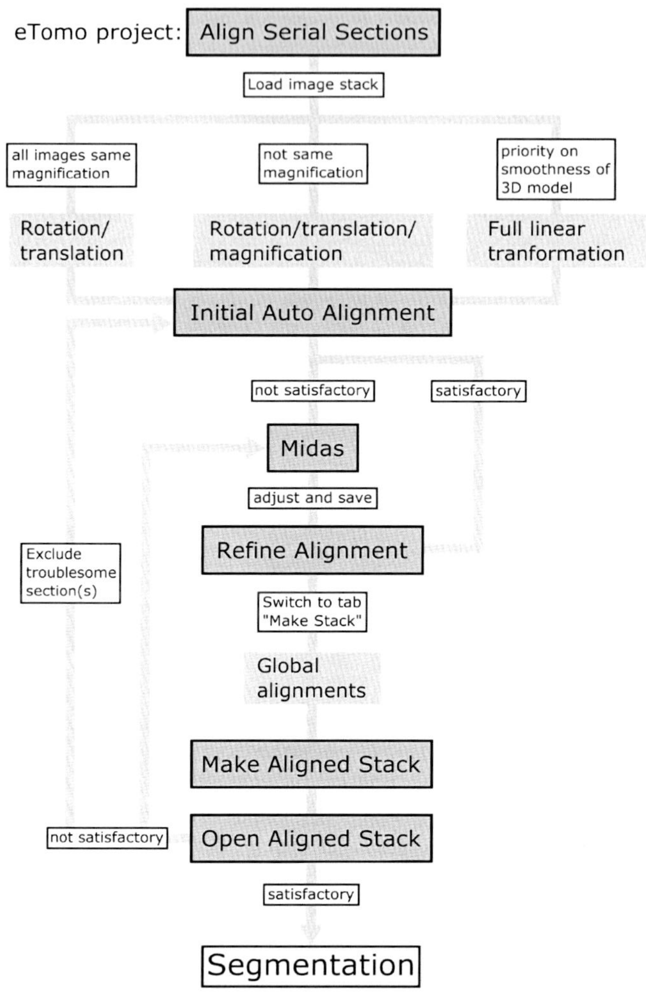

FIGURE 5

Flow chart of the steps necessary to align an image stack of serial scanning electron microscopy sections with eTomo (IMOD).

1.9.2 Fiji and AMIRA

We use AMIRA 6.0 (FEI, Visualization Sciences Group) for the reconstruction of nervous tissue of social insects. In this case the SEM images are aligned as a separate stack using the TrackEM plugin of Fiji (Schindelin et al., 2012). Afterward, the image stack of correlated SIM and SEM images is created similarly as

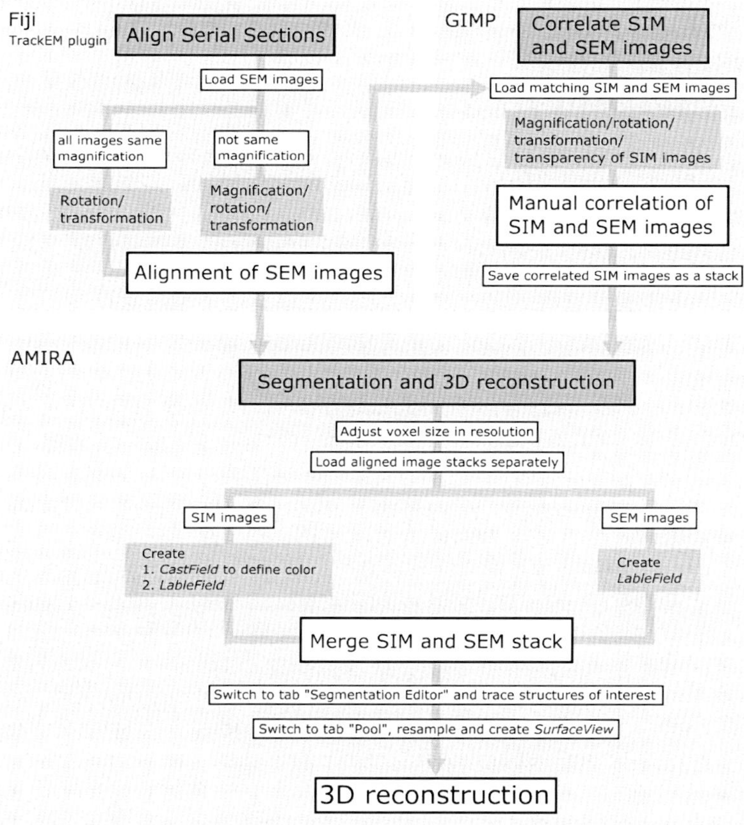

FIGURE 6

Flow chart of the workflow for alignment, correlation, and 3D reconstruction of scanning electron microscopy and structured illumination microscopy imaging data using a combination of Fiji, GIMP, and AMIRA.

described in Section 1.8 using GIMP. Both stacks are then loaded to AMIRA separately. For segmentation, attach "LableField" modules (use "CastField" function for fluorescence images) to the image stack modules and switch to the "Segmentation Editor" tab. Fluorescent signals in the SIM image stack as well as the ultrastructure in the SEM image stack can be traced by assigning voxels to individual "Materials." These materials in turn can be used for 3D visualizations and quantifications (Fig. 6).

2. METHODS—MODEL-SPECIFIC ADAPTATIONS AND CONSIDERATIONS

2.1 *CAENORHABDITIS ELEGANS*

The young adult hermaphrodite of *C. elegans* is only ~80 μm in diameter and thus can be high-pressure frozen living and intact. Make sure that the worm population is well fed for at least three generations and select young adults for freezing. If unsure, pick L4-larvae the day before on a separate plate. Any other stage, including *dauer*, can also be frozen with good results. We place the 100 μm recess platelet on the agar of the worm plate, just outside of the bacterial lawn. By using a stereomicroscope with light sources from below and above, it is possible to see the worms on the plate as well as in the platelet. We overfill the platelet with freeze protectant/filler solution. For *C. elegans* 10% BSA solution in M9 buffer works well. A bacterial paste made from the worms' feeding bacteria (just resuspend a pellet of bacteria in a very small volume of 10% BSA solution) can also be recommended. For an overview of fillers and freeze protectants for high-pressure freezing see McDonald et al. (2010).

Usually 10–30 worms are placed into the solution with a worm pick. The lid (0 μm recess) is placed on top. A little bit of liquid should flow out during lid placement. This insures that the platelet is filled completely. It is important that there are no air bubbles in the platelet cavity or freezing quality will deteriorate. The sandwich is then immediately transferred to the high-pressure freezing machine for cryoimmobilization. *C. elegans* samples do not require any special treatment during freeze substitution and embedding. However, since the worms are small, it is advisable to use a stereomicroscope whenever possible to prevent loss of specimens. We usually embed the whole high-pressure frozen and substituted pellet or, if it broke during the process, all pieces of it. If desired, the worms can also be removed from the pellet using very thin needles. Some will break, but if they are handled very carefully, individual intact worms can be freed completely from the surrounding material and embedded separately.

For correlation, DNA staining is especially useful in *C. elegans*, since nuclei are present in almost every section and many of them show distinct patterns of heterochromatin that can be precisely matched in EM and light microscopic images. Depending on the tissue of interest, mitochondria, microtubules, or lipid droplets are also valuable candidates for correlation.

Two application examples of superresolution AT applied to *C. elegans* young adult hermaphrodites are shown in Fig. 7. The first example (Fig. 7 A–E) showcases a nuclear staining. It becomes apparent, how precisely the superresolved heterochromatin signal of a Hoechst staining can be correlated to the ultrastructure. The staining of the nuclear lamina, as well as the nuclear pore complexes, fits perfectly onto the nuclear membrane and shows the expected alternating pattern, where the lamina is discontinued around the pores.

The second example (Fig. 7 F–K) illustrates how a staining against a fluorescent protein tag can be a valuable alternative in absence of good direct antibodies. By using a standard antibody against GFP in an already established worm line expressing a UNC-7::GFP fusion protein, a gap junction in the ventral nerve cord containing the innexin UNC-7 can be readily identified by CLEM. Due to the high resolution of SIM and the precision of the heterochromatin-assisted correlation, potentially all such gap junctions in the worm can be mapped with high confidence (Markert et al., 2016). Because of the ease of sample handling and preparation for high-pressure freezing and the vast genetic toolbox available for this model, *C. elegans* is well suited for our described CLEM approach, using superresolution AT.

2.2 *TRYPANOSOMA BRUCEI*

African trypanosomes, the causative agents of the deadly sleeping sickness, are unicellular blood parasites. The flagellate protozoa are an interesting cell biological model system because many basic cellular processes, ranging from gene expression to cell division are influenced by the parasitic lifestyle. For microscopists, *Trypanosoma brucei* is a very attractive specimen, simply due to its tidy cell structure. The cell is highly polarized, with endocytosis and membrane recycling restricted to a small invagination at the posterior pole. Most major organelles are present in single copies, and their location within the cell is well conserved. A dense subpellicular microtubule corset supports the plasma membrane, which is covered with an impervious layer of mainly one protein, i.e., the variant surface glycoprotein (VSG). The mitochondrial genome is condensed and forms the characteristic kinetoplast. Furthermore, a complete molecular genetics toolbox is available and all kinds of "omics" have been conducted (Alsford et al., 2012; Dejung et al., 2016; Mony et al., 2014). A project to label each trypanosome protein with a fluorescent tag is well advanced (http://tryptag.org). In addition, the ultrastructure of the parasites has been described in detail (Hughes, Borrett, Towers, Starborg, & Vaughan, 2017). Thus, trypanosomes are an ideal object for establishing CLEM protocols.

For high-pressure freezing of *T. brucei* it is important to obtain a dense pellet of cells. Cells are harvested from a suspension culture. We centrifuge at least 2×10^7 cells at room temperature for 3 min at $750 \times g$. The parasites are carefully resuspended in 10 mL of HMI9-medium, containing 50% fetal calf serum [(FCS) fetal calf serum, from Sigma-Aldrich in this case] for freeze protection. Following centrifugation at $750 \times g$ for 3 min, the trypanosomes are resuspended in 200 μL of HMI9-medium, containing 50% FCS, and transferred to a 200 μL PCR tube. The sample is centrifuged in a microfuge (labnet) at $2{,}000 \times g$ for 5 s. The supernatant is removed and 2 μL of the cell pellet is transferred into the 100 μm recess platelet, which is covered with a 0 μm recess lid and immediately processed. All subsequent steps follow the core protocol.

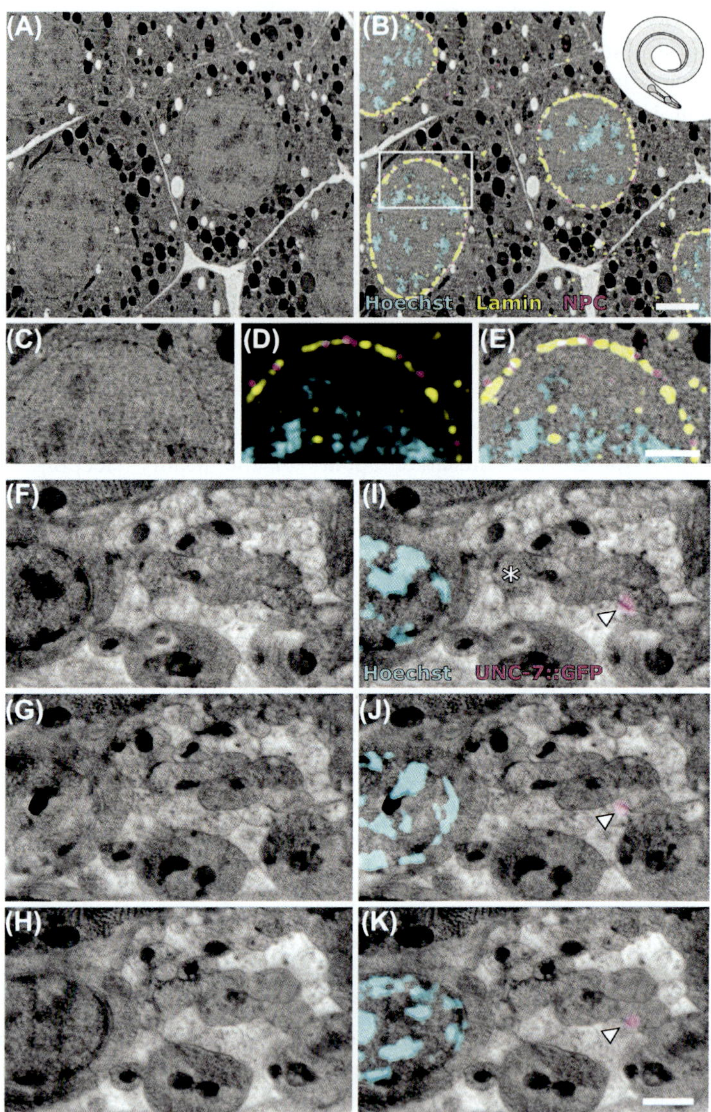

FIGURE 7

Superresolution Array Tomography on ultrathin sections of *Caenorhabditis elegans* samples. (A) 100 nm-LR White section (thermally cured) of an early embryo in utero imaged with the LABE detector of a JEOL JSM-7500F scanning electron microscope (SEM). (B) Same images as in (A) overlaid with three structured illumination microscopy (SIM) fluorescence channels. Lamin (yellow), nuclear pore complex (NPC) (magenta), and heterochromatin (cyan) are stained. Scale bar: 2 μm. (C–E) Detail indicated in (B) with SEM and fluorescence channels shown separately (C and D), as well as overlaid (E). Scale bar: 1 μm. (F–H) Consecutive 100 nm-LR White sections (UV cured) of the *C. elegans* ventral nerve cord of a young adult

Unlike tissues and larger organisms, trypanosomes are small (20 × 4 μm) and grow in suspension. Thus, the cells are randomly positioned in the sample and present in all possible orientations. Therefore, it is important during imaging to choose an area of the resin sample that contains distinctive features that can easily be identified in the SEM.

The use of SIM and *d*STORM for trypanosome CLEM is possible and rewarding. As a proof-of-principle example, a costaining of the VSG surface coat and the underlying microtubule cytoskeleton is shown in Fig. 8. SIM clearly reveals the expected location of fluorescent signals, with tubulin on the cytoplasmic site and VSG on the plasma membrane. As the endosomal system in trypanosomes is largely one extensive structure, CLEM is needed to resolve subcompartments. Thus, African trypanosomes are excellent model cells for CLEM, and CLEM is the perfect technique for studying the structure–function relationships of organelles in trypanosomes.

2.3 SOCIAL INSECTS (*APIS MELLIFERA* AND *CATAGLYPHIS FORTIS*)

The following describes model-specific protocols and considerations for neuronal tissue of the European honeybee, *Apis mellifera carnica*, and the desert ant, *Cataglyphis fortis*. These protocols should be applicable for most (social) insect species. Social insects exhibit a remarkable neuronal plasticity associated with development, maturation, division of labor, aging, as well as learning and memory processes. Using social insects as models to investigate mechanisms underlying neuronal plasticity, CLEM offers fantastic new opportunities to analyze structural changes within the synaptic network of various brain centers and, at the same time, for precise localization of involved molecular components.

Since most social insects are too large for high-pressure freezing as a whole, neuronal tissue needs to be dissected first. For dissection, *A. mellifera* or *C. fortis* are immobilized on ice, harnessed (e.g., in plastic tubes or custom-made acryl glass holders) and the head is fixed using soft dental wax (Surgident, Sigma Dental Systems). To gain easy access to the brain tissue the antennae are removed and a rectangular-shaped window is cut into the head capsule between the base of the antennae, the ocelli, and the complex eyes. Hereafter, all glands, muscles, and trachea are removed and the brain can be detached from the head capsule. If necessary, ant or bee physiological saline solution is applied during dissection (for ants see Stieb, Muenz, Wehner, & Rössler, 2010; for bees see Groh, Lu, Meinertzhagen, & Rössler, 2012). *C. fortis* brains are transferred immediately after

◀

hermaphrodite imaged with the same SEM as (A). (I–K) Same images as in (F–H) overlaid with two SIM fluorescence channels. A staining for UNC-7::GFP (magenta) reveals the location of a gap junction (arrowhead). Additionally, heterochromatin (cyan) is stained. Asterisk in (I) indicates a chemical synapse with dense projections. Scale bar: 500 nm.

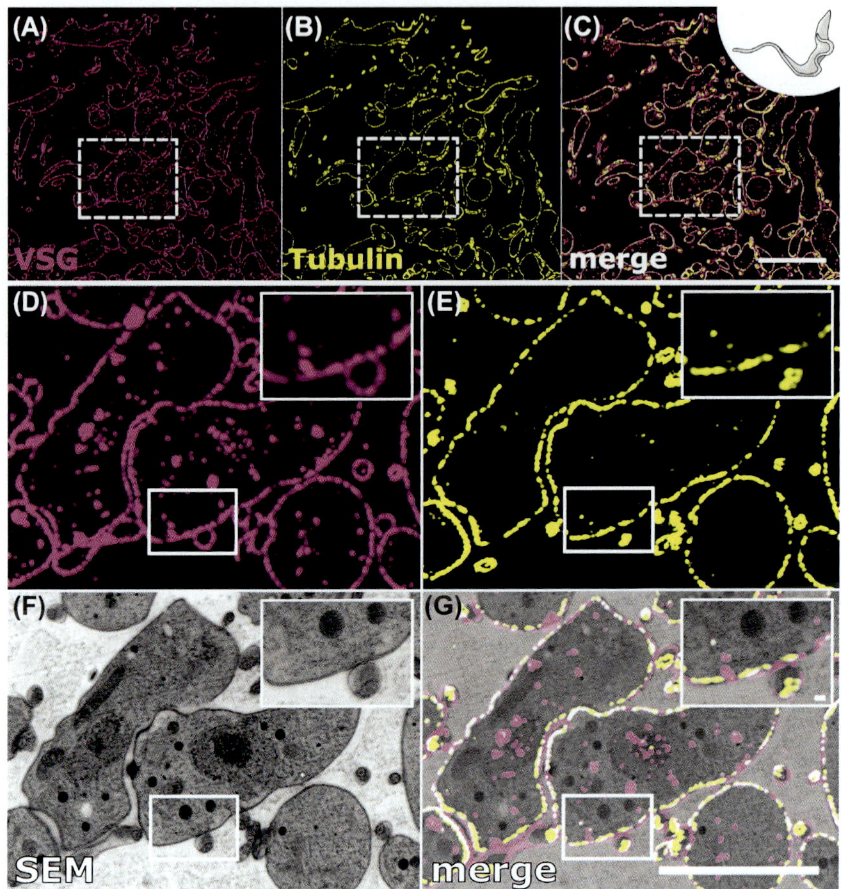

FIGURE 8

Immunohistochemistry staining against the VSG MITat1.1 and tubulin in *Trypanosoma brucei* embedded in LR White. The *T. brucei* strain Lister 427 expressing the variant surface glycoprotein (VSG) MITat1.1 was stained for VSG and tubulin and imaged with structured illumination microscopy (A–E). (A) VSG localization shown in magenta. (B) Tubulin staining shown in yellow. (C) Merged image of (A) and (B). (D, E) Magnification of boxed area shown in (A) and (B), respectively. Top right insets show enlargements of the boxed area in the respective images. (F) The same sections were imaged with a JEOL JSM-7500F scanning electron microscope with LABE detector and (G) correlated with both light microscopy images (D, E). Scale bars in (A), (B), and (C): 10 μm; in (D–G): 5 μm; in the inlays in (D–G): 200 nm.

dissection as a whole to the lecithin-coated platelets (200 μm depth) as illustrated in the core protocol, using hexadecene as filler. In contrast, the larger *A. mellifera* brain tissue needs further processing to fit the size of the high-pressure freezing platelets. Therefore, a 1% formaldehyde solution (methanol free, 28908, Fisher

Scientific) in phosphate-buffered saline (PBS, pH 7.2) is applied for 20 min onto the brain after opening the head capsule and before removing the brain to facilitate a chemical prefixation process. The fixed brain is now suitable for embedding in a droplet of low-melting point agarose (Agarose type II, Amresco) on a precooled metal slide. Make sure to work quickly when adjusting the position of the brain in the rapidly curing agarose. Afterward, the brain tissue in the cured agarose droplet is trimmed on ice to the region of interest and adjusted to the size of the platelets. The resulting tissue block can be glued to a metal disc and cut in 90 μm-thick sections in ice-cold PBS using a vibrating microtome (Leica VT 1000S, Leica Microsystems). Individual sections containing regions of interest are selected and carefully transferred to the platelets (100 μm depth), again using hexadecene as filler. When closing the "freezing sandwich" it is important to take special care that the tissue is not squeezed into the platelets or lifts off the lid. For LR White embedding, ultramicrotomy, fluorescent labeling, and both imaging steps no special adaptations to the core protocol are necessary.

The combination of AT and superresolution microscopy is a very promising tool for the use in (social) insect (nervous) tissues. Particularly the use of chemical prefixation and agarose sectioning prior to high-pressure freezing opens this technique to a huge variety of different tissue applications and possible questions to be addressed. For example, it will be possible to analyze different subregions of the brain (or other tissue) under experimental conditions reflecting the in vivo status of individuals. The increased resolution of protein detection combined with simultaneous detection of ultrastructural features will be key to gain further access to molecular mechanisms underlying synaptic plasticity and changes in characteristic neuronal microcircuits, such as the highly redundant microglomerular synaptic complexes in the mushroom bodies, a prominent multimodal sensory integration center in the insect brain involved in learning, spatial orientation, and memory processes (for review see Fahrbach, 2006) (Fig. 9). Transgenic tools are not yet available in social insects, but the broad access to working antibodies offers a great potential for future studies, even though each antibody has to be tested and adapted for the use on LR White-embedded ultrathin sections. In addition, fluorescently labeled neuronal tracers and other molecular markers can be tested in the near future. Depending on the question (e.g., detailed characterization of active zones or analyzing individual vesicles) the degree of structure preservation and resolution obtained from SEM imaging for ultrastructural morphology emerged as a crucial factor that needs further adjustments. However, the enormous potential of CLEM to combine quantitative ultrastructural analyses with superresolution localization of specific molecules such as multi protein localization and colocalization studies will significantly advance our understanding of mechanisms underlying pre- and postsynaptic neuronal plasticity, their role in functional adjustments in neuronal circuits, and their role in behavioral plasticity.

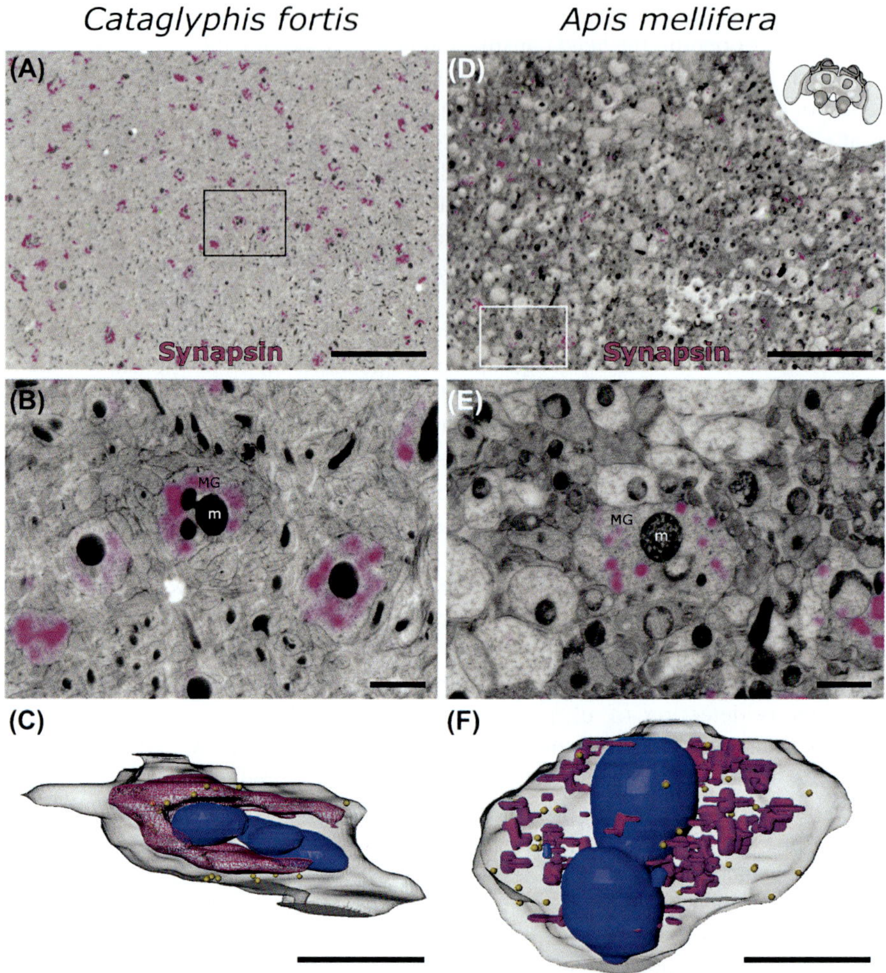

FIGURE 9

Microglomeruli-based synaptic architecture of the mushroombody calyx in *Cataglyphis fortis* (A–C) and *Apis mellifera* (D–F). LR White-embedded mushroom body tissue was stained for the synaptic vesicle–associated protein synapsin (magenta) and imaged with superresolution structured illumination microscopy. The same sections were imaged with scanning electron microscopy and correlated with the light microscopy images. Correlated composite images show presynaptic boutons of individual microglomeruli (MG), which can be clearly identified by morphology and anti-synapsin fluorescence labeling (A, B, D, E). Higher magnifications reveal large mitochondria (m) inside the bouton surrounded by synapsin positive labeling, electron dense active zones on the cell membrane, and few dense core vesicles (B, E). Reconstructions based on information derived from both ultrastructure and fluorescence images depict the 3D architecture of a microglomerulus complex synapse including precise synapsin localization (C, F). Scale bars in (A) and (D): 10 μm, in (B) and (E): 1 μm, in (C) and (F): 750 nm.

3. INSTRUMENTATION AND MATERIALS

3.1 HIGH-PRESSURE FREEZING

3.1.1 General

Instrumentation: High-pressure freezing machine EM HPM100 (Leica Microsystems).
Materials: Freezing platelets type A and type B (Leica Microsystems).
Reagents: Lecithin, chloroform, ethanol (absolute) for the HPM100.

3.1.2 Caenorhabditis elegans

Instrumentation: Stereomicroscope, worm pick made from platinum wire mounted on a Pasteur pipet, alcohol burner, and centrifuge for 50 mL plastic tubes.
Materials: *C. elegans* culture, overnight culture of *Escherichia coli* in LB medium, 50 mL plastic tubes, 1.5 mL plastic tubes, cryoprotectant bacteria paste (resuspend a pellet of *E. coli* bacteria in a very small volume of 10% BSA solution).
Reagents: LB medium, worm buffer M9 (22 mM potassium phosphate monobasic (KH_2PO_4), 19 mM NH_4Cl, 48 mM sodium phosphate dibasic (Na_2HPO_4), 9 mM NaCl), 10% BSA solution in M9.

3.1.3 Trypanosoma brucei

Instrumentation: Centrifuge for 50 mL plastic tubes, microfuge (labnet).
Materials: *T. brucei* suspension culture, 50 mL plastic tubes, 200 μL plastic tubes.
Reagents: FCS (from Sigma-Aldrich in this case), HMI9-medium (Hirumi & Hirumi, 1989).

3.1.4 Cataglyphis fortis *and* Apis mellifera

Instrumentation: Vibrating microtome (Leica VT 1000S, Leica Microsystems), dissecting microscope.
Materials: Ice, harness for bees or ants (e.g., plastic tubes or custom made acryl glass holders), dental wax (Surgident, Sigma Dental Systems), scalpel, heater for melting agarose.
Reagents: PBS (phosphate-buffered saline, pH 7.2), physiological saline (for ants see Stieb et al., 2010; for bees see Groh et al., 2012), 1% formaldehyde solution (methanol free, 28908, Fisher Scientific), low-melting point agarose (Agarose type II, Amresco), hexadecene as filler for platelets.

3.2 FREEZE SUBSTITUTION

Instrumentation: Automated freeze substitution system EM AFS2 (Leica Microsystems), Bunsen burner.
Materials: Metal washing containers with bottom discs and flow-through plastic capsules (Leica Microsystems), thin needle or syringe mounted with tape or glue to a Pasteur pipet, small glass vials with lid, Pasteur pipets.

Reagents: $KMnO_4$, anhydrous acetone (EM grade), ethanol (96% is sufficient), LR White Medium Grade Acrylic Resin, (London Resin Company Ltd.) including accelerator.

3.3 EMBEDDING

Instrumentation: For UV curing: UV lamp attachment to AFS2 or separate UV lamp for LR White polymerization, for thermal curing: incubator at 48–52°C.
Materials: Gelatin embedding capsules, LR White Medium Grade Acrylic Resin, (London Resin Company Ltd.).

3.4 ULTRAMICROTOMY

Instrumentation: Ultramicrotome Leica EM UC7 (Leica Microsystems), histo Jumbo diamond knife (DiATOME), incubator at 50°C.
Materials: Glue (Pattex Gel Compact), black pigment (Spinell Black 47400, Kremer Pigmente, Aichstetten, Germany), thin mounted needles or syringes, poly-L-lysine coated slides (Polysine, Thermo Fisher).

3.5 LIGHT MICROSCOPY

Instrumentation: Structured illumination microscope Elyra S.1 (Zeiss), table top centrifuge.
Materials: Hydrophobic pen (PAP pen) or PDMS polymer chamber, humid box such as a StainTray staining system (Sigma-Aldrich), high-precision coverslips (Carl Roth).
Reagents: Tris buffer (50 mM Tris in ddH_2O, pH 7.6), glycine solution (50 mM glycine in 50 mM Tris buffer), blocking solution (0.1% BSA and 0.05% Tween 20 in Tris buffer), Live Hoechst 33342 (Sigma-Aldrich), Mowiol.

3.5.1 Primary Antibodies

- *C. elegans*: Lamin: polyclonal, guinea pig, kind gift from Georg Krohne, University of Würzburg, Germany; nuclear pore complexes: Mab414, monoclonal, mouse, Abcam, product number: ab24609.
- *T. brucei*: VSG: anti-MITat1.1 [Molteno Institute Trypanozoon Antigen Type 1.1 (Cross, 1975)], polyclonal, rabbit, kind gift from M. Carrington, Cambridge, UK; tubulin: anti-α-tubulin, monoclonal, mouse, Sigma–Aldrich, product number: T5168.
- *C. fortis* and *A. mellifera*: Synapsin: SYNORF1, monoclonal, mouse, kind gift from E. Buchner, University of Würzburg, Germany.

3.6 CONTRASTING AND CARBON COATING

Instrumentation: Carbon coater Med 010 (Balzers Union).
Materials: Diamond pen, tweezers, silver paint, SEM specimen holder stubs, carbon stickers for specimen holders.

Reagents: decocted ddH_2O, 2.5% uranyl acetate in ethanol (96%), 50% Reynolds' lead citrate (Reynolds, 1963) in decocted ddH_2O, sodium hydroxide pellets for CO_2 absorption.

3.7 SCANNING ELECTRON MICROSCOPY

Instrumentation: Field emission scanning electron microscope JSM-7500F (JEOL) with LABE detector.
Imaging parameters: 5 kV acceleration voltage, 300 pA probe current, 6–8 mm working distance.

3.8 IMAGE PROCESSING AND CORRELATION OF STRUCTURED ILLUMINATION MICROSCOPY AND SCANNING ELECTRON MICROSCOPY IMAGES

Software: Vector graphics editor Inkscape (version 0.91; http://www.inkscape.org), image editing software GIMP (http://www.gimp.org), ec-CLEM plugin for the software Icy (http://icy.bioimageanalysis.org/plugin/ec-CLEM#documentation), Fiji (ImageJ) (Schindelin et al., 2012).

3.9 ALIGNMENT AND 3D RECONSTRUCTION

Software: 3D reconstruction software IMOD (Kremer et al., 1996; details can be retrieved from http://bio3d.colorado.edu), Fiji (ImageJ) (Schindelin et al., 2012), AMIRA 6.0 (FEI, Visualization Sciences Group).

CONCLUSIONS

With this superresolution AT protocol we provide a valuable and versatile tool for answering challenging biological questions, such as the superresolved localization of proteins of interest in their ultrastructural context in 3D. We show some application examples for four models, but superresolution AT should be applicable to almost any model and tissue. One of the most crucial parts is achieving good ultrastructural preservation. High-pressure freezing and freeze substitution offers superior preservation in many systems (McDonald & Auer, 2006), but it is time-consuming and might not work well for certain large samples. However, high-pressure freezing might not be necessary for answering particular biological questions.

The potentially most crucial part of our approach is to obtain proper staining for light microscopy. Good antibodies are hard to come by and might not work on plastic sections due to potential modifications of the epitope. However, testing antibodies on a few sections beforehand is easy and fast, so feasibility of a study will become apparent very quickly. Alternatively, a protein of interest might be tagged with fluorescent proteins or other tags, such as His-tags, for which commercial and well established antibodies exist. And of course, this method is not restricted to antibodies. Notably, we recently also established a superresolution AT/CLEM protocol for RNA in situ hybridization (Jahn et al., 2016).

With the AT protocol repetitive rounds of staining and destaining can be incorporated (Micheva & Smith, 2007). For the application examples that we present here with our superresolution AT protocol a single round of staining was sufficient, since we used at most three different fluorophores at a time. We tried additional destaining and restaining steps according to the original AT publication (Micheva & Smith, 2007). This worked also for our superresolution AT protocol, but as reported by Collman et al. (2015), we observed that the quality of the ultrastructure is affected. Therefore, if the biological application allows, we try to avoid these antibody elution steps.

Here, we apply SIM as the superresolution technique of choice, since it is easy to use and does not require any specific sample preparation nor specially adapted fluorophores. If even higher lateral resolution is necessary, superresolution AT can also be combined with direct stochastic optical reconstruction microscopy (*d*STORM) (Markert et al., 2016), which offers a spatial resolution of ~20 nm (Galbraith & Galbraith, 2011; van de Linde et al., 2011).

A higher resolved fluorescence signal also requires a more precise correlation with the ultrastructure. To ensure this, some rely on special fiducials, such as quantum dots (Kukulski et al., 2011; Nisman, Dellaire, Ren, Li, & Bazett-Jones, 2004) or special beads such as gold nanoparticles (Watanabe et al., 2011). However, such fiducials might be expensive, difficult to apply, offering low contrast in EM, or they might become dislocated in between imaging steps (Watanabe et al., 2011). As an alternative, we use intrinsic landmarks for the correlation to avoid such problems (Löschberger, Franke, Krohne, Linde, & Sauer, 2014; Markert et al., 2016). Any structure that can be stained with fluorophores for light microscopy and visualized in the SEM image potentially can be used as a landmark for correlation of light microscopic and EM images. Since the landmark signals are superresolved too, they can be matched to their ultrastructure very precisely, thus allowing for an accurate correlation of the signal of interest (Markert et al., 2016).

For correlation and 3D reconstruction a plethora of software exists. We generally would recommend to use free and open source software such as Fiji and Inkscape, as their functions are transparent and reproducible. IMOD in particular has a very active community of users and program updates and new features are published regularly. The correlation workflow using Inkscape and GIMP presented here works well for small datasets, but for bigger projects we recommend trying semiautomated correlation software. Such software is still emerging, but the ec-CLEM plugin for Icy that we mentioned already works very well, although it is still in beta version as of now.

We want to encourage the interested reader to try out different approaches and to speak with a local EM expert, if results are not satisfactory right away. A very helpful collection of further hints for high-pressure freezing is published in McDonald et al. (2010). With these additional resources at hand, it should be possible to funnel even very challenging samples into the here presented superresolution AT workflow.

ACKNOWLEDGMENTS

This work was supported by the Bundesministerium für Bildung und Forschung (BMBF) Grant No. 13N12781 (MS), by the German Research Foundation (DFG), Collaborative Research Center SFB1047 "Insect Timing" (Project B6 to WR), by a "Messreise" grant of the Deutsche Gesellschaft für Elektronenmikroskopie (SB), and by a PhD grant from the Studienstiftung des Deutschen Volkes (SMM). We cordially thank H. Schwarz, E. Meyer-Natus, M. Soiza-Reilly, M. Lang, J.-L. Bessereau, C. Luccardini, H. Zhan, S. Proppert, G. Krohne, M. Zhen, D. Holmyard, B. Mulcahy, D. Witvliet, and M. Behringer for experimental support and/or fruitful discussions throughout the project. For antibodies we thank M. Carrington, G. Krohne, M.-C. Dabauvalle, and E. Buchner.

We further thank C. Gehrig, B. Trost, and D. Bunsen for excellent technical support.

REFERENCES

Albrecht, U., Seulberger, H., Schwarz, H., & Risau, W. (1990). Correlation of blood−brain barrier function and HT7 protein distribution in chick brain circumventricular organs. *Brain Research, 535*(1), 49−61.

Alsford, S., Eckert, S., Baker, N., Glover, L., Sanchez-Flores, A., Leung, K. F., … Horn, D. (2012). High-throughput decoding of antitrypanosomal drug efficacy and resistance. *Nature, 482*(7384), 232−236.

Camilli, P. D., Cameron, R., & Greengard, P. (1983). Synapsin I (protein I), a nerve terminal-specific phosphoprotein. I. Its general distribution in synapses of the central and peripheral nervous system demonstrated by immunofluorescence in frozen and plastic sections. *The Journal of Cell Biology, 96*(5), 1337−1354.

Collman, F., Buchanan, J., Phend, K. D., Micheva, K. D., Weinberg, R. J., & Smith, S. J. (2015). Mapping synapses by conjugate light-electron array tomography. *The Journal of Neuroscience, 35*(14), 5792−5807.

Cross, G. A. M. (1975). Identification, purification and properties of clone-specific glycoprotein antigens constituting the surface coat of *Trypanosoma brucei. Parasitology, 71*(3), 393−417.

Dejung, M., Subota, I., Bucerius, F., Dindar, G., Freiwald, A., Engstler, M., … Janzen, C. J. (2016). Quantitative proteomics uncovers novel factors involved in developmental differentiation of *Trypanosoma brucei. PLoS Pathogens, 12*(2), e1005439.

Fahrbach, S. (2006). Structure of the mushroom bodies of the insect brain. *Annual Review of Entomology, 51*(1), 209−232.

Fialka, I., Schwarz, H., Reichmann, E., Oft, M., Busslinger, M., & Beug, H. (1996). The estrogen-dependent c-JunER protein causes a reversible loss of mammary epithelial cell polarity involving a destabilization of adherens junctions. *The Journal of Cell Biology, 132*(6), 1115−1132.

Galbraith, C. G., & Galbraith, J. A. (2011). Super-resolution microscopy at a glance. *Journal of Cell Science, 124*(10), 1607–1611.

Groh, C., Lu, Z., Meinertzhagen, I. A., & Rössler, W. (2012). Age-related plasticity in the synaptic ultrastructure of neurons in the mushroom body calyx of the adult honeybee *Apis mellifera. The Journal of Comparative Neurology, 520*(15), 3509–3527.

Gustafsson, M. G. L. (2000). Surpassing the lateral resolution limit by a factor of two using structured illumination microscopy. *Journal of Microscopy, 198*(2), 82–87.

Haraguchi, C. M., & Yokota, S. (2002). Immunofluorescence technique for 100-nm-thick semithin sections of Epon-embedded tissues. *Histochemistry and Cell Biology, 117*(1), 81–85.

Herken, R., Fussek, M., Barth, S., & Götz, W. (1988). LR-White and LR-Gold resins for post-embedding immunofluorescence staining of laminin in mouse kidney. *The Histochemical Journal, 20*(8), 427–432.

Hirumi, H., & Hirumi, K. (1989). Continuous cultivation of *Trypanosoma brucei* blood stream forms in a medium containing a low concentration of serum protein without feeder cell layers. *The Journal of Parasitology, 75*(6), 985–989.

Hughes, L., Borrett, S., Towers, K., Starborg, T., & Vaughan, S. (2017). Patterns of organelle ontogeny through a cell cycle revealed by whole cell reconstructions using 3D electron microscopy. *Journal of Cell Science, 130*(3), 637–647.

Jahn, M. T., Markert, S. M., Ryu, T., Ravasi, T., Stigloher, C., Hentschel, U., & Moitinho-Silva, L. (2016). Shedding light on cell compartmentation in the candidate phylum Poribacteria by high resolution visualisation and transcriptional profiling. *Scientific Reports, 6*, 35860.

Kremer, J. R., Mastronarde, D. N., & McIntosh, J. R. (1996). Computer visualization of three-dimensional image data using IMOD. *Journal of Structural Biology, 116*(1), 71–76.

Kukulski, W., Schorb, M., Welsch, S., Picco, A., Kaksonen, M., & Briggs, J. A. G. (2011). Correlated fluorescence and 3D electron microscopy with high sensitivity and spatial precision. *The Journal of Cell Biology, 192*(1), 111–119.

Kurth, T., Schwarz, H., Schneider, S., & Hausen, P. (1996). Fine structural immunocytochemistry of catenins in amphibian and mammalian muscle. *Cell and Tissue Research, 286*(1), 1–12.

van de Linde, S., Löschberger, A., Klein, T., Heidbreder, M., Wolter, S., Heilemann, M., & Sauer, M. (2011). Direct stochastic optical reconstruction microscopy with standard fluorescent probes. *Nature Protocols, 6*(7), 991–1009.

Löschberger, A., Franke, C., Krohne, G., van de Linde, S., & Sauer, M. (2014). Correlative super-resolution fluorescence and electron microscopy of the nuclear pore complex with molecular resolution. *Journal of Cell Science, 127*(20), 4351–4355.

Markert, S. M., Britz, S., Proppert, S., Lang, M., Witvliet, D., Mulcahy, B., … Stigloher, C. (2016). Filling the gap: Adding super-resolution to array tomography for correlated ultrastructural and molecular identification of electrical synapses at the *C. elegans* connectome. *Neurophotonics, 3*(4), 041802.

McDonald, K., Schwarz, H., Müller-Reichert, T., Webb, R., Buser, C., & Morphew, M. (2010). Chapter 28-"Tips and Tricks" for high-pressure freezing of model systems. In T. Müller-Reichert (Ed.), *Methods in cell biology: Vol. 96. Electron microscopy of model systems* (pp. 671–693). Academic Press.

McDonald, K. L., & Auer, M. (2006). High-pressure freezing, cellular tomography, and structural cell biology. *Biotechniques, 41*(2), 137–139.

Micheva, K. D., & Smith, S. J. (2007). Array tomography: A new tool for imaging the molecular architecture and ultrastructure of neural circuits. *Neuron, 55*(1), 25–36.

Mony, B. M., MacGregor, P., Ivens, A., Rojas, F., Cowton, A., Young, J., … Matthews, K. (2014). Genome-wide dissection of the quorum sensing signalling pathway in *Trypanosoma brucei. Nature, 505*(7485), 681–685.

Nanguneri, S., Flottmann, B., Horstmann, H., Heilemann, M., & Kuner, T. (2012). Three-dimensional, tomographic super-resolution fluorescence imaging of serially sectioned thick samples. *PLoS One, 7*(5), e38098.

Nisman, R., Dellaire, G., Ren, Y., Li, R., & Bazett-Jones, D. P. (2004). Application of quantum dots as probes for correlative fluorescence, conventional, and energy-filtered transmission electron microscopy. *Journal of Histochemistry & Cytochemistry, 52*(1), 13–18.

Ojeda, J. L., Ros, M. A., & Icardo, J. M. (1989). A technique for fluorescence microscopy in semithin sections. *Stain Technology, 64*(5), 243–248.

Perkovic, M., Kunz, M., Endesfelder, U., Bunse, S., Wigge, C., Yu, Z., … Frangakis, A. S. (2014). Correlative light- and electron microscopy with chemical tags. *Journal of Structural Biology, 186*(2), 205–213.

Reynolds, E. S. (1963). The use of lead citrate at high ph as an electron-opaque stain in electron microscopy. *The Journal of Cell Biology, 17*(1), 208–212.

Schindelin, J., Arganda-Carreras, I., Frise, E., Kaynig, V., Longair, M., Pietzsch, T., … Cardona, A. (2012). Fiji: An open-source platform for biological-image analysis. *Nature Methods, 9*(7), 676–682.

Schwarz, H., & Humbel, B. (2014). Correlative light and electron microscopy using immunolabeled sections. In J. Kuo (Ed.), *Electron microscopy, methods in molecular biology* (pp. 559–592). Humana Press.

Stieb, S. M., Muenz, T. S., Wehner, R., & Rössler, W. (2010). Visual experience and age affect synaptic organization in the mushroom bodies of the desert ant *Cataglyphis fortis. Developmental Neurobiology, 70*(6), 408–423.

Watanabe, S., Punge, A., Hollopeter, G., Willig, K. I., Hobson, R. J., Davis, M. W., … Jorgensen, E. M. (2011). Protein localization in electron micrographs using fluorescence nanoscopy. *Nature Methods, 8*(1), 80–84.

Weimer, R. (2006). Preservation of *C. elegans* tissue via high-pressure freezing and freeze-substitution for ultrastructural analysis and immunocytochemistry. In K. Strange (Ed.), C. elegans, *methods in molecular biology* (pp. 203–221). Humana Press.

CHAPTER

3

Preserving the photoswitching ability of standard fluorescent proteins for correlative in-resin super-resolution and electron microscopy

Errin Johnson[1], Rainer Kaufmann

University of Oxford, Oxford, United Kingdom

[1]*Corresponding author: E-mail: errin.johnson@path.ox.ac.uk*

CHAPTER OUTLINE

Methods in Cell Biology, Volume 140, ISSN 0091-679X, http://dx.doi.org/10.1016/bs.mcb.2017.04.001

Abstract

There are many different correlative light and electron microscopy (CLEM) techniques available. The use of super-resolution microscopy in CLEM is an emerging application and while offering the obvious advantages of improved resolution in the fluorescence image, and therefore more precise correlation to electron microscopy (EM) ultrastructure, it also presents new challenges. Choice of fluorophore, method of fixation, and timing of the fluorescence imaging are critical to the success of super-resolution CLEM and the relative importance, and technical difficulty, of each of these factors depends on the type of super-resolution microscopy being employed. This chapter details the method we developed for in-resin super-resolution CLEM using single molecule localization microscopy (SMLM) with standard fluorescent proteins (e.g., GFP and mVenus). The key to this approach is being able to preserve not only the fluorescence, but also, and more importantly, the photoswitching ability of the fluorescent proteins throughout the EM sample preparation procedure. Cells are cryofixed using high pressure freezing for optimal structural preservation and then freeze substituted in tannic acid, which preserves the photoswitching ability of the fluorescent proteins and is essential for high-quality SMLM imaging. Resin sections are then imaged using SMLM, achieving a structural resolution of 40–50 nm and a localization precision of ~17 nm, followed by transmission electron microscopy. This produces high quality correlative images without the use of specialized fluorescent proteins or antibodies.

INTRODUCTION

Although by no means a new concept (see Hayat, 1987), correlative microscopy is becoming an increasingly powerful tool in biological research, as technological and methodological advances enable a wider range of imaging modalities (e.g., X-ray, light, electron, and atom force microscopy) to be combined and applied to a greater range of samples (reviewed by Caplan, Niethammer, Taylor, & Czymmek, 2011; for a recent example see Karreman et al., 2016). Correlative light and electron microscopy (CLEM) is arguably the most widely used version of the approach, due to the highly complementary features of fluorescence microscopy and electron microscopy (EM), but also the relative accessibility of these techniques and huge array of fluorescent probes available (see Brown & Verkade, 2010; Giepmans, Adams, Ellisman, & Tsien, 2006). CLEM has two distinct applications: (1) to visualize rare events or a specific subset of cells within a larger population using fluorescence microscopy, so that they can be accurately pinpointed at the EM level (e.g., Kobayashi et al., 2012), saving a great deal of microscope time and (2) to localize fluorescently labeled proteins of interest and place them into the ultrastructural context of the EM image (e.g., Peddie et al., 2014). However, the resolution gap of ~200 nm between light microscopy and EM reduces the precision and information content of this correlation, which can often complicate the interpretation of data. Super-resolution microscopy (SRM) overcomes this limitation and can narrow the gap to the 10 nm range (see Wegel et al., 2016 for a practical comparison of super-resolution techniques), turning CLEM into a truly powerful tool for molecular and cellular biology.

Indeed, there are a growing number of studies focusing on super-resolution CLEM, each utilizing a different approach to address the unique challenges associated with combining these two distinct imaging modalities without sacrificing the image quality of either (reviewed by Hauser et al., 2017). The key factors to consider for successful super-resolution CLEM are the mode of fixation and how this affects fluorescence and/or ultrastructure, the choice of fluorophore, which is particularly important for SRM as it requires a high signal to noise ratio, and the stage at which the SRM is performed (i.e., prior to, or post, EM sample preparation).

Chemical fixation has been used in correlative studies with direct stochastic optical reconstruction microscopy (dSTORM), stimulated emission depletion and photoactivated localization microscopy on samples expressing fluorescent proteins optimized for SRM (e.g., Betzig et al., 2006; Kopek, Shtengel, Xu, Clayton, & Hess, 2012) or labeled with Alexa dye–tagged antibodies (e.g., Kim et al., 2015; Löschberger, Franke, Krohne, van de Linde, & Sauer, 2014). However, the alterations to ultrastructure which occur due to chemical fixation, and subsequent dehydration with solvents (Bleck et al., 2010; Kellenberger et al., 1992), are of much greater significance with super-resolution CLEM where the structural resolution at the fluorescence level is up to 10× better than conventional CLEM. As such, the usefulness and quality of the correlation may be improved by instead using cryopreparation techniques, where cells are preserved as close as possible to their native state (McDonald, 2009). For instance, samples cryofixed using plunge freezing with liquid ethane or high pressure freezing (HPF) with liquid nitrogen can either be imaged directly with cryo-SRM followed by cryo-EM (e.g., Chang et al., 2014; Liu et al., 2015), or processed into resin for SRM imaging post-EM processing (Watanabe et al., 2011).

There are several different ways to fluorescently label proteins of interest for super-resolution CLEM. For antibody labeling and subsequent imaging using dSTORM, samples can be permeabilized and immunolabeled prior to EM processing (Kim et al., 2015) or the EM preparation can be modified so that the antigenicity is preserved for postembedding immunolabeling. Another way to avoid permeabilizing the sample to allow the antibody access to the antigen is to use the Tokuyasu cryosectioning approach with chemically fixed samples (e.g., Suleiman et al., 2013), which, though more technically challenging, avoids the use of both detergents and resin embedding altogether, increasing the chance of successful labeling. However, the quality and specificity of the primary antibody is critically important, because low labeling efficiency and high background levels significantly decrease the structural resolution of SRM. As an alternative, genetically encoded fluorescent proteins specifically modified for SRM can be used. These include citrine (Watanabe et al., 2011), mEos2 (Kopek et al., 2012), Dronpa

(Liu et al., 2015), and mEos4a (Paez-Segala et al., 2015). The latter has been engineered to be more tolerant to osmium tetroxide, such that a standard transmission electron microscopy (TEM) prep can be used, and structural preservation and contrast can be improved, without sacrificing fluorescence. A potential drawback of using fluorescent protein fusions is that tagging them to your protein of interest without disrupting its function and/or localization can be challenging.

The timing of SRM imaging in the CLEM procedure is important to consider. With conventional CLEM, the sample is often first imaged with LM and subsequently prepared for EM. This may be technically more straightforward, but the resulting changes, such as extraction and shrinkage (Kopek et al., 2012; Peddie et al., 2014), which occur before imaging the sample with EM can significantly affect the quality of the correlation. This is particularly relevant with SRM, where the structural resolution is well below 100 nm. Therefore, depending on the application, it can be advantageous to perform both the SRM and EM postsample processing. If using fluorescent proteins to label the protein of interest, the challenge then is to maintain not only the fluorescence itself throughout the TEM sample preparation procedure, but also the photoswitching capability of the fluorophore to enable high quality SRM of resin sections, while simultaneously preserving cellular ultrastructure and introducing sufficient contrast for TEM imaging.

Standard fluorescent proteins (e.g., GFP and YFP) have been shown to possess sufficient photoswitching capabilities for single molecule localization microscopy (SMLM; Lemmer et al., 2008) and our aim was to exploit this ability for super-resolution CLEM. Since in-resin GFP and RFP fluorescence can be preserved for CLEM using HPF and freeze substitution (Kukulski et al., 2011; Peddie et al., 2014), we sought to optimize these cryopreparation procedures to preserve the photoswitching of standard fluorescent proteins expressed in mammalian cells for high quality SMLM, followed by TEM for super-resolution CLEM. While choice of cryoprotectant, duration of freeze substitution, and mounting medium for SMLM imaging all affected the quality of in-resin SMLM imaging, the most critical factor was the composition of the freeze substitution medium. We found that the addition of tannic acid to the freeze substitution medium was vital for high-quality SMLM imaging, significantly improving both the single molecule localization accuracy and the photoswitching of the FPs (Johnson et al., 2015). Using this SMLM-optimized sample preparation protocol, it is possible to achieve true super-resolution in the fluorescent images (17 nm average single molecule localization accuracy; 40–50 nm structural resolution) while preserving ultrastructure at the EM level in resin-embedded mammalian cells (Fig. 1).

1. RATIONALE

Our goals were to establish a method for super-resolution CLEM that uses standard fluorescent proteins, so that the technique is widely accessible to the cell biology community, cryofixation for optimal ultrastructural preservation and postembedding

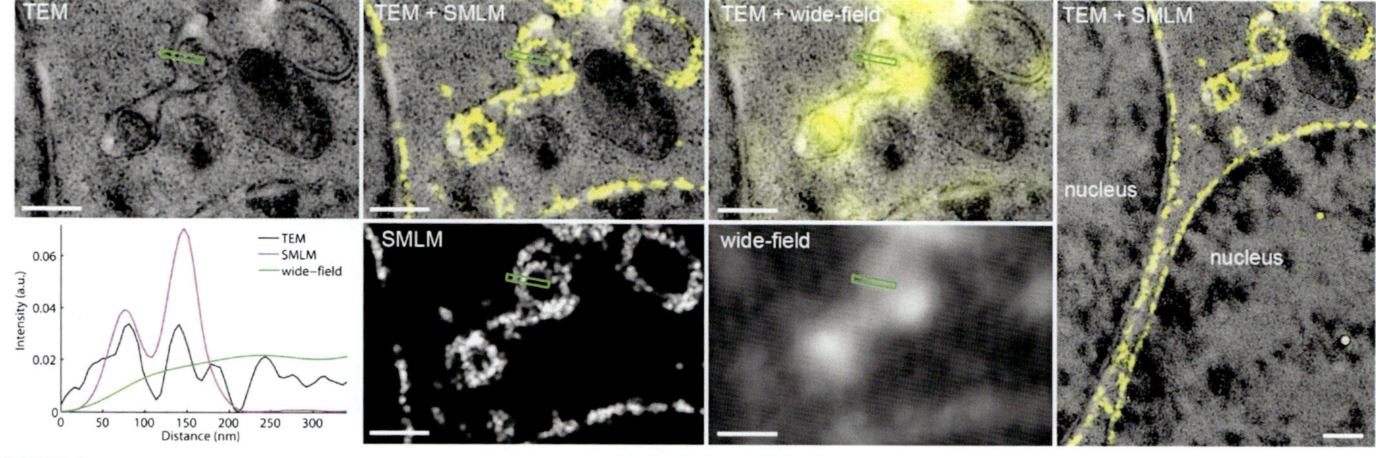

FIGURE 1

Resolution comparison of in-resin fluorescence imaging. ER-localized EphA2-mVenus in HEK293T cells was imaged with SMLM and conventional wide-field fluorescence microscopy than compared to ultrastructure in the corresponding TEM image. Unlike the wide-field image, in the corresponding SMLM image the two membranes (which are ~60 nm apart) are clearly distinguishable and the distribution of fluorescent molecules matches well with the EM ultrastructure. Conventional fluorescence microscopy does not allow to discriminate whether the fluorescent molecules are located on both membranes or not. Scale bars are 500 nm. *EM*, electron microscopy; *ER*, endoplasmic reticulum; *SMLM*, single molecule localization microscopy; *TEM*, transmission electron microscopy.

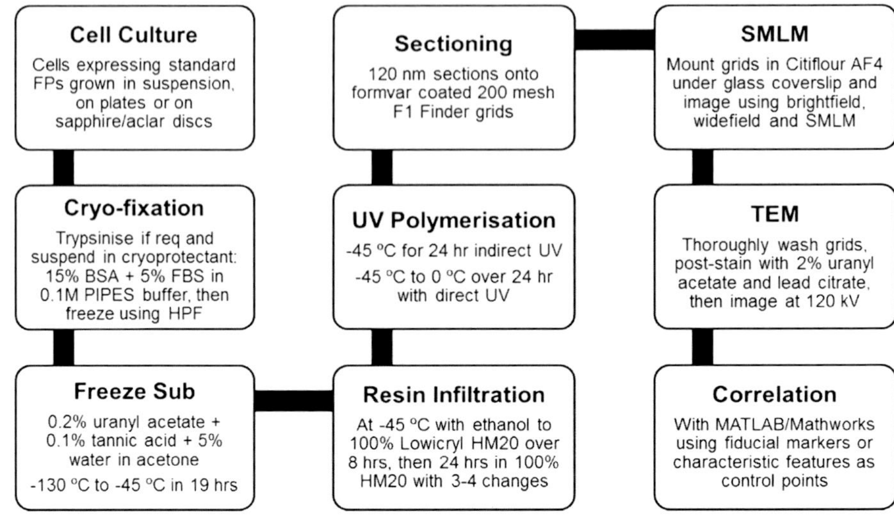

FIGURE 2

Overview of the workflow for correlative in-resin SMLM and TEM. The entire procedure takes about 1.5 weeks to complete. *SMLM*, single molecule localization microscopy; *TEM*, transmission electron microscopy.

SRM for highly precise correlations (see Fig. 2 for an overview of the technique). This chapter provides a detailed description of the method we developed for in-resin correlative SRM and EM using standard fluorescent proteins (Johnson et al., 2015). We highlight the key steps and discuss the applications and limitations of the technique.

2. MATERIALS

2.1 INSTRUMENTATION

1. High pressure freezer (we use the Leica EM PACT2 and Leica EM ICE)
2. Inverted epifluorescence microscope
3. Freeze substitution unit with UV light attachment, hereafter referred to as the AFS2 (Leica)
4. Ultramicrotome
5. Diamond knife (we use a Diatome 45° knife)
6. SMLM microscope capable of acquiring low magnification maps. We use an OMX V2 microscope with a UPlanSApo to 100× 1.4 NA oil objective and an EMCCD camera (Photometrics Evolve 512 Delta) modified for SMLM imaging according to Lemmer et al. (2008), for details see Johnson et al. (2015).
7. TEM. We use an FEI T12 120 kV TEM equipped with a Gatan OneView CMOS camera for digital imaging.
8. Workstation equipped with MATLAB (Mathworks)

2.2 MATERIALS

1. Culture cells expressing GFP, YFP/mVenus, mRuby2, or similar FP (but not mCherry)
2. Leica 1.5 × 0.1 mm membrane carriers (for EM PACT2)
3. Leica aluminum Type A and Type B 3 mm carriers (for EM ice)
4. Consumables for Leica AFS2 (glass bottles, reagent baths, and flow-through rings)
5. Aluminum foil
6. 1.5 mL or 2 mL cryotubes
7. Glass slides
8. Diamond knife
9. 200 mesh copper F1 alphanumeric finder grids (Agar Scientific) with a 0.6% formvar film
10. Perfect loop (Agar Scientific)
11. High precision coverslips no. 1.5 (170 μm thickness)
12. Tweezers: Antimagnetic no. 7, type 5 angled at 45 degrees and flat end
13. Parafilm, 4 in × 250 ft

2.3 CHEMICALS

1. Albumin bovine serum (BSA, Fraction V, ≥98%)
2. Fetal bovine serum (FBS, Sigma)
3. PIPES buffer (Sigma)
4. Trypsin (TryplE Express, Gibco)
5. Acetone (99.6% ACS)
6. Low molecular weight tannic acid (Electron Microscopy Sciences)
7. Uranyl acetate
8. Methanol
9. Pure ethanol
10. Lowicryl HM20 resin kit (Polysciences)
11. Nail polish
12. Ultrapure water
13. Citifluor AF4
14. Lead nitrate
15. Sodium citrate
16. Sodium hydroxide pellets

3. METHODS

3.1 REAGENT PREPARATION

1. *0.2 M PIPES buffer, pH 7.2*: Add 300 mL ultrapure water and 302.37 g PIPES to a 1 L beaker containing a magnetic flea and place on a magnetic stirrer at high

speed. Slowly add 10 M NaOH until the solution turns from cloudy to clear. Adjust the pH to 7.2 and make up to 500 mL final volume with ultrapure water. Store the buffer at 4°C.

2. *Cryoprotectant*: 15% (w/v) BSA + 5% (v/v) FBS in 0.1M PIPES buffer, pH 7.2. Add 2.25 g BSA to a 50 mL centrifuge tube, tap the base to settle the BSA, then add 0.75 mL FBS and make up to 15 mL with 0.1 M PIPES. Tap the base of the tube again and add more PIPES buffer if required. Hold the tube under running warm water for a few minutes, gently inverting until the BSA is dissolved.
3. *10% (w/v) Tannic acid (TA)*: Immediately before use, dissolve 0.025 g low molecular weight TA in 0.25 mL acetone.
4. *5% Uranyl acetate (UA)*: Add 0.5 g UA to 10 mL methanol in a foil wrapped 15 mL centrifuge tube and mix. The solution is stable for months when stored at −20°C.
5. *Freeze substitution medium*: 0.1% TA (v/v) + 5% water (v/v) + 0.2% UA in acetone. For 10 mL, add 100 μL freshly prepared 10% TA, 400 μL 5% UA in methanol stock, 500 μL water, and 9 mL acetone to a 15 mL plastic centrifuge tube and mix well. The solution will be a reddish brown due to the reaction between UA and TA, which may form a precipitate and would not affect the freeze substitution.
6. *Lowicryl HM20 resin*: Transfer a fine balance to the fumehood and weigh the HM20 components into a 50 mL plastic centrifuge tube in the following order: 5.96 g Crosslinker D, 34.04 g Monomer E, and 0.2 g Initiator C. Cap and vigorously shake the resin for approximately 30 s to dissolve the initiator and then store at −20°C until use.
7. *Reynolds lead citrate*: Prepare according to Reynolds (1963). Add 30 mL degassed water (boiled for a few minutes in the microwave and then cooled), 1.33 g lead nitrate ($PbNO_3$) and 1.76 g sodium citrate ($Na_3C_6H_5O_7 \cdot 2H_2O$) to a 50 mL tube. Cap and vigorously shake the tube for 1 min, then shake every 5−10 min for 30 min. During this time, prepare a fresh 1 M sodium hydroxide solution by dissolving 1 g pellets in 25 mL degassed water and inverting gently to dissolve. Add 8 mL of freshly prepared 1 M sodium hydroxide to the lead solution and invert slowly; the solution should turn from milky to clear. Add 12 mL degassed water and gently mix, then warp the tube in foil to protect it from light and store in at 4°C. If a precipitate develops over time, discard it and prepare a fresh solution.
8. *2% UA (aqueous)*: Add 1 g UA and 50 mL ultrapure water to a brown glass bottle, then mix on a magnetic stirrer for 30 min and let it rest overnight before use. Filter the stock through a 0.22 μm syringe filter and store at 4°C. Centrifuge briefly before use to pellet any precipitates that form over time.

3.2 HIGH PRESSURE FREEZING AND FREEZE SUBSTITUTION

Always check the fluorescence in your sample(s) before starting. If the transfection efficiency is lower than expected, the fluorescence is fainter than it should be and/or there is an abnormally high proportion of dead/dying cells, then abort the experiment and start new cultures, as these factors significantly decrease the chance of a successful CLEM experiment.

The protocol we describe here is for suspension cells and for adherent cells grown in a six-well culture plate. For the latter, we lift the cells using trypsin just prior to freezing and pool the cells from up to three wells (depending on cell density) for cryofixation, keeping at least one well as a backup in case there are any problems during the HPF procedure. However, if your fusion protein could potentially be perturbed by trypsinization (i.e., if localized to the plasma membrane), cells can instead be grown on sapphire/aclar discs, or transwell membranes carefully cut to size (1.2, 3, or 6 mm depending on the carrier you are using), which can be directly transferred to the HPF carrier and frozen (Jimenez, Humbel, Van Donselaar, Verkleij, & Burger, 2006; Morphew & McIntosh, 2003; Reipert, Fischer, & Wiche, 2004). When using the latter option, we add a drop of 2% low melting point agarose (in cell culture medium supplemented with 15% BSA) on top of the membrane/disc before freezing.

Weakly adherent cells, such as HEK293T cells, can be lifted from the plate without the use of trypsin. However, despite gentle pipetting, the cells tend to clump together, leading to incomplete resin infiltration in some places. We achieve more consistent polymerization and distribution of cells within the block by treating HEK293T cells with trypsin as for adherent cells.

It is important to know the molecular weight of the tannic acid you are using. High molecular weight tannic acid ($C_{76}H_{52}O_{46}$; MW: 1701) is readily available, but does not penetrate well into cells and can contribute to protein extraction (Simionescu & Simionescu, 1976). We use low molecular weight tannic acid $(C_{14}H_{10}O_9)_n$ (MW: 1000–1500) from Electron Microscopy Sciences.

The freeze substitution protocol is adapted from Hawes, Netherton, Mueller, Wileman, and Monaghan (2007). We found that longer durations of the freeze substitution (e.g., Kukulski et al., 2011), which nicely preserved the in-resin fluorescence, negatively affected the photoswitching ability of the fluorophores. The super quick freeze substitution method of McDonald and Webb (2011), which has been successfully applied to preserving in-resin fluorescence for conventional CLEM by Peddie et al. (2014), produced comparable results to the overnight freeze substitution we use here and can be used instead, if preferred.

1. Place the cell culture plate on an inverted epifluorescence microscope to check that the cells are fluorescent (if transiently transfected, the efficiency should ideally be >50%) and look healthy, then return the plate to the 37°C incubator until you are ready to proceed with the HPF.
2. Warm the trypsin, a 15 mL tube of 0.1 M PIPES and the cryoprotectant to 37°C.
3. Set up the HPF and AFS2. Once filled with liquid nitrogen, set the AFS2 to −130°C.
4. Prepare the freeze substitution solution. Fill a 1.5 mL cryotube or Leica reagent bath (+flow through rings) with the freeze substitution medium and transfer to the AFS2 chamber to cool.

5. *Trypsinization of adherent cells*: Withdraw the culture media from each well, leaving a thin layer covering the cells so that they do not dry out, and then briefly wash the cells with 1 mL warm PIPES buffer. Withdraw the buffer and apply 0.5 mL warm trypsin to the each well, then return to the incubator for 4 min. Lightly tap the plate to release the cells, add 1 mL warm cryoprotectant to the well and use a plastic transfer pipette to gently resuspend the cells. Transfer the suspensions to a 15 mL centrifuge tube, spin for 2 min at 1000 rpm and go directly to the HPF.
   Suspension culture cells: Transfer 5–10 mL of the culture to a 15 mL centrifuge tube and spin for 1 min at 500 rpm. Resuspend cells in 0.1–0.5 mL warm cryoprotectant solution, transfer to a 1.5 mL tube, and spin for 30 s at 10,000 rpm. Proceed directly to the HPF.
6. Carefully withdraw the supernatant and load the cell slurry into the HPF carrier, so that it is only very slightly overfilled (see also Vanhecke, Zuber, Brugger, & Studer, 2012), then immediately load the carrier into the HPF and freeze the sample. We recommend three replicate carriers per sample.
7. Repeat Steps 5 and 6 for the remaining samples.
8. From this point on, great care should be taken not to warm the carriers during handling. Always precool tweezers, etc., before they come into contact with the sample, to prevent ice crystal formation and therefore damage to the sample.
9. Under liquid nitrogen, place the carriers in the cap of a 50 mL centrifuge tube and transfer this to the AFS2 chamber. Quickly and carefully transfer the carriers to the prepared cryotubes/reagent bath and start the freeze substitution program (Table 1).

3.3 RESIN INFILTRATION AND POLYMERIZATION

1. When the freeze substitution is finished, hold the temperature of the AFS2 at −45°C and wash the samples with pure acetone (precooled to −45°C) for 10–15 min, then 3× with pure ethanol (pre-cooled to −45°C) for 10–15 min each. If required, transfer the carriers to the reagent bath + flow-through rings and ensure that each carrier is sitting flush with the base of the tube.
2. Infiltrate with resin as outlined in Table 1 using pure ethanol for the resin dilutions steps.
3. Immediately prior to UV polymerization, cover the top of the reagent bath with a small square of aluminum foil so that the samples are exposed to indirect UV light for the first 24 h of polymerization, which results in more evenly polymerized resin (Schwarz & Humbel, 2007).
4. Attach the UV head to the AFS2 chamber and start the UV polymerization program (Table 1). Remove the foil cover half way through the program.
5. Transfer the samples to the fumehood and leave at room temperature for 1–2 days. The resin might turn pink as it warms up, but this will fade after a few days and will not affect the fluorescence.

Table 1 Outline of the Sample Preparation Steps Undertaken in the Automated Freeze Substitution (AFS) Unit

Program	Step	Temperature 1	Temperature 2	Slope	Duration	Reagent	UV
Freeze substitution	1	−130°C	−90°C	20°C/h	2 h	Freeze substitution medium	No
	2	−90°C	−90°C	–	6 h		
	3	−90°C	−45°C	5°C/h	9 h		
	4	−45°C	−45°C	–	3 h		
Washing	1	−45°C	−45°C	–	~1 h	Acetone/ ethanol	
Resin infiltration	1	−45°C	−45°C	–	2–3 h	25% HM20	
	2				2–3 h	50% HM20	
	3				2–3 h	75% HM20	
	4				12–14 h	100% HM20	
	5				2–3 h	100% HM20	
	6				2–3 h	100% HM20	
	7				2–3 h	100% HM20	
UV polymerization	1	−45°C	−45°C	–	24 h[a]	100% HM20	Yes
	2	−45°C	0°C	3.7°C/h	12 h[b]		
	3	0°C	0°C	–	12 h[b]		

Samples cryofixed using HPF are transferred to the AFS2 unit and freeze substituted in 0.2% uranyl acetate, 0.1% tannic acid, 5% water in acetone and warmed to −45°C over a period of 20 h, washed with acetone/ethanol and infiltrated with HM20 acrylic resin over ~32 h before polymerization with UV light.

[a] *With foil cover for indirect UV polymerization.*

[b] *Without foil cover for direct UV polymerization.*

6. Carefully cut the polymerized blocks out of the tubes. Remove the carriers by dipping only the very tip of the block in liquid nitrogen and then immediately scraping off the resin around the top and sides of the carrier. Use a pair of tweezers to pop off the carrier and expose the pellet.
7. Store the blocks at room temperature in the dark. The fluorescence and photo-switching ability of the fluorescent proteins will be stable for many months.

3.4 ULTRAMICROTOMY

We initially used bare finder grids, but switched to formvar-coated grids to mitigate loss of sections from the grids following SMLM imaging and increase section stability under the electron beam.

1. Take 100−150 nm sections on an ultramicrotome using a diamond knife. Use a perfect loop to transfer the sections onto 200 mesh Finder grids coated with 0.6% formvar.
2. Proceed immediately to the SMLM, as the in-resin fluorescence and blinking of the sections will significantly decline after 24 h.

3.5 SINGLE MOLECULE LOCALIZATION MICROSCOPY

For SMLM imaging, the grids were mounted between a glass slide and coverslip for imaging with an oil immersion objective. The signal from dry mounted sections was comparable to those mounted in PBS and glycerol-based antifade reagents, but we recommend using the latter as sections were less likely to adhere to the coverslip when unmounting the grids for EM imaging. Fiducial markers can also be applied at this stage as for Kukulski et al. (2011) to facilitate the correlation.

1. Apply one droplet of Citiflour AF4 mounting medium (total volume for all droplets: 20 μL) in the center of the coverslip and one in the center of the glass slide. Add four more droplets to the coverslip around the center droplet, which reduces the chance of the grid drifting toward the edge when applying the coverslip. Place the grid section side up on the droplet on the glass slide. Carefully lower the coverslip on top and secure it with nail polish once the mounting medium has spread.
2. Use an epifluorescence microscope to canvas the grid at low magnification and identify areas of interest (Fig. 3A).
3. Transfer the slide to the SMLM and use the 100× objective to acquire a map of the areas selected in Step 2, with both transmitted light and fluorescence to record both the finder grids reference and position of the cell of interest within the grid square. This enables the same cell to be tracked back on the TEM (Fig. 3B and C).
4. Set the microscope parameters to match the fluorophore in use and acquire the SMLM data set. Use the 488 nm laser line for yellow and green fluorescent proteins and the 593 nm laser line (on other setups the 561/568 nm laser) for red

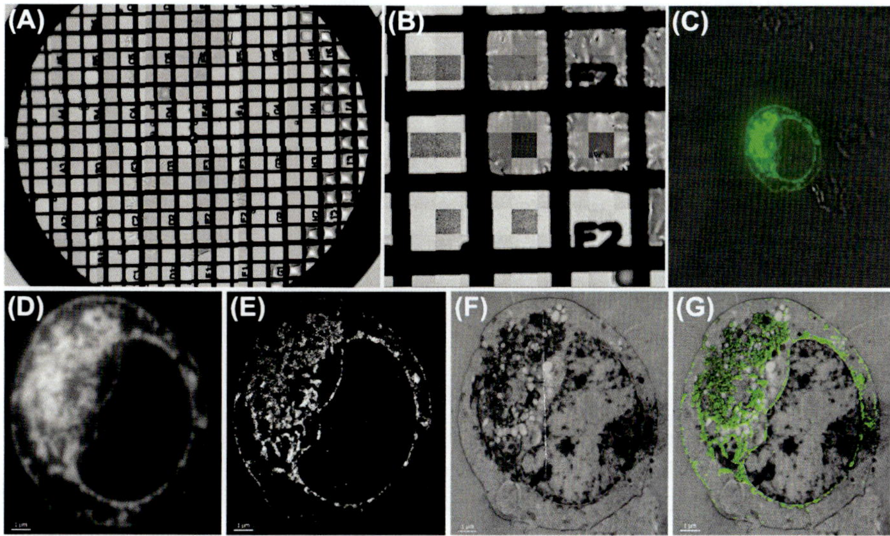

FIGURE 3

The image acquisition process for correlative SMLM and TEM. Ultrathin sections on formvar-coated finder grids are mounted onto glass slides in glycerol-based antifade and sealed under a coverslip, then imaged at low magnification on an epifluorescence microscope to identify areas of interest (A). On a microscope configured for SMLM, the 100× oil objective is used to take a mosaic of these areas with brightfield (B) and fluorescence (C). Noting its grid reference, the cell of interest is then imaged [shown here with 4× magnification compared to (C)] (D), followed by the SMLM data acquisition and reconstruction (E), which takes about 5 min. This process is repeated for several cells on the grid. The grids are then carefully unmounted and washed to remove the antifade and poststained. Using the reference and features recorded at (B) and (C), the same cell is identified and imaged using TEM (F). The SMLM image is then overlaid onto the TEM image (G), either by manual registration using characteristic features as control points or fiducial markers. *SMLM*, single molecule localization microscopy; *TEM*, transmission electron microscopy.

fluorescent proteins, at an intensity of 2–10 kW/cm^2 in the object plane for SMLM imaging (high laser intensity required for driving fluorophores to long-lived dark state).

5. Record the stochastic recovery of fluorescence (blinking) using the EMCCD with an integration time of 50 ms and a frame rate as fast as possible (in our experience the second fastest pixel readout rate gives the best compromise between speed and noise).
6. Run the single molecule localization procedure (with this protocol typically 4000 frames are sufficient). Because there can be increased background fluorescence from the resin and formvar, we use the maximum likelihood–based *fastSPDM*

algorithm (Grull et al., 2011), which has a sliding window to subtract background.

7. Generate super-resolution images based on the SMLM position data (software for visualization of SMLM data typically is part of the software of determining the single molecule positions, Fig. 3E).
8. Apply a few droplets of warm PBS to the nail polish, to help dislodge the coverslip, and very carefully lift the coverslip and remove the grid. Pass it over three droplets of ultrapure water, as residual mounting media on the section can lead to staining artifacts.

3.6 TRANSMISSION ELECTRON MICROSCOPY

1. Poststain the sections for 10 min on 30 μL droplets 2% uranyl acetate (protect from light), wash by passing the grids over warm ultrapure water droplets for 5 × 2 min, then stain with Reynolds lead citrate for 10 min (surround droplets with sodium hydroxide pellets to reduce exposure to CO_2) and wash as before. Take care not to breathe on the lead stain during the procedure as this will lead to the formation of lead carbonate precipitates on your section and could obscure the ultrastructure in your cell of interest.
2. Image the grids in the TEM, using the transmitted light and fluorescence images to locate the cell(s) of interest imaged with SMLM (Fig. 3F).

3.7 CORRELATION OF SINGLE MOLECULE LOCALIZATION MICROSCOPY AND TRANSMISSION ELECTRON MICROSCOPY IMAGES

The following describes an example of a basic procedure that can easily be implemented in other workflows for automation. Alternatively, ec-CLEM (Paul-Gilloteaux et al., 2017) provides a dedicated registration suite for CLEM.

1. Use the *Control Point Selection Tool* of MATLAB (Mathworks) to define control points that are visible in both the SMLM and TEM images (i.e., fiducial markers or clearly identifiable organelles).
2. Use these points to determine the coordinate system transformation between the SMLM and TEM images. A linear conformal transformation may be applied if there was minimal shrinkage due to electron beam damage during TEM imaging. Otherwise (for instance, if the cell of interest was near a hole in the resin), apply an affine transformation to improve the correlation accuracy.
3. Apply the coordinate transformation to the SMLM image and overlay it onto the TEM image (Fig. 3G). Adjust the dynamic range of the TEM image so that it fills the lower half of the range (e.g., 0–127 for an eight-bit image) so that both images use a maximum and equal range.

4. DISCUSSION

This chapter provides a detailed, step-by-step guide to performing super-resolution CLEM using standard fluorescent proteins. Because it does not require the use of antibodies or specialized fluorophores, this procedure can readily be applied across a wide range of cell biology research. By avoiding chemical fixation and by preserving the ability of the fluorescent proteins to photoswitch for high-quality SMLM post-EM processing, changes to sample ultrastructure, and their impact on the final correlation precision, are minimized.

4.1 PRESERVATION OF FLUOROPHORE PHOTOSWITCHING

It was surprising to find that while the fluorescence of a variety of standard fluorescent proteins could survive the cryopreparation and resin embedding procedures, the same was not true for their photoswitching ability (Johnson et al., 2015). We established that freeze substitution with tannic acid was crucial for preserving the photoswitching of fluorescent proteins for high-quality in-resin SMLM. The mechanisms behind this remain unclear. It is possible that the tannic acid and uranyl acetate, which are known to interact (Hayat, 1993), form complexes that bind to the fluorescent proteins and protect them during the solvent dehydration process. Alternatively, tannic acid may alter the local pH to conditions optimal for photoswitching.

Aside from tannic acid, the other factor in preserving fluorescent protein photoswitching was the duration of the freeze substitution. While fluorescence preservation for wide-field imaging was marginally reduced when increasing the length of the freeze substitution from several hours up to 60–80 h, only samples processed with the quick or intermediate length (up to 20 h) freeze substitution protocols retained sufficient photoswitching ability for SMLM (Johnson et al., 2015).

When imaging the resin sections, we found that no special mounting medium, such as switching buffers, is required for SMLM. We used a glycerol-based medium as this prevented the sections from attaching to the coverslip during unmounting of the sample after the SMLM imaging. The AF4 medium also contains an antifade reagent, which seemed to improve fluorescence imaging in general compared to pure glycerol. The advantage of using glycerol-based over water-based media for the photoswitching is probably because the higher viscosity of the former reduces reactions of the fluorescent proteins with oxygen. Similarly, we observed very good photoswitching when sections were attached to the coverslip without any medium present. Sealing the section on one side by the glass seems to have a similar effect to using high viscosity media.

4.2 APPLICATIONS OF THE TECHNIQUE

This technique works very well with overexpressed YFP, GFP, mVenus, mGFP, and mRuby2 fusion proteins in mammalian culture cells (we have so far tested it with

HeLa, HEK, and T-cells). We have used this technique to correlate fluorescence to endoplasmic reticulum, mitochondria, the nucleus, and lysosomes. It offers a powerful alternative to immunolabeling, particularly if there is no appropriate antibody available or if the background labeling is too high. The strength of the technique is that because tagging proteins with GFP or YFP is now a common practice, it can therefore be readily applied to cell biology research without the need for extra cloning steps.

4.3 LIMITATIONS

There are, however, some limitations to the technique which should be taken into consideration before using it. While it works well when fusion proteins are overexpressed, we have had variable success with proteins expressed under their native promoter, where fluorescence is often quite faint. This is problematic because a high density of fluorescent signal is required to generate sufficient data for SMLM and because background autofluorescence from resin section and/or the sample itself (e.g., yeast) can obscure the true signal. In these cases, it is useful to consider the live-cell imaging conditions. If long exposures are required for sufficient signal there, then there is little chance of retaining enough signal for in-resin SMLM.

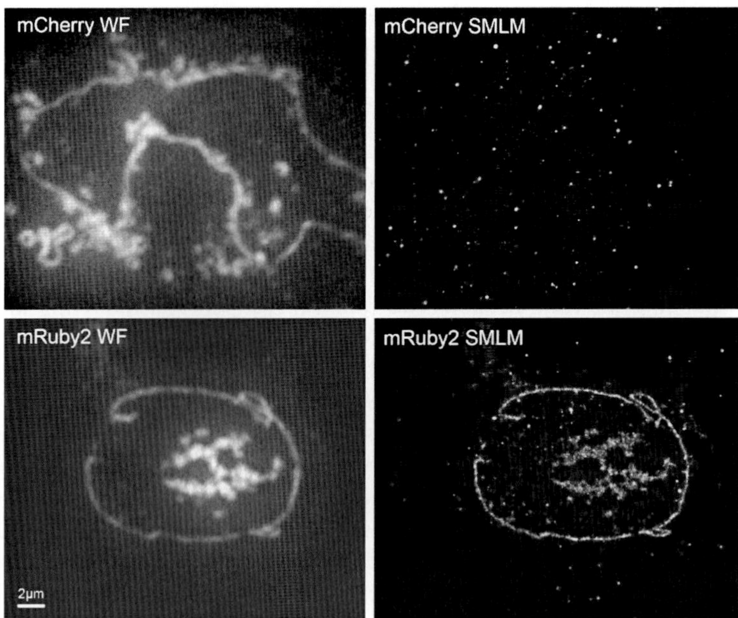

FIGURE 4

Comparison of mCherry and mRuby2 in-resin fluorescence with wide-field microscopy and SMLM. HEK cells expressing an mCherry fusion protein localized to the plasma membrane (top row) or mRuby2 fused with an ER/plasma membrane–localized protein (bottom row) and imaged with wide-field (left) or SMLM (right). *EM*, endoplasmic reticulum; *SMLM*, single molecule localization microscopy.

Despite being successfully optimized for GFP and YFP variants, the procedure is not well suited to the red fluorescent proteins mCherry and RFP. It does, however, work very well with mRuby2, which was originally engineered for FRET imaging (Lam et al., 2012). Fig. 4 illustrates how both mCherry and mRuby2 fluorescence is well preserved for in-resin wide-field fluorescence imaging, but that only mRuby2 retains the ability to photoswitch for SMLM. It is unclear why the red fluorescent proteins are more sensitive to the procedure than the green and yellow proteins.

4.4 FUTURE PROSPECTS

We are currently working on applying this technique to a wider range of fluorescent proteins, including mNeonGreen (Shaner et al., 2013), which is substantially brighter than GFP, and other fluorophores specialized for SRM, to determine the maximum structural resolution achievable with in-resin SMLM. We are also extending the technique to multicolor applications (e.g., GFP and mRuby2 cotransfections) and to tissue, where preliminary experiments show that the longer resin infiltration times required for tissue (e.g. *Drosophila* larvae) does not detrimentally affect the blinking. Finally, we are looking to make the leap to 3D super-resolution CLEM by using serial sections with EM tomography and array tomography, together with in-resin SMLM in the hope of adding a useful 3D tool to the growing CLEM toolbox.

ACKNOWLEDGMENTS

We thank Elena Seiradake, Marek Drozdz, and Lei Song for culture and transfection of cells, together with Christoph Hagen, Richard Parton, Ilan Davis, Jordan Raff, Kay Grünewald, E Yvonne Jones, and Ian Dobbie for their support and scientific discussions. This work was carried out in the Dunn School EM Facility and Micron Advanced Bioimaging Unit and was supported by the Wellcome Trust core award to the Wellcome Trust Centre of Human Genetics (090532/Z/09/Z), Wellcome Truse Senior Research Fellowships (090895/Z/09/Z; 096144/Z/11/Z), the Wellcome Trust Strategic Award to Micron (091911) and a Cancer Research UK programme grant (A10976).

REFERENCES

Betzig, E., Patterson, G. H., Sougrat, R., Lindwasser, O. W., Olenych, S., Bonifacino, J. S., … Hess, H. F. (2006). Imaging intracellular fluorescent proteins at nanometer resolution. *Science, 313*(5793), 1642–1645.

Bleck, C. K. E., Merz, A., Gutierrez, M. G., Walther, P., Dubochet, J., Zuber, B., & Griffiths, G. (2010). Comparison of different methods for thin section EM analysis of *Mycobacterium smegmatis*. *Journal of Microscopy, 237*(1), 23–38.

Brown, E., & Verkade, P. (2010). The use of markers for correlative light electron microscopy. *Protoplasma, 244*(1–4), 91–97.

Caplan, J., Niethammer, M., Taylor, R. M., & Czymmek, K. J. (2011). The power of correlative microscopy: Multi-modal, multi-scale, multi-dimensional. *Current Opinion in Structural Biology, 21*(5), 686–693.

Chang, Y. W., Chen, S., Tocheva, E. I., Treuner-Lange, A., Löbach, S., Søgaard-Andersen, L., & Jensen, G. J. (2014). Correlated cryogenic photoactivated localization microscopy and cryo-electron tomography. *Nature Methods, 11*(7), 737–739.

Giepmans, B. N., Adams, S. R., Ellisman, M. H., & Tsien, R. Y. (2006). The fluorescent toolbox for assessing protein location and function. *Science, 312*(5771), 217–224.

Grull, F., Kirchgessner, M., Kaufmann, R., Hausmann, M., & Kebschull, U. (September 2011). Accelerating image analysis for localization microscopy with FPGAs. In *Field programmable logic and applications (FPL), 2011 International Conference* (pp. 1–5). IEEE.

Hauser, M., Wojcik, M., Kim, D., Mahmoudi, M., Li, W., & Xu, K. (2017). Correlative super-resolution microscopy: New dimensions and new opportunities. *Chemical Reviews*. http://dx.doi.org/10.1021/acs.chemrev.6b00604.

Hawes, P., Netherton, C. L., Mueller, M., Wileman, T., & Monaghan, P. (2007). Rapid freeze-substitution preserves membranes in high-pressure frozen tissue culture cells. *Journal of Microscopy, 226*(2), 182–189.

Hayat, M. A. (1987). *Correlative microscopy in biology. Instrumentation and methods*. Orlando: Academic Press.

Hayat, M. A. (1993). *Stains and cytochemical methods*. Springer Science & Business Media.

Jimenez, N., Humbel, B. M., Van Donselaar, E., Verkleij, A. J., & Burger, K. N. J. (2006). Aclar discs: A versatile substrate for routine high-pressure freezing of mammalian cell monolayers. *Journal of Microscopy, 221*(3), 216–223.

Johnson, E., Seiradake, E., Jones, E. Y., Davis, I., Grünewald, K., & Kaufmann, R. (2015). Correlative in-resin super-resolution and electron microscopy using standard fluorescent proteins. *Scientific Reports, 5*, 9583.

Karreman, M. A., Mercier, L., Schieber, N. L., Solecki, G., Allio, G., Winkler, F., … Schwab, Y. (2016). Fast and precise targeting of single tumor cells in vivo by multimodal correlative microscopy. *Journal of Cell Science, 129*(2), 444–456.

Kellenberger, E., Johansen, R., Maeder, M., Bohrmann, B., Stauffer, E., & Villiger, W. (1992). Artefacts and morphological changes during chemical fixation. *Journal of Microscopy, 168*(2), 181–201.

Kim, D., Deerinck, T. J., Sigal, Y. M., Babcock, H. P., Ellisman, M. H., & Zhuang, X. (2015). Correlative stochastic optical reconstruction microscopy and electron microscopy. *PLoS One, 10*(4), e0124581.

Kobayashi, K., Cheng, D., Huynh, M., Ratinac, K. R., Thordarson, P., & Braet, F. (2012). Imaging fluorescently labeled complexes by means of multidimensional correlative light and transmission electron microscopy: Practical considerations. *Methods in Cell Biology, 111*, 1–20.

Kopek, B. G., Shtengel, G., Xu, C. S., Clayton, D. A., & Hess, H. F. (2012). Correlative 3D superresolution fluorescence and electron microscopy reveal the relationship of mitochondrial nucleoids to membranes. *Proceedings of the National Academy of Sciences of the United States of America, 109*(16), 6136–6141.

Kukulski, W., Schorb, M., Welsch, S., Picco, A., Kaksonen, M., & Briggs, J. A. (2011). Correlated fluorescence and 3D electron microscopy with high sensitivity and spatial precision. *The Journal of Cell Biology, 192*(1), 111–119.

Lam, A. J., St-Pierre, F., Gong, Y., Marshall, J. D., Cranfill, P. J., Baird, M. A., … Tsien, R. Y. (2012). Improving FRET dynamic range with bright green and red fluorescent proteins. *Nature Methods, 9*(10), 1005–1012.

Lemmer, P., Gunkel, M., Baddeley, D., Kaufmann, R., Urich, A., Weiland, Y., … Cremer, C. (2008). SPDM: Light microscopy with single-molecule resolution at the nanoscale. *Applied Physics B, 93*(1), 1.

Liu, B., Xue, Y., Zhao, W., Chen, Y., Fan, C., Gu, L., … Ding, W. (2015). Three-dimensional super-resolution protein localization correlated with vitrified cellular context. *Scientific Reports, 5*, 13017.

Löschberger, A., Franke, C., Krohne, G., van de Linde, S., & Sauer, M. (2014). Correlative super-resolution fluorescence and electron microscopy of the nuclear pore complex with molecular resolution. *Journal of Cell Science, 127*(20), 4351–4355.

McDonald, K. L. (2009). A review of high-pressure freezing preparation techniques for correlative light and electron microscopy of the same cells and tissues. *Journal of Microscopy, 235*(3), 273–281.

McDonald, K. L., & Webb, R. I. (2011). Freeze substitution in 3 hours or less. *Journal of Microscopy, 243*(3), 227–233.

Morphew, M. K., & McIntosh, J. R. (2003). The use of filter membranes for high-pressure freezing of cell monolayers. *Journal of Microscopy, 212*(1), 21–25.

Paez-Segala, M. G., Sun, M. G., Shtengel, G., Viswanathan, S., Baird, M. A., Macklin, J. J., … Hess, H. F. (2015). Fixation-resistant photoactivatable fluorescent proteins for CLEM. *Nature Methods, 12*(3), 215–218.

Paul-Gilloteaux, P., Heiligenstein, X., Belle, M., Domart, M. C., Larijani, B., Collinson, L., … Salamero, J. (2017). eC-CLEM: Flexible multidimensional registration software for correlative microscopies. *Nature Methods, 14*(2), 102–103.

Peddie, C. J., Blight, K., Wilson, E., Melia, C., Marrison, J., Carzaniga, R., … Collinson, L. M. (2014). Correlative and integrated light and electron microscopy of in-resin GFP fluorescence, used to localise diacylglycerol in mammalian cells. *Ultramicroscopy, 143*, 3–14.

Reipert, S., Fischer, I., & Wiche, G. (2004). High-pressure freezing of epithelial cells on sapphire coverslips. *Journal of Microscopy, 213*(1), 81–85.

Reynolds, E. S. (1963). The use of lead citrate at high pH as an electron-opaque stain in electron microscopy. *The Journal of Cell Biology, 17*(1), 208.

Schwarz, H., & Humbel, B. M. (2007). Correlative light and electron microscopy using immunolabeled resin sections. *Electron Microscopy: Methods and Protocols*, 229–256.

Shaner, N. C., Lambert, G. G., Chammas, A., Ni, Y., Cranfill, P. J., Baird, M. A., … Davidson, M. W. (2013). A bright monomeric green fluorescent protein derived from *Branchiostoma lanceolatum*. *Nature Methods, 10*(5), 407–409.

Simionescu, N., & Simionescu, M. (1976). Galloylglucoses of low molecular weight as mordant in electron microscopy. I. Procedure, and evidence for mordanting effect. *The Journal of Cell Biology, 70*(3), 608–621.

Suleiman, H., Zhang, L., Roth, R., Heuser, J. E., Miner, J. H., Shaw, A. S., & Dani, A. (2013). Nanoscale protein architecture of the kidney glomerular basement membrane. *eLife, 2*, e01149.

Vanhecke, D., Zuber, B., Brugger, S. D., & Studer, D. (2012). Safe high-pressure freezing of infectious micro-organisms. *Journal of Microscopy, 246*(2), 124–128.

Watanabe, S., Punge, A., Hollopeter, G., Willig, K. I., Hobson, R. J., Davis, M. W., … Jorgensen, E. M. (2011). Protein localization in electron micrographs using fluorescence nanoscopy. *Nature Methods, 8*(1), 80–84.

Wegel, E., Göhler, A., Lagerholm, B. C., Wainman, A., Uphoff, S., Kaufmann, R., & Dobbie, I. M. (2016). Imaging cellular structures in super-resolution with SIM, STED and localisation microscopy: A practical comparison. *Scientific Reports, 6*.

CHAPTER

Minimal resin embedding of multicellular specimens for targeted FIB-SEM imaging

4

Nicole L. Schieber*, Pedro Machado*, Sebastian M. Markert§, Christian Stigloher§, Yannick Schwab*,[1], Anna M. Steyer*,[1]

**European Molecular Biology Laboratory, Heidelberg, Germany*

§University of Würzburg, Würzburg, Germany

[1]Corresponding authors: E-mail: schwab@embl.de; steyer@embl.de

CHAPTER OUTLINE

Abstract

Correlative light and electron microscopy (CLEM) is a powerful tool to perform ultrastructural analysis of targeted tissues or cells. The large field of view of the light microscope (LM) enables quick and efficient surveys of the whole specimen. It is also compatible with live imaging, giving access to functional assays. CLEM protocols take advantage of the features to efficiently retrace the position of targeted sites when

Methods in Cell Biology, Volume 140, ISSN 0091-679X, http://dx.doi.org/10.1016/bs.mcb.2017.03.005

switching from one modality to the other. They more often rely on anatomical cues that are visible both by light and electron microscopy. We present here a simple workflow where multicellular specimens are embedded in minimal amounts of resin, exposing their surface topology that can be imaged by scanning electron microscopy (SEM). LM and SEM both benefit from a large field of view that can cover whole model organisms. As a result, targeting specific anatomic locations by focused ion beam−SEM (FIB-SEM) tomography becomes straightforward. We illustrate this application on three different model organisms, used in our laboratory: the zebrafish embryo *Danio rerio*, the marine worm *Platynereis dumerilii*, and the *dauer* larva of the nematode *Caenorhabditis elegans*. Here we focus on the experimental steps to reduce the amount of resin covering the samples and to image the specimens inside an FIB-SEM. We expect this approach to have widespread applications for volume electron microscopy on multiple model organisms.

INTRODUCTION

Correlative light and electron microscopy (CLEM) aims at imaging the same specimen with multiple modalities, namely light or fluorescence microscopy and electron microscopy. CLEM is very efficient for targeted subcellular studies, on heterogeneous and large multicellular specimens. One obvious application is the study of specific cells or organs in animal models. The fluorescence microscope, thanks to its large field of view and the possibility to directly visualize fluorescent molecules, is a powerful tool to precisely identify the site of expression of a given protein, or the position of organs or cell types. We have developed strategies in the past to target cells in nematodes (Kolotuev, Bumbarger, Labouesse, & Schwab, 2012; Kolotuev, Hyenne, Schwab, Rodriguez, & Labouesse, 2013; Kolotuev, Schwab, & Labouesse, 2010), in zebrafish embryos (Durdu et al., 2014; Goetz et al., 2014; Nixon et al., 2009; Schieber, Nixon, Webb, Oorschot, & Parton, 2010), and in mice (Karreman et al., 2014; Karreman, Mercier, et al., 2016) where the fluorescence, as seen in the living specimen or after preparation for electron microscopy (EM), is used to precisely depict the position of the region of interest (ROI) in the resin block. With this information, the sample is traditionally imaged by transmission electron microscopy (TEM) by collecting sections at the predicted ROI through very precise trimming. Alternative EM methods for imaging the ultrastructure are automated serial imaging in scanning electron microscopy (ASI-SEM). ASI-SEM is the method of choice for acquiring three-dimensional (3D) data on voluminous samples (Peddie & Collinson, 2014; Titze & Genoud, 2016). In particular, focused ion beam−SEM (FIB-SEM) is a powerful way to generate volume images at isotropic resolutions in the range of a few nanometers. Seen as a "quiet revolution in biology" (Narayan & Subramaniam, 2015), FIB-SEM enables fine ultrastructural measurements for rather large volumes covering multiple cells. Even more critical than for TEM, the volume to be imaged has to be exposed very close to the surface of the resin block. When working on adherent cultured cells, imaging can be achieved by accessing the cells from the attachment surface side (after removing the culture substrate). CLEM is then enabled by the footprint left by coordinates present at the surface of the coverslip. For bulkier specimens though, precise trimming is mandatory to reach the ROI and CLEM has been a way to perform this task efficiently (Armer et al., 2009; Karreman, Mercier, et al., 2016; Maco et al., 2013). Interestingly, when embedded in a very thin layer of resin, cultured cells can be imaged in FIB-SEM while still attached to their substrate (Kizilyaprak, Bittermann, Daraspe, & Humbel, 2014). The minimal amount of resin does not mask the cell topology that can be visualized using the secondary electrons in the SEM (Belu et al., 2016). As a result, our approach allows specific cells to be selected and even a sub-ROI at the cell surface. Similarly, we found out that minimizing the amount of resin surrounding the sample enables fast and precise targeting of regions within multicellular organisms based on their topology as seen in the SEM.

1. RATIONALE

Here we outline the methods developed to minimize resin on three model organisms, the zebrafish embryo *Danio rerio*, the marine worm *Platynereis dumerilii*, and the nematode *Caenorhabditis elegans* that were prepared by either high-pressure freezing (HPF) or chemical fixation. After standard EM processing and resin infiltration, we blot away the excess resin and then allow it to polymerize. With this method, only a very thin layer of plastic is coating the sample's surface, while its core, fully infiltrated, is ready for serial imaging with the FIB-SEM. This method enabled us to directly target a single neuromast (a group of sensory cells) from the lateral line of a zebrafish; the ventral nerve cord within the anterior body region of *C. elegans dauer* larvae; and a ciliated region below the palpi of the *Platynereis's* head.

2. METHODS

In the following part we describe the detailed workflow for minimal resin embedding for multicellular specimens, exposing their surface topology that can be imaged by SEM. The focus of this section is to describe the handling of the specimens after they have been imaged by fluorescence microscopy. We do not detail here the method to correlate the fluorescence data to the FIB-SEM, but one would simply overlay the fluorescence information obtained either before or after fixation, to the bulk surface anatomy of the specimen as seen in the SEM. Previous work describes such image registration (Armer et al., 2009; Durdu et al., 2014; Kolotuev et al., 2010).

2.1 FROM IMMOBILIZATION TO INFILTRATION

2.1.1 Chemical fixation

2.1.1.1 Zebrafish embryos (*Danio rerio*)

For live confocal imaging, embryos were mounted in 0.8% low melting agarose in glass-bottom dishes and imaged at 28°C (Durdu et al., 2014). After live imaging, anesthetized embryos (0.01% tricaine) were removed from agarose. Tails were removed by cutting after the yolk extension, and the bodies were immediately fixed with 2.5% glutaraldehyde (GA) and 4% formaldehyde (FA) in 0.1 M PHEM buffer for 14 min in the microwave (100-W cycling intervals of 2 min on and off under vacuum). A Pelco Biowave microwave containing ColdSpot was used for all processing steps with slight modifications to Schieber et al. (2010). To process several fishes together while keeping track of each specimen, they were processed in flow-through chambers as described previously (Fig. 1A insert, Goetz, Monduc, Schwab, & Vermot, 2015). After fixation, they were rinsed with PHEM buffer and postfixed with 1% osmium tetroxide (OsO_4) (in water) and then 1% OsO_4 with 1.5% potassium ferrocyanide ($K_3Fe(CN)_6$), each for 14 min (100-W cycling intervals of 2 min on and off under vacuum). Samples were then rinsed with water, stained with 1% uranyl acetate (UA) (in water) for 7 min (150-W cycling intervals of 1 min on—off—on under vacuum), and rinsed again with water. The embryos were then taken through a series of dehydration steps with ethanol (25%, 50%, 75%, 90%), each step 40 s (250-W, no vacuum). Samples were further dehydrated once with 95% ethanol and twice with 100% dried ethanol. Finally, they were infiltrated through a series of Durcupan resin (seven steps including two changes of 100% resin each for 3 min at 250 W under vacuum). All processing steps were performed at room temperature (RT). Although we have seen that the use of the microwave improves the ultrastructure significantly (Schieber et al., 2010), this particular protocol is not essential for minimal resin. It would also be feasible to use a different protocol that is finished by the gentle removal of excess resin.

2.1.1.2 *Platynereis* (*Platynereis dumerilii*)

Anesthetized *Platynereis* larvae (late nectochaete stage) were immersed in 2.5% GA in 0.15 M cacodylate buffer with 2 mM calcium chloride at RT for 15 min followed by 4°C incubation for 4 days (adapted from Deerinck, Bushong, Thor, & Ellisman, 2010). After fixation, the samples were washed in 0.15 M cacodylate buffer with 2 mM calcium chloride and then immersed in freshly prepared 1.5% $K_3Fe(CN)_6$ in the same buffer. Samples were immersed in 2% aqueous OsO_4 on ice for 1 h, filtered thiocarbohydrazide (TCH) solution for 20 min, and then in 2% OsO_4 in ddH_2O for 30 min at RT. Between each of the previous three steps, samples were rinsed in ddH_2O. The larvae were placed in 1% UA at 4°C overnight. The following day, after rinsing in ddH_2O, they were immersed in lead aspartate solution in a 60°C oven for 30 min. The embryos were rinsed again in ddH_2O, followed by dehydration in a graded series of ethanol and finally in glass distilled pure acetone. The embryos were infiltrated in Durcupan resin following a graded series up to an overnight infiltration step in pure resin.

2.1.2 High-pressure freezing

2.1.2.1 *Caenorhabditis elegans dauer* larvae

C. elegans dauer larvae were collected either by using some M9 buffer to wash them off the plates or were picked individually and then transferred to the HPF

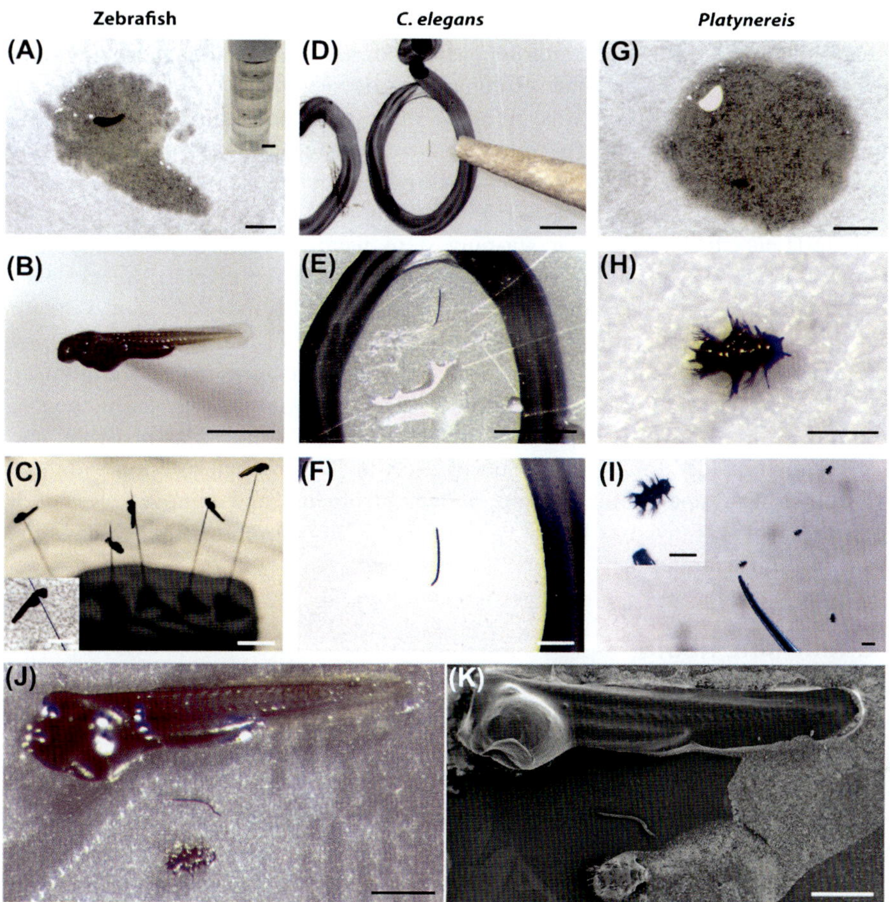

FIGURE 1

Summary of preparation for minimal resin embedding. (A–C) Zebrafish. (A) After processing in multiplex basket (insert), the resin-infiltrated specimens are placed on Whatman filter paper. Scale bar 2 mm. (B) Samples after resin removal are mounted onto a pin to ensure there is no resin pooling and reduce the attached surface, which helps with their clean removal from the pin. Scale bar 1 mm. (C) Samples after polymerization on pins. Scale bar 2 mm. (D–F) *Caenorhabditis elegans*. (D) Infiltrated *dauer* larvae are placed on Aclar film. The excess of resin is removed by gently moving the worms across the surface using a toothpick. Scale bar 1 mm. (E) Samples after resin removal. Scale bar 1 mm. (F) Samples polymerized on Aclar. Scale bar 250 μm. (G–I) *Platynereis*. (G) Infiltrated samples are placed on paper (scale bar 1 mm) and moved on the filter surface until the resin is completely drained (H, scale bar 250 μm). (I) They are transferred on Aclar films for polymerization, scale bar 250 μm. (J) Stereoscopic view of all three organisms on one SEM stub. (K) SEM view of all three organisms on SEM stub. Scale bars 500 μm. *SEM*, scanning electron microscopy.

carriers containing 20% bovine serum albumin (BSA). After HPF, the worms were freeze-substituted following the protocol from Stigloher and colleagues with modifications (Stigloher, Zhan, Zhen, Richmond, & Bessereau, 2011). A 0.1% tannic acid and 0.5% GA solution was prepared in acetone, filled in aluminum vials, and frozen down with LN_2. The samples were transferred on top of the solution making sure that the carriers were open to allow the chemicals access to the worms. After incubating the samples for 96 h at −90°C in an automatic freeze substitution machine (Leica EM AFS), four washing steps were performed with anhydrous acetone. OsO_4 (2% in anhydrous acetone) was added for 28 h to the samples. Next, the temperature was raised over the course of 14 h to −20°C and kept for 16 h. Finally, the temperature was raised over 4 h to 4°C, and the OsO_4 solution was removed by washing four times with anhydrous acetone. After warming up to RT, the worms were transferred into 50% Durcupan in acetone and incubated for 5 h. For the following steps, we removed as much solution as possible, and fresh resin was added for 2 h each. Successively 90% Durcupan solution and three times 100% Durcupan solution were added, incubated, and removed. Residues from BSA were removed as best as possible with an entomology pin to have the worms nicely separated.

2.2 MINIMAL RESIN

To minimize the resin, there were slight variations of the method between the three model organisms. For all, once the samples were taken through the EM processing, they were gently placed with a toothpick or a pin on either absorbent paper, filter paper, or Aclar, depending on the size of the organism. In each case the samples were gently moved around on their substrate until there was no more resin surrounding them.

For zebrafish we used both Whatman #1 filter paper and standard kitchen absorbent paper (Fig. 1A). To further the draining of the resin, they were pierced between the head and yolk sac with entomology pins to suspend with a small point of contact (Fig. 1B and C). For *C. elegans*, they were left on the piece of thick Aclar after being stripped of excess resin using a toothpick or pin (Fig. 1D–F). For *Platynereis* we used Whatman #1 filter paper (Fig. 1G and H) and then placed individual organisms onto thin Aclar sheets (Fig. 1I). All three samples were polymerized in a 60°C oven for 48 h. Zebrafish samples were removed from the pin with a razor blade and added to an SEM stub with a conductive carbon sticker. *Platynereis* samples were gently removed from the Aclar film by means of fine forceps and also added to an SEM stub. *C. elegans* were left on the Aclar film that was cut out to fit and put on an SEM stub. All were sputter-coated with gold for 180 s at 30 mA (Quorum, Q150RS). Silver paint was added to some samples to help with sample stability and to avoid charging. For the sake of presentation, all three models were mounted together on an SEM stub (Fig. 1J and K), but would be treated separately otherwise.

3. RESULTS: FIB-SEM TARGETING AND IMAGING

The samples were targeted and imaged inside an FIB-SEM (Auriga 60 or to the Crossbeam 540, Carl Zeiss Company). ATLAS 3D, being part of Atlas5 software from Fibics, was used to prepare the sample for acquiring image stacks.

Since the samples were not embedded in a block of resin, inside the FIB-SEM the topology of the specimen could be directly examined with the secondary electrons secondary ions (SESI) detector, which is using the secondary electrons to visualize surface information (Fig. 2). This revealed the overall structure of the specimen (Fig. 2A, E and I) including its length, height, and orientation (posterior/anterior), as well as finer structural details. In the zebrafish, it was possible to see the underlying muscle chevrons and even the scales (Fig. 2A–C). In *C. elegans dauer* larva structures such as the cuticle with alae, a very distinctive pattern of a set of raised cuticular ridges that extend along both sides of the animal was visible (Fig. 2F). Within the specimen of the *Platynereis*, the different segments of the animal became visible especially the chaetae, anal cirri, and cilia in the posterior part (Fig. 2I–K). All of those different features can be directly registered to images acquired in the light microscope; therefore no additional intermediate steps of added landmarks such as laser brandings are necessary.

Utilizing the structural features revealed by the minimal resin embedding technique, we could easily target an ROI. The areas that were acquired were dictated only by the questions we had for each of the different model organisms (Fig. 2C, G and K), but were not limited by the technique since the whole organism was accessible.

To protect the imaging region from the ion beam as well as avoiding curtaining on the cross section, the surface of the ROI was coated with a 1-μm-thick platinum coat. To open up the surface a coarse trench in front of the ROI is created using a 15-nA FIB current, followed by a polishing step using 3 nA. For the final acquisition of images the FIB was operated at 1.5 nA with the SEM and the FIB operating simultaneously (Narayan et al., 2014). The images are acquired with the SEM at 1.5 kV with the energy-selective back-scattered electron (EsB) detector with a grid voltage of 1100 V, analytical mode at a 700 pA current, setting the dwell time and line average to add up to about 1.5 min per image and an x/y pixel size of 5–8 nm and a slice thickness of 8 nm. Parameters chosen for these samples were specific for our Crossbeam 540; however, this sample preparation did not require extraordinary measures.

After exposing a polished cross section through the specimen, the different structures of interest could be imaged. In the zebrafish, the different cells within the neuromast were visible (Figs. 2D and 3A). In the *C. elegans dauer* larva, the neurons in the ventral nerve cord, as well as the surrounding muscles, were imaged (Figs. 2H and 3B). In *Platynereis*, the dorsal ciliated cells with motile cilia were imaged (Figs. 2L and 3C). The presence of cilia increases the surface area leading to a thicker layer of resin over these structures (Fig. 3C). Since in general there is only a very thin layer of resin left covering the specimen, the FIB-SEM acquisition can be performed with common parameters, for example, cutting slices of 8 nm thickness over 10's of μm.

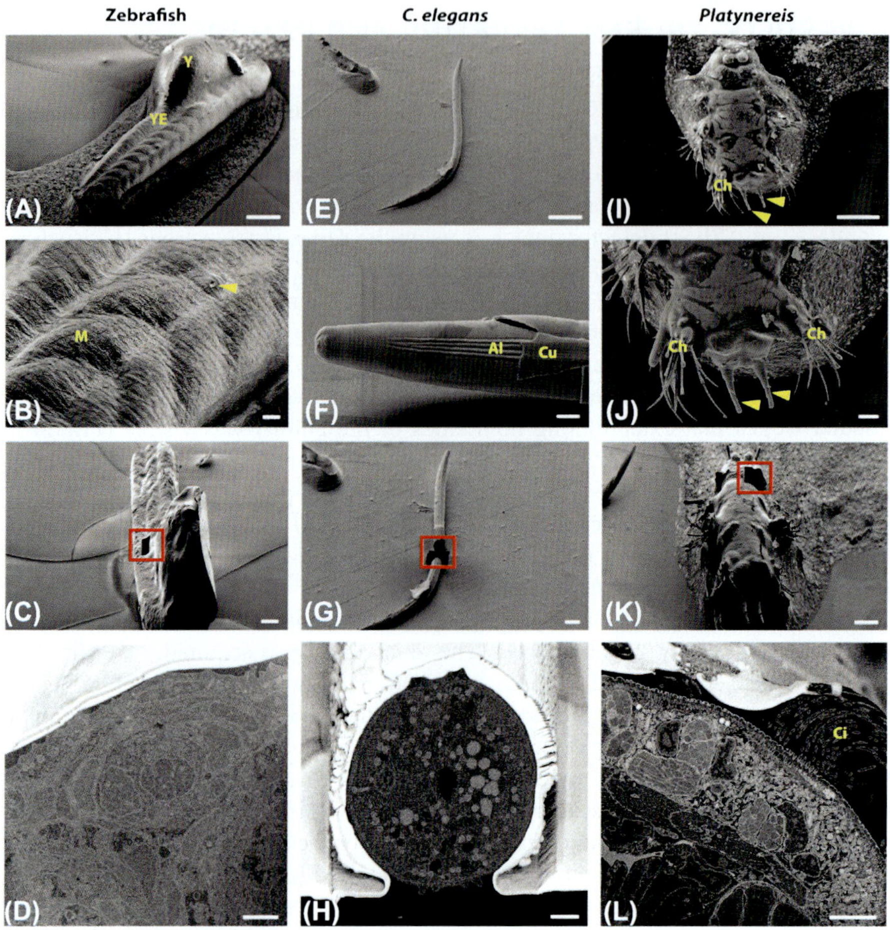

FIGURE 2

Overview of polymerized samples and FIB-SEM acquisition setup. (A–D) Zebrafish. (A) SEM (SESI) overview showing many anatomical features of the sample including the large yolk (Y) and yolk extension (YE). Scale bar 200 μm. (B) Higher magnification SEM (SESI) illustrating the level of details that remain visible after blotting the resin away like individual muscle chevrons (M) in their distinct pattern. The neuromast of the lateral line is bulging underneath the epidermis (*arrow head*). Scale bar 20 μm. (C) SEM (SESI) image of the sample after milling a trench (*highlighted in red*) to expose the imaging surface that is shown in (D). Scale bar 100 μm. (D) SEM (EsB) high magnification imaging of the exposed cross section. Scale bar 5 μm. (E–H) *Caenorhabditis elegans*. (E) SEM (SESI) overview image. Scale bar 50 μm. (F) Detail SEM (SESI) image showing anatomical features of the *C. elegans dauer* larva including the cuticle (Cu) and the alae (Al) in their distinct pattern. Scale bar 5 μm. (G) Trench *highlighted in red*. Scale bar 20 μm. (H) SEM (EsB) high magnification of cross section. Scale bar 2 μm. (I–L) *Platynereis*. (I) Dorsal view of a late nectochaete larva, anterior

4. INSTRUMENTATION AND MATERIALS

4.1 CHEMICAL FIXATION

Instrumentation: Pelco Biowave microwave with ColdSpot (Ted Pella Inc.), oven (INCU-Line, VWR), sputter coater (Quorum, Q150RS), Auriga 60/Crossbeam 540 (Carl Zeiss Company).
Material: Kitchen paper, forceps, pins (entomology pins 0.1 × 12 mm, Bioform Cat# B12c), Aclar embedding film (2 mil thickness, EMS Cat# 50426-25), Whatman #1 filter paper, SEM stub (6 mm length, Agar Scientific, Cat# G301F), conductive carbon sticker (12 mm, Plano GmbH, Cat# G3347), silver paint (Colloidal Silver Liquid, Ted Pella Inc., Cat# 16031).
Reagents: tricaine: 4 g ethyl 3-amino benzoate methanesulfonate salt (Sigma Cat# A5040), 10 g Na_2HPO_4 (Merck Cat# 1.06580.1000); GA (EM grade EMS Cat# 16220), paraformaldehyde (16% EM grade, EMS Cat# 15710), PHEM buffer pH 6.9: 240 mM PIPES (Sigma Cat# P6757-100G), 100 mM Hepes (Biomol Cat# 05288.100), 8 mM $MgCl_2$ (Merck Cat# 1.05833.1000), 40 mM EGTA (Sigma Cat# E3389-100G); cacodylate buffer pH 7.2 (sodium cacodylate trihydrate in H_2O, Sigma Cat# C0250-100G), calcium chloride (Merck Cat# 1.02382.1000), osmium tetroxide (100 mg, Serva Cat# 31251), potassium ferrocyanide $K_3[Fe(CN)_6]$ (Merck Cat# 1115305), thiocarbohydrazide (Sigma Cat# 88535-5G), UA (Serva Cat# 77870.01), lead aspartate, final pH 5.5: L-aspartic acid (Sigma Cat# A-9256), lead nitrate (Sigma Cat# L-6258), ethanol (Merck Cat# 1.000983.2500), acetone (EMSURE Millipore Cat# 100014), Durcupan resin (Sigma Cat# 44610).

4.2 HIGH-PRESSURE FREEZING

Instrumentation: Leica EM HPM100 and HPM010 (ABRA Fluid), Leica EM AFS/AFS2, oven (INCU-Line, VWR), sputter coater (Quorum, Q150RS), Auriga 60/Crossbeam 540 (Carl Zeiss Company).
Material: Carriers [Ø 3.0 × 0.5 mm type A (100 μm side) and B, middle plate and half cylinder for HPM100, Leica], Aclar embedding film (7.8 mil thickness, EMS Cat# 50425-10), toothpicks, conductive carbon sticker (12 mm, Plano GmbH, Cat# G3347), SEM stub (6 mm length, Agar Scientific, Cat# G301F), silver paint (Colloidal Silver Liquid, Ted Pella Inc., Cat# 16031).

side up. Scale bar 100 μm. (J) Detail of the posterior side where it is possible to visualize chaetae (Ch), anal cirri (*arrow head*). Scale bar 20 μm. (K) Milled trench on the larva first segment. Scale bar 50 μm. (L) High magnification cross section of the first segment. The resin surrounding the cilia (Ci) was not drained away. Scale bar 5 μm. *EsB*, energy-selective back-scattered electron; *FIB-SEM*, focused ion beam—SEM; *SEM*, scanning electron microscopy; *SESI*, secondary electrons secondary ions.

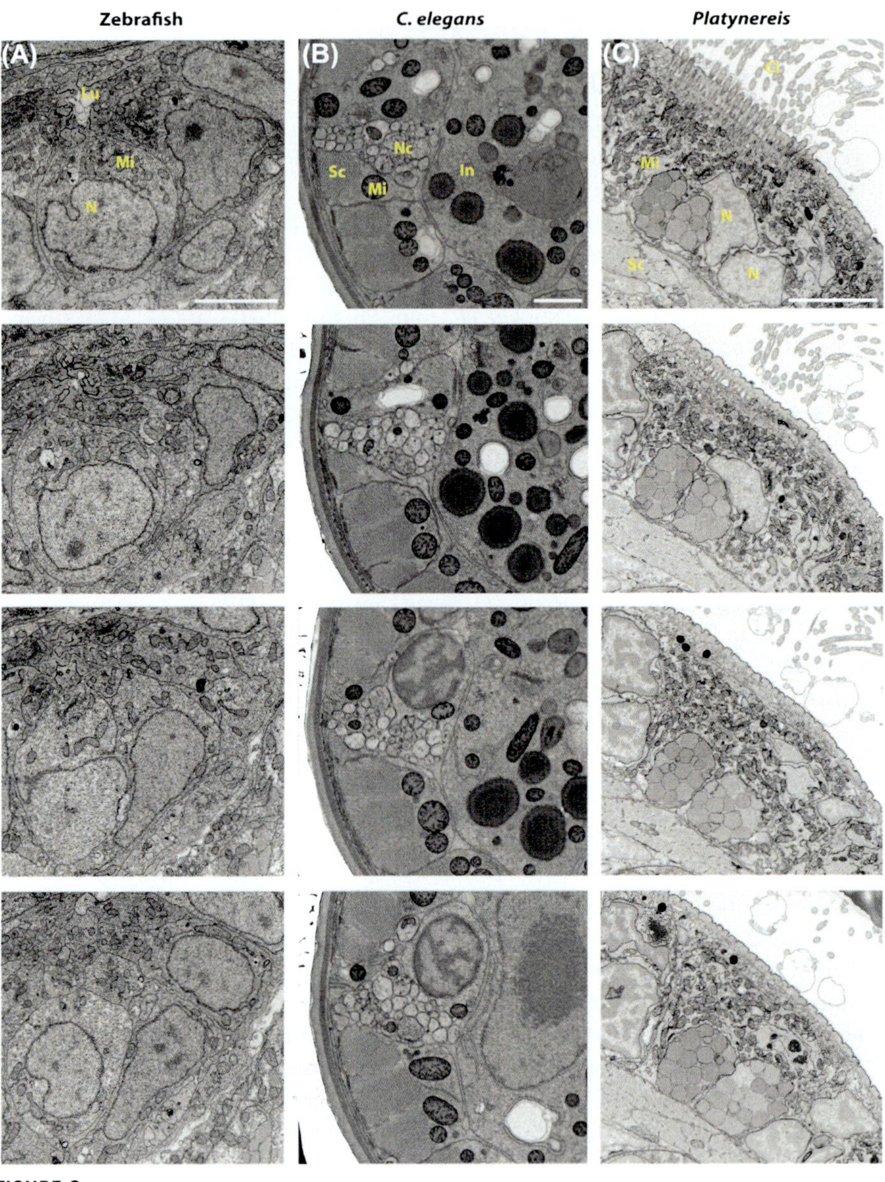

FIGURE 3

FIB-SEM imaging of the samples cross sections at different milling depth. (A) Zebrafish. Approximately 2 μm between slices, pixel size x/y 8 nm. Scale bar 10 μm. The targeting efficiently exposed the neuromast in which the cells are organized as a rosette, forming an apical lumen (Lu) clearly visible. (B) *Caenorhabditis elegans.* Approximately 2 μm between slices, pixel size x/y 5 nm. Scale bar 1 μm. The FIB-SEM cross section is exposing the ventral nerve chord of *C. elegans dauer* larva (Nc), the sarcomeres (Sc) of the body

Reagents: M9 buffer: 3 g KH_2PO_4 (Merck Cat# 1.04873.1000), 6 g Na_2HPO_4 (Merck Cat# 1.06580.1000), 5 g NaCl (Sigma Cat# S9888-1KG), 1 mL 1 M $MgSO_4$ (Merck Cat# A950386), H_2O (1 L, autoclave), BSA (Sigma Cat# A2153-50G), tannic acid ($C_{76}H_{52}O_{46}$, EMS Cat# 21700), GA (25% EM grade EMS Cat# 16220), acetone (EMSURE Millipore Cat# 100014), osmium tetroxide (Serva Cat# 31251), Durcupan resin (Sigma Cat# 44610).

5. DISCUSSION

When applied to model organisms, CLEM is a powerful technique that combines functional to ultrastructural information (Karreman, Hyenne, et al., 2016). Even though recent developments significantly improved their throughput, methods can still be tedious and difficult to implement across a majority of laboratories or service facilities. The minimal resin method described in this chapter is a simple technique that enables to target surface-associated structures within multicellular organisms. As long as similar anatomical features are recognizable in both LM and SEM, they can be used as references to register and target with precision the ROIs, whether previously determined by fluorescence imaging or simply by transmitted light.

In an earlier work, we have used targeted ultramicrotomy to image the forming sensory organs of the lateral line in the developing zebrafish (Durdu et al., 2014). Even though the targeting was efficient, the unprecise registration of the LM onto the sample, as seen inside the resin block, required the production of serial sections through a depth necessarily larger than the ROI. This precaution was meant to compensate for registration offsets, but introduced delays and potential material losses. A typical experiment would take about 1 day for the sectioning (including 1 h for the laser etching, 30 min for targeted trimming), 1.5 days for the serial imaging, and 2 weeks for image processing (alignment, modeling). Using the minimal resin method, the sensory organs are directly visible in the SEM, as they form a mass bulging underneath the epidermis. Their imaging by 3DEM is therefore direct and does not rely on estimates of their position. We estimate the gain in time to be in the order of 1 week (a few minutes for the targeting and 2 days for FIB-SEM imaging of a $12 \times 12 \times 20\ \mu m^3$ volume). Similarly, the research on transient blood vessel fusion events in zebrafish could benefit from the method (Armer et al., 2009). It could be done by targeting specific body segments as identifiable by the topology formed by the underlying muscle chevrons (Fig. 2A). Topological information has

wall muscles and the intestinal cells (In). (C) *Platynereis*. Approximately 1 μm between slices, x/y 5 nm. Scale bar 5 μm. A multiciliated area was exposed, where you can visualize the microtubule organization of motile axonemes and some details of adjacent muscle. *Ci*, cilia; *FIB-SEM*, focused ion beam–scanning electron microscopy; *Mi*, mitochondria; *N*, nucleus; *Sc*, sarcomere.

also been utilized to target ROIs in resin-free, vitrified samples for the making of thin lamellas with cryo-FIB-SEMs. Such preparations allowed cryo-EM tomography on various samples such as cells (Arnold et al., 2016) and nematodes (Harapin et al., 2015).

It is very important to stress that reducing the resin as much as possible does not compromise the ultrastructure and the more that can be removed the better. Previously published work has mainly used three methods to remove or reduce as much resin as possible: (1) draining of the resin by gravity, (2) temperature, and (3) centrifugation (Kizilyaprak et al., 2014). In principle, the resin must be removed before going into the oven for polymerization. The Durcupan resin, which we use for good milling quality in the FIB-SEM (Hayworth et al., 2015), is too viscous to be efficiently removed by gravity or centrifugation, even with increased temperature. Fine manipulation of the specimens is therefore necessary to achieve proper draining of the excess of the resin.

The advantage of the technique relies on the direct visualization of the samples morphology, which precludes the need for intermediate steps for targeting a specific ROI. Because the imaging is restricted to the ROI, multiple regions can be analyzed from the same specimen. Surprisingly, volume imaging of such samples with the FIB-SEM is very stable, even in the absence of a flat top surface. This is mainly made possible when a thick protective layer of platinum is deposited above the ROI. The absence of empty resin around the sample also allows direct access to the ROI. Trimming of the block on an ultramicrotome or long FIB milling is therefore not necessary, which saves a lot of time and labor. FIB-SEMs have limited milling capacities and would not allow direct access to structures located deep inside the sample. In our experience, while milling a deep trench might not be the limiting factor, the position of the imaging window on the exposed surface should stay as close as possible to the top surface (5–40 μm). When working on large specimens and for internal ROIs, trimming outside the microscope is therefore mandatory. After minimal resin embedding though, the specimens can have reduced contact surfaces with the supporting SEM stub, which would preclude their trimming in an ultramicrotome as they would either break or detach. Similarly, samples with complex morphology (e.g., *Platynereis*) will have protruding body parts. These would easily detach on slice and view imaging in the FIB-SEM. This would be the case when imaging the anterior part of a nematode larva that is not fully lying on the support (see, for example, Fig. 2F) as well. An alternative would consist in consolidating the specimen attachment to the substrate by adding resin or conductive glue.

We have developed the technique for FIB-SEM imaging, because it is powerful to obtain isotropic volumes at high resolution, but we also foresee applications with serial block face SEM (SBF-SEM). Following the sample's morphology while slicing would also allow precise targeting. Moreover, the absence of empty resin around the sample is expected to dramatically reduce the charging issues often encountered when imaging large block surfaces (Peddie & Collinson, 2014; Wanner, Kirschmann, & Genoud, 2015). One main advantage of SBF-SEM is access to deeper structures in larger specimens than what can be achieved within the FIB-SEM (Titze & Genoud, 2016).

The presented method can be applied across a wide range of multicellular organisms as demonstrated in this chapter with zebrafish, *Platynereis*, and *C. elegans*. These organisms are routinely imaged live by fluorescence microscopy. In a CLEM workflow, the topology of the specimen is used to bridge the LM and the EM observation. Because the sample preparation that we have described here involves the use of heavy metals and embedding in a resin, the fluorescence is not observed on the embedded samples but rather overlaid to the topology image. It would also be interesting to try other sample preparation methods, where methacrylate resins are used, leaving the potential to exploit in-resin fluorescence (Kukulski et al., 2011; Nixon et al., 2009) for precise targeting by image registration or using integrated fluorescence and scanning electron microscopes (Brama et al., 2016).

The power of the minimal resin embedding relies on the straightforward targeting of precise regions of interest on multicellular organisms. Leading to an enhanced throughput, we expect this method to reach routine applications for volume EM imaging in a large number of laboratories.

ACKNOWLEDGMENTS

We would like to thank Sevi Durdu (Gilmour lab, EMBL Heidelberg) for providing the zebrafish. Darren Gilmour for giving valuable feedback to the project. Hernando Martinez Vergara (Arendt lab; EMBL Heidelberg) for providing the *Platynereis* and Detlev Arendt for his support in the project. We would like to thank the Electron Microscopy Core Facility (EMCF, EMBL Heidelberg) for support.

REFERENCES

Armer, H. E., Mariggi, G., Png, K. M., Genoud, C., Monteith, A. G., Bushby, A. J., ... Collinson, L. M. (2009). Imaging transient blood vessel fusion events in zebrafish by correlative volume electron microscopy. *PLoS One, 4*(11), e7716. http://dx.doi.org/10.1371/journal.pone.0007716.

Arnold, J., Mahamid, J., Lucic, V., de Marco, A., Fernandez, J. J., Laugks, T., ... Plitzko, J. M. (2016). Site-specific cryo-focused ion beam sample preparation guided by 3D correlative microscopy. *Biophysical Journal*. http://dx.doi.org/10.1016/j.bpj.2015.10.053.

Belu, A., Schnitker, J., Bertazzo, S., Neumann, E., Mayer, D., Offenhausser, A., & Santoro, F. (2016). Ultra-thin resin embedding method for scanning electron microscopy of individual cells on high and low aspect ratio 3D nanostructures. *Journal of Microscopy, 263*(1), 78–86. http://dx.doi.org/10.1111/jmi.12378.

Brama, E., Peddie, C. J., Wilkes, G., Gu, Y., Collinson, L. M., & Jones, M. L. (2016). ultraLM and miniLM: Locator tools for smart tracking of fluorescent cells in correlative light and electron microscopy. *Wellcome Open Research, 1*, 26. http://dx.doi.org/10.12688/wellcomeopenres.10299.1.

Deerinck, T. J., Bushong, E. A., Thor, A., & Ellisman, M. H. (2010). *NCMIR methods for 3D EM: A new protocol for preparation of biological specimens for serial block face scanning electron microscopy − SBEM Protocol v7_01_10*. Retrieved from https://www.ncmir.ucsd.edu/sbem-protocol.

Durdu, S., Iskar, M., Revenu, C., Schieber, N., Kunze, A., Bork, P., … Gilmour, D. (2014). Luminal signalling links cell communication to tissue architecture during organogenesis. *Nature, 515*(7525), 120−124. http://dx.doi.org/10.1038/nature13852.

Goetz, J. G., Monduc, F., Schwab, Y., & Vermot, J. (2015). Using correlative light and electron microscopy to study zebrafish vascular morphogenesis. *Springer Science + Business Media New York, 1189*. http://dx.doi.org/10.1007/978-1-4939-1164-6_3.

Goetz, J. G., Steed, E., Ferreira, R. R., Roth, S., Ramspacher, C., Boselli, F., … Vermot, J. (2014). Endothelial cilia mediate low flow sensing during zebrafish vascular development. *Cell Reports, 6*(5), 799−808. http://dx.doi.org/10.1016/j.celrep.2014.01.032.

Harapin, J., Bormel, M., Sapra, K. T., Brunner, D., Kaech, A., & Medalia, O. (2015). Structural analysis of multicellular organisms with cryo-electron tomography. *Nature Methods, 12*(7), 634−636. http://dx.doi.org/10.1038/nmeth.3401.

Hayworth, K. J., Xu, C. S., Lu, Z., Knott, G. W., Fetter, R. D., Tapia, J. C., … Hess, H. F. (2015). Ultrastructurally smooth thick partitioning and volume stitching for large-scale connectomics. *Nature Methods, 12*(4), 319−322. http://dx.doi.org/10.1038/nmeth.3292.

Karreman, M. A., Hyenne, V., Schwab, Y., & Goetz, J. G. (2016). Intravital correlative microscopy: Imaging life at the nanoscale. *Trends in Cell Biology*. http://dx.doi.org/10.1016/j.tcb.2016.07.003.

Karreman, M. A., Mercier, L., Schieber, N. L., Shibue, T., Schwab, Y., & Goetz, J. G. (2014). Correlating intravital multi-photon microscopy to 3D electron microscopy of invading tumor cells using anatomical reference points. *PLoS One, 9*(12), e114448. http://dx.doi.org/10.1371/journal.pone.0114448.

Karreman, M. A., Mercier, L., Schieber, N. L., Solecki, G., Allio, G., Winkler, F., … Schwab, Y. (2016). Fast and precise targeting of single tumor cells in vivo by multimodal correlative microscopy. *Journal of Cell Science, 129*(2), 444−456. http://dx.doi.org/10.1242/jcs.181842.

Kizilyaprak, C., Bittermann, A. G., Daraspe, J., & Humbel, B. M. (2014). FIB-SEM tomography in biology. *Methods in Molecular Biology, 1117*, 541−558. http://dx.doi.org/10.1007/978-1-62703-776-1_24.

Kolotuev, I., Bumbarger, D. J., Labouesse, M., & Schwab, Y. (2012). Targeted ultramicrotomy: A valuable tool for correlated light and electron microscopy of small model organisms. *Methods in Cell Biology, 111*, 203−222. http://dx.doi.org/10.1016/B978-0-12-416026-2.00011-X.

Kolotuev, I., Hyenne, V., Schwab, Y., Rodriguez, D., & Labouesse, M. (2013). A pathway for unicellular tube extension depending on the lymphatic vessel determinant Prox1 and on osmoregulation. *Nature Cell Biology, 15*(2), 157−168. http://dx.doi.org/10.1038/ncb2662.

Kolotuev, I., Schwab, Y., & Labouesse, M. (2010). A precise and rapid mapping protocol for correlative light and electron microscopy of small invertebrate organisms. *Biology of the Cell, 102*(2), 121−132. http://dx.doi.org/10.1042/BC20090096.

Kukulski, W., Schorb, M., Welsch, S., Picco, A., Kaksonen, M., & Briggs, J. A. (2011). Correlated fluorescence and 3D electron microscopy with high sensitivity and spatial precision. *Journal of Cell Biology, 192*(1), 111−119. http://dx.doi.org/10.1083/jcb.201009037.

Maco, B., Holtmaat, A., Cantoni, M., Kreshuk, A., Straehle, C. N., Hamprecht, F. A., & Knott, G. W. (2013). Correlative in vivo 2 photon and focused ion beam scanning electron microscopy of cortical neurons. *PLoS One, 8*(2), e57405. http://dx.doi.org/10.1371/journal.pone.0057405.

Narayan, K., Danielson, C. M., Lagarec, K., Lowekamp, B. C., Coffman, P., Laquerre, A., … Subramaniam, S. (2014). Multi-resolution correlative focused ion beam scanning electron microscopy: Applications to cell biology. *Journal of Structural Biology, 185*(3), 278–284. http://dx.doi.org/10.1016/j.jsb.2013.11.008.

Narayan, K., & Subramaniam, S. (2015). Focused ion beams in biology. *Nature Methods, 12*(11), 1022–1031. http://dx.doi.org/10.1038/NMETH.3623.

Nixon, S. J., Webb, R. I., Floetenmeyer, M., Schieber, N., Lo, H. P., & Parton, R. G. (2009). A single method for cryofixation and correlative light, electron microscopy and tomography of zebrafish embryos. *Traffic, 10*(2), 131–136. http://dx.doi.org/10.1111/j.1600-0854.2008.00859.x.

Peddie, C. J., & Collinson, L. M. (2014). Exploring the third dimension: Volume electron microscopy comes of age. *Micron, 61*, 9–19. http://dx.doi.org/10.1016/j.micron.2014.01.009.

Schieber, N. L., Nixon, S. J., Webb, R. I., Oorschot, V. M., & Parton, R. G. (2010). Modern approaches for ultrastructural analysis of the zebrafish embryo. *Methods in Cell Biology, 96*, 425–442. http://dx.doi.org/10.1016/S0091-679X(10)96018-4.

Stigloher, C., Zhan, H., Zhen, M., Richmond, J., & Bessereau, J. L. (2011). The presynaptic dense projection of the *Caenorhabditis elegans* cholinergic neuromuscular junction localizes synaptic vesicles at the active zone through SYD-2/liprin and UNC-10/RIM-dependent interactions. *Journal of Neuroscience, 31*(12), 4388–4396. http://dx.doi.org/10.1523/JNEUROSCI.6164-10.2011.

Titze, B., & Genoud, C. (2016). Volume scanning electron microscopy for imaging biological ultrastructure. *Biology of the Cell*, 1–17. http://dx.doi.org/10.1111/boc.201600024.

Wanner, A. A., Kirschmann, M. A., & Genoud, C. (2015). Challenges of microtome-based serial block-face scanning electron microscopy in neuroscience. *Journal of Microscopy, 259*(2), 137–142. http://dx.doi.org/10.1111/jmi.12244.

CHAPTER

A new method for cryo-sectioning cell monolayers using a correlative workflow

5

Androniki Kolovou, Martin Schorb, Abul Tarafder, Carsten Sachse, Yannick Schwab[1], Rachel Santarella-Mellwig[1]

European Molecular Biology Laboratory, Heidelberg, Germany

[1]*Corresponding authors: E-mail: schwab@embl.de; santarel@embl.de*

CHAPTER OUTLINE

Methods in Cell Biology, Volume 140, ISSN 0091-679X, http://dx.doi.org/10.1016/bs.mcb.2017.03.011

Abstract

Cryo-electron microscopy (cryo-EM) techniques have made a huge advancement recently, providing close to atomic resolution of the structure of protein complexes. Interestingly, this imaging technique can be performed in cells, giving access to the molecular machines in their natural context, therefore bridging structural and cell biology. However, in situ structural electron microscopy faces one major challenge, which is the ability to focus on specific subcellular regions to capture the objects of interest. Correlative light and electron microscopy (CLEM) is one very efficient solution for this. Here we present a sample preparation technique that enables cryo-sections of vitrified cell monolayers in an orientation that places the cryo-section parallel to the fluorescence imaging plane. The main advantage of this approach is that it exploits the potentials of CLEM for cryo-EM investigation, for selecting specific cells of interest in a heterogeneous population, or for finding identified subcellular regions on sections.

INTRODUCTION

We recently celebrated the 70th anniversary of the birth of cellular electron microscopy (EM) pioneered by Porter, Claude, and Fullam who were the first to image cultured fibroblasts in a transmission electron microscope (TEM) (Porter, Claude, & Fullam, 1945). Cells were grown and fixed on films that were then transferred onto EM grids for direct, whole-mount visualization, which strikingly revealed subcellular features such as mitochondria or different parts of the cytoskeleton. Already then, the authors realized the limitations of the TEM when studying whole-mount cells. When operated at voltages around 100 kV, electron scattering fully hampers the imaging of thicker regions of the cells limiting studies to thinner portions like the cell periphery. From this seminal work stemmed the development of thinning procedures such as ultramicrotomy, conditioning the need for dedicated sample preparation methods for cell biology [fixation, resin embedding; see Griffiths for an excellent historical overview (Griffiths, 1993)].

Routine methods in EM are performed at room temperature, which has many drawbacks (see Kellenberger in Steinbrecht & Zierold, 1987) precluding the preservation of the native cellular architecture and the visualization of subcellular contents at high resolution. Cryo-EM was thus expected to fulfill these needs. The revolution of cryo-EM started in the 1980s with the work of Dubochet [see Dubochet's historical perspective (Dubochet, 2011)] and was followed by decades of development in sample preparation and in instrumentation. Recently coming of age, the technique has been elected the method of the year in 2015 (Anon, 2015; Callaway, 2015) and is expected to yield many ground-breaking results in biology for the years to come.

High-resolution cryo-EM can be performed on whole cells when grown and vitrified directly on EM grids (Medalia et al., 2002; Resch, Brandstetter, Wonesch, & Urban, 2011; Sartori et al., 2007); however, the visualization of structures is restricted to the thin cell periphery or to cell extensions (e.g., neurites, lamellipodia). Developing efficient means of cryo-sectioning through vitrified specimens required great efforts (Fernandez-Moran, 1953; McDowall et al., 1983) finally leading to the development of the cryo-electron microscopy of vitrified sections (CEMOVIS) technique (Al-Amoudi et al., 2004; Dubochet et al., 1988), which uniquely opened access to high-resolution imaging within the core of cells or tissues. In recent years an alternative to CEMOVIS has been developed with the use of a focused ion beam (FIB) inside a scanning electron microscope. When operated under cryo-conditions, this technique is able to generate thin lamellas through vitrified cells (Marko, Hsieh, Schalek, Frank, & Mannella, 2007), exposing the fine cellular ultrastructure (Mahamid et al., 2016) with fewer artifacts than with CEMOVIS. When the sample is homogenous and when the macromolecular structures to be resolved are dense enough, any random section can provide similar chances to expose the object of interest. However, when a rare event is to be observed in a cell, targeted strategies are crucial. For this, cryo-correlative light and electron microscopy (cryo-CLEM) has been developed where light microscopy is used to target specific regions on the EM grid. Cryo-CLEM has been used on whole-mount cells (Sartori et al., 2007), on cryo-sections (Nolin et al., 2012; Schorb et al., 2017) and in conjunction with cryo-FIB lamellas (Arnold et al., 2016). CLEM is routine on resin-embedded cell monolayers; however, it can be a real challenge under cryo-conditions. Solutions were found for cryo-sectioning chemically fixed and frozen monolayers (van Rijnsoever, Oorschot, & Klumperman, 2008), but the challenge when doing cryo-EM is that the cells are grown and vitrified directly on EM grids. Therefore, the targeting and the sectioning of the cell of interest have to be performed directly on this grid. RJ Mesman developed a very elegant way to cryo-section a monolayer of cells while cutting perpendicular to the culture substrate surface (Mesman, 2013); however, straightforward correlative light and cryo-EM methods for targeting specific cells in a monolayer are still missing.

1. RATIONALE

Here we have developed a method for cryo-sectioning a monolayer of cells parallel to the cell substrate, while incorporating several correlative options for targeting the cell of interest (Fig. 1). First, we image and localize cells of interest growing in a monolayer on the surface of a culture substrate. Thus, we can identify a unique phenotype among a heterogeneous cell population. This first correlation is performed either by live-cell imaging (option A), by imaging high-pressure frozen cells in a cryo-light microscope (option B) or by visualizing the mounted fluorescent cells directly in the chamber of the cryo-ultramicrotome (option C). The cell position is recorded relative to the topology of the finder grid. We then trim and section the high-pressure frozen specimens precisely to the site of the cell of interest. In a second on-section correlative step, high-accuracy cryo-CLEM (option D) can be utilized to spot the subcellular structure of interest on series of sections collected on EM grids. In addition, we show how to perform this procedure using two sectioning techniques: either perpendicular or parallel to the culture substrate.

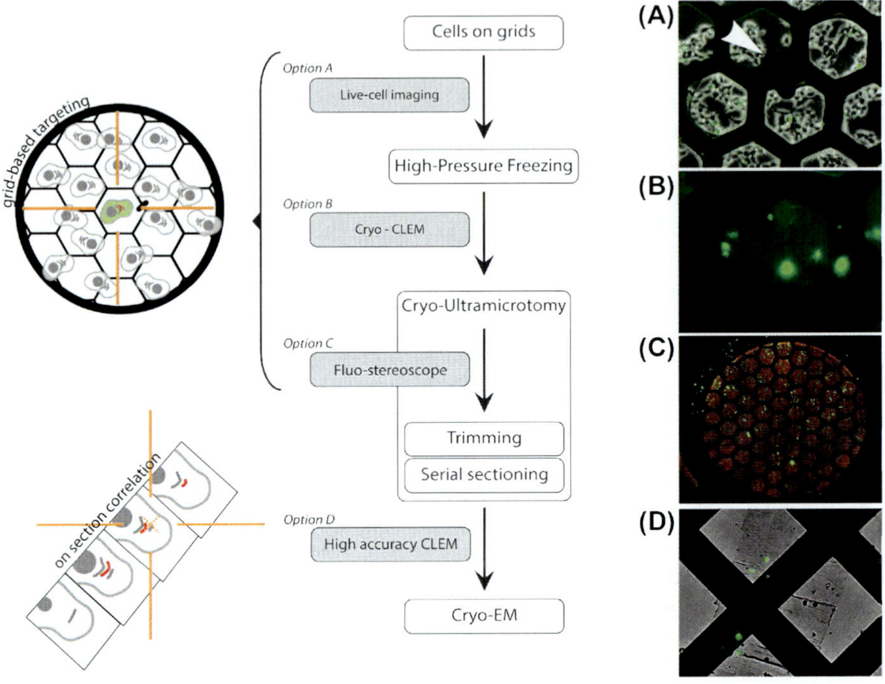

FIGURE 1

Correlative light and electron microscopy (CLEM) workflows. The position of the cell of interest on the finder grid is determined by light microscopy either by live-cell imaging (Panel A, Option A), by imaging the vitrified cells in a cryo-light microscope (Panel B, Option B) or by imaging the cell fluorescence directly in the chamber of the cryo-ultramicrotome (Panel C, Option C). The position of the cell of interest is recorded relative to a central mark on the finder grid (Panel A, *arrowhead*). Note that the grids are observed with an epifluorescence microscope. Cells growing above a grid bar are therefore visible, but shall not be selected. Ribbons of vitrified sections are then collected on carbon-coated electron microscopy grids for inspection either in a cryo-light microscope (for high-accuracy CLEM) (Panel D, Option 4) or directly in the cryo-electron microscope.

2. METHODS

2.1 CELL CULTURE ON FINDER GRIDS

This part of the workflow was adapted from a protocol published by RJ Mesman (2013). The aim here is threefold: (1) seeding cells on a finder grid will provide an appropriate coordinate system to record the position of a given cell of interest, (2) using carbon–Formvar-coated grids provides a convenient substrate for manipulating the cells and for vitrifying them by high-pressure freezing (HPF) and (3) the use of Matrigel prevents the grids from floating in the growing medium and secures them to the bottom of the dish (see also Jiménez et al., 2010).

2.1.1 Preparation of the finder grids

1. For this preparation, gold, hexagonal, 135 mesh, finder grids were used with a diameter of 3.05 mm. Before use wash them briefly in acetone and let them air dry on a filter paper.
2. Next, prepare Aclar strips for picking up the grids at a later point. Aclar sheets 51 μm thick are used, cut into $3 \times 3\ cm^2$ squares, and washed briefly with acetone and water. They can be placed on a filter paper to dry.
3. To generate Formvar films, use a 1% Formvar solution in chloroform as described previously (Peters & Pierson, 2008), float them on water, and place the finder grids shiny side up on the Formvar film.
4. Using the Aclar strip prepared above, slowly pick up the floating Formvar film so that it falls flat onto the Aclar strip (Fig. 2A).
5. Let the Formvar-coated grids air-dry overnight in a closed Petri dish.
6. On the next day, sputter-coat the grids with a thin layer of carbon (2 nm).
7. For preparing the Matrigel, precool culture dishes on ice, thaw the Matrigel, and keep it on ice for the remainder of the time. Pipette droplets of Matrigel (total volume: 5 μL) on an area of the culture dish smaller than the piece of Aclar to be attached (Fig. 2B). Place the Aclar strip onto the Matrigel droplets, the finder grids facing dull side up (Fig. 2C and D). The droplets should spread without overflowing onto the Aclar surface.
8. To ensure the gelling of Matrigel, transfer the culture dishes onto a warm plate and incubate for 30 min at 37°C.
9. Sterilize dishes by UV irradiation in a tissue culture hood for 1 h.

2.1.2 Cell culture for whole-mount correlation

Setting up this technique was done with three cell lines: HEK293, Hela K-EMBL, and BV2 microglial cells. The medium to use is the following: low glucose Dulbecco's modified Eagle's medium (DMEM) supplemented with 1% Pen/Strep, 1% L-Glutamine, 10% foetal calf serum, and 25 mM Hepes.

Seed the cells onto the dishes with the grids one day before HPF (Fig. 2D). Cell confluency should be around 80%. Note that for higher confluency, the cells have a tendency to detach from the grids when handled for HPF.

2.2 CORRELATIVE LIGHT AND ELECTRON MICROSCOPY OPTION A: LIVE-CELL FLUORESCENCE MICROSCOPY BEFORE HIGH-PRESSURE FREEZING

If the fluorescent signal is very dim, it is preferable to image the living cells before the freezing (Fig. 1, Option A), as at this stage of the workflow, high-end light microscopes can be utilized. By doing so, the operator can follow dynamic processes and record with precision the position of the cell of interest within the grid. The region of interest (the hexagon or grid square that will be trimmed later) should be around the mark in the middle of the grid (Fig. 1A). It is not wise to choose areas

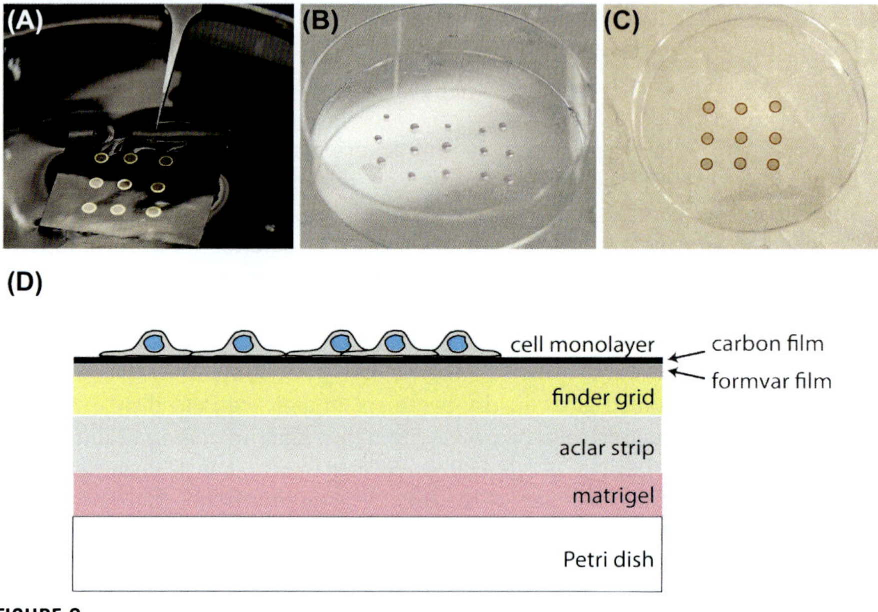

FIGURE 2

Seeding cells onto the finder grid. (A) Picking up of the grids on the Formvar film with a strip of Aclar. (B) Image of the Petri dish with drops of Matrigel covering a surface equivalent to the Aclar strip. (C) Image of the Aclar strip with the finder grids on top of the Matrigel. (D) Schematic side view of this set up once the cells have been seeded onto the grids.

that are close to the edge of the grid; first, they are difficult to target, and second these areas are very fragile. Grid bars close to the rim can come out while trimming.

2.3 HIGH-PRESSURE FREEZING

HPF is performed with the HPM010 (Abra Fluid) using the adapted carriers. If another system is to be utilized, some modification of the method might be necessary, especially for the choice of carriers in which the cell-bearing finder grids have to be frozen.

1. Two types of carriers are necessary for HPF (Fig. 3). The cell-bearing grid is sandwiched between a gold-coated copper type A carrier (0.1/0.2 mm) and the flat side of a type B aluminum carrier with the cells facing the 0.1 mm deep side of the type A carrier.
2. Before freezing, coat the carriers with hexadecane. Place a Whatman No. 1 filter paper inside a glass Petri dish and add hexadecane until the paper is completely covered. Place the carriers onto the Whatman paper (the 0.1 mm side of the type A carrier and the flat side of the type B carrier touching the filter paper) so they are coated with hexadecane before freezing. This helps to break the two carriers

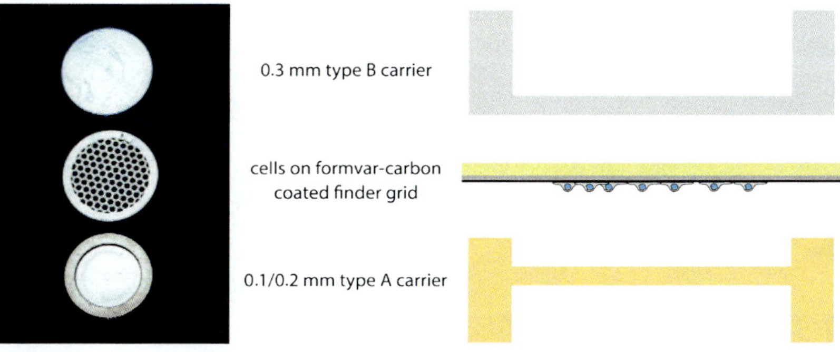

FIGURE 3

High-pressure freezing. The gold, type A carrier (bottom) has the 0.1 mm side facing up. The cells are laying on a Formvar and carbon-coated finder grid and the grid is placed with the cells facing down on the gold carrier (middle). The aluminum, type B carrier (top) is placed on top of the sandwich with the flat side facing the grid.

apart under liquid nitrogen after freezing. Note: make sure there is not an excess of hexadecane floating around in the bottom of the Petri dish. This could interfere with loading the sample in the next steps.

3. Dilute 20% dextran in culture medium to be used as a cryoprotectant for HPF.
4. Pipette 1 μL of cryoprotectant onto the type A carrier.
5. Lift a grid out of the Petri dish, dip it into the cryoprotectant, and place it with the cells facing down, onto the type A carrier. Finally, place the type B carrier on top with the flat surface in contact with the grid (Fig. 3).
6. This montage is placed in the HPF holder and frozen (2100 bars, −196°C) in the HPM010 machine.
7. After the freezing and under liquid nitrogen immersion, the B carrier is discarded and the cells, mounted on the A carrier, are stored in LN2 until further use.

2.4 CORRELATIVE LIGHT AND ELECTRON MICROSCOPY OPTION B: CRYO-FLUORESCENCE MICROSCOPY AFTER HIGH-PRESSURE FREEZING

After HPF, bright fluorescent signals can be located on a cryo-light microscope (Fig. 1, Option B). As mentioned before, we favor selecting an area located near the center of the grid.

2.5 MOUNTING OF THE CARRIER INTO THE CRYO-ULTRAMICROTOME

Cryo-ultramicrotomy is performed with a Leica UC6/FC6 cryo-ultramicrotome. Depending on the sectioning orientation, two types of sample holders are used: the cryo-atomic force microscope (AFM) holder for sectioning perpendicular to the finder grid (Fig. 4D–F—see also Mesman, 2013), and the universal sample

FIGURE 4

Mounting the carrier into the microtome. (A) Image of the surface of the grid where the grid is not bent or broken. (B) Image of the surface of the grid where one piece of the grid is slightly bent (*white arrowhead*). (C) Image of the surface of the grid where one side is broken and/or bent (*white arrowhead*). (D) Top view of the carrier inserted into the cryo-atomic force microscope holder for sectioning perpendicular to the culture substrate. (E) 91 degrees tilt of the holder before trimming. (F) Side view showing how deep the carrier is inserted in the cryo-holder. (G) Top view of the carrier inserted into the chuck for sectioning parallel to the culture substrate. (H) Low-magnification view of (G) when mounted into the cryo-ultramicrotome holder. (I) Side view of (G) and (H).

holder with special insert called a chuck for sectioning parallel to the grid (Fig. 4G–I). To avoid devitrification of the samples, all steps described here are performed either under liquid nitrogen immersion or in the chamber of the cryo-ultramicrotome at a temperature set to −150°C.

1. Before mounting the sample into the microtome, the quality of the carrier/grid sandwich should be assessed (e.g., for bubbles, cracks, and/or bending,

Fig. 4A–C). The width of the grid is 3.05 mm, while the width of the carrier is 3.0 mm, which could cause bending of the grid during freezing and/or opening of the sandwich after HPF (Fig. 4B). Make sure the grid is not strongly bent (Fig. 4C). A slightly bent grid rim can still be used (Fig. 4B).

2. For sectioning perpendicular to the grid, insert the carrier into a cryo-AFM holder making sure the location of the cell of interest is accessible for ultramicrotomy (Figs. 4D–F and 5). Most of the carrier should be held tight to avoid vibrations during trimming and sectioning.
3. For sectioning parallel to the grid, insert the chuck (Fig. 4G) into sample holder, place the carrier/grid on top (Fig. 4G–I), and fix it in the microtome chamber.

2.6 CORRELATIVE LIGHT AND ELECTRON MICROSCOPY OPTION C: FLUORESCENCE TARGETING AT THE MICROTOME

Cells can also be located directly in the chamber of the cryo-ultramicrotome with the help of a fluorescence device mounted on the stereoscope (Fig. 1, Option C). We have used a Fluorescence Module GFP LP adapted to the M-series stereomicroscope (Leica Microsystems). Nevertheless, other solutions can be adapted to the ultramicrotome (Leforestier et al., 2014). For such direct correlation and targeting, finder grids are not necessary anymore and can be replaced by regular 100 mesh grids.

Note that imaging cells after freezing (Options B and C above) efficiently ascertains if the cells detached during the manipulation (handling the grid from the culture dish to the cryo-protection, then to the high-pressure freezer), which can occur frequently. If many of the cells do detach, discard the sample, and move on to the next one. Furthermore, the use of dry objective lens with rather low magnification and limited numerical aperture would not allow the visualizing of dim fluorescent signals.

2.7 TRIMMING

A 45-degree cryo-trim diamond knife is used for trimming. Both trimming and sectioning are done at −150°C. The grid hexagon containing the cell of interest is carefully spotted using one or a combination of the methods described in Section 2.1.2 (Fig. 1) to trim away the surrounding material.

When sectioning perpendicular to the grid (Fig. 5; see also Mesman, 2013), the approach is performed stepwise. Specific care should be given not to trim while the cells are facing up. The upward cutting movement will lead to the detachment of the grid and breaking of the vitrified ice.

1. Mount the carrier/grid sandwich horizontally, grid facing up, to localize the grid hexagon where the cell of interest is located (Fig. 5A).
2. Rotate the holder 91 degrees anticlockwise (Fig. 5B) and trim the front surface to expose the vitrified sample. The 1 degree angle makes sure that the grid is not

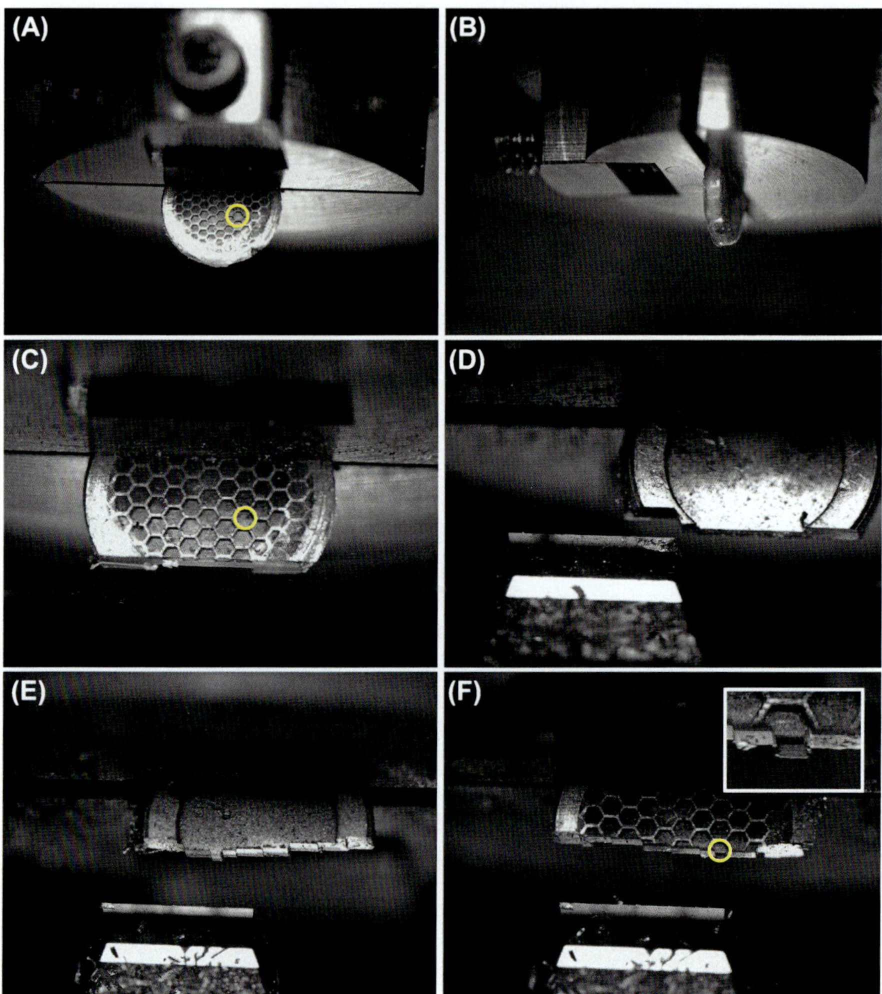

FIGURE 5

Trimming method for sectioning perpendicular to the cell substrate. (A) Top view of the carrier inserted into the cryo-atomic force microscope holder, with the grid exposed grid-side up. (B) The frontal view of (A) when the holder is rotated 91 degrees. (C) Top view of the carrier after the initial trimming step. (D and E) The carrier is positioned grid-side down and trimmed further using a step-wise trimming approach. (F) The carrier is rotated back to its original orientation to check the trimming respective to the targeted area. The final pyramid excludes the grid bars (inset). The *yellow circle* in (A), (C), and (F) indicates the region of interest to be targeted.

being pushed out of the sandwich while trimming. The trimming feed is set to 100 nm and the speed to 80 mm/s.

3. Turn the holder to the horizontal position to trim the sides stepwise. Flipping the carrier back and forth and up and down ascertains accurate targeting of the proper position on the grid (Fig. 5C–F).
4. The size of the block face can be chosen such that grid bars are excluded (Fig. 5F—insert).
5. After trimming the sides, turn the holder to the vertical position and trim the bottom of the carrier away.

When sectioning parallel to the grid (Fig. 6).

1. Here the trimming is performed stepwise taking care not to remove widths larger than 400 μm. The trimming feed is first set to 100 nm and the speed to 80 mm/s. The central mark of the finder grid (Fig. 6C and D) helps to keep track of the position of the region of interest (ROI).
2. Gradually reduce the trimming depth while progressing toward the region of interest. The first depth is set to about 100 μm and the last one, making the final pyramid, is set to 30 μm (Fig. 6B–D and F).
3. The trimmed pyramid should have a width of about 100 μm (Fig. 6F).
4. Leave a small piece of the gold grid on one corner of the trimmed square (Fig. 6F—inset). This is crucial to later find the plane that contains the cells (Figs. 6E,F and 7).

2.8 SECTIONING

Mounting the HPF carrier in the cryo-ultramicrotome first exposes the finder grids to sectioning. A significant amount of material thus needs to be removed before reaching the cells (Fig. 7A). The first layer to be trimmed is the cryoprotectant that intercalates between the finder grid and the type B carrier (Fig. 7A, Layer 1, and Fig. 3). Trimming further thins the small parts of the finder grid that were left during the trimming (Fig. 7A, Layer 2). As the grid depth is known, tracking the presence of this metal corner helps predicting the depth at which sectioning of the cells will start (Fig. 7B). Typically, we section a depth of 15 μm before starting to collect ribbons. Note that the diamond knife will be damaged when sectioning through the grid. For this reason, a dedicated part of the knife shall be used for fine trimming of this metal corner. When reaching the cell level, another part of the knife shall be used for collecting cryo-sections.

1. A 35-degree cryo-immuno diamond knife and micromanipulator from Diatome (Studer, Klein, Iacovache, Gnaegi, & Zuber, 2014) (Fig. 7C) are used for cryo-sectioning.
2. To cut ultrathin sections, set the sectioning feed to 50 nm and the speed to 0.4 mm/s.
3. Collect the ribbons on C-flat holey carbon grids.

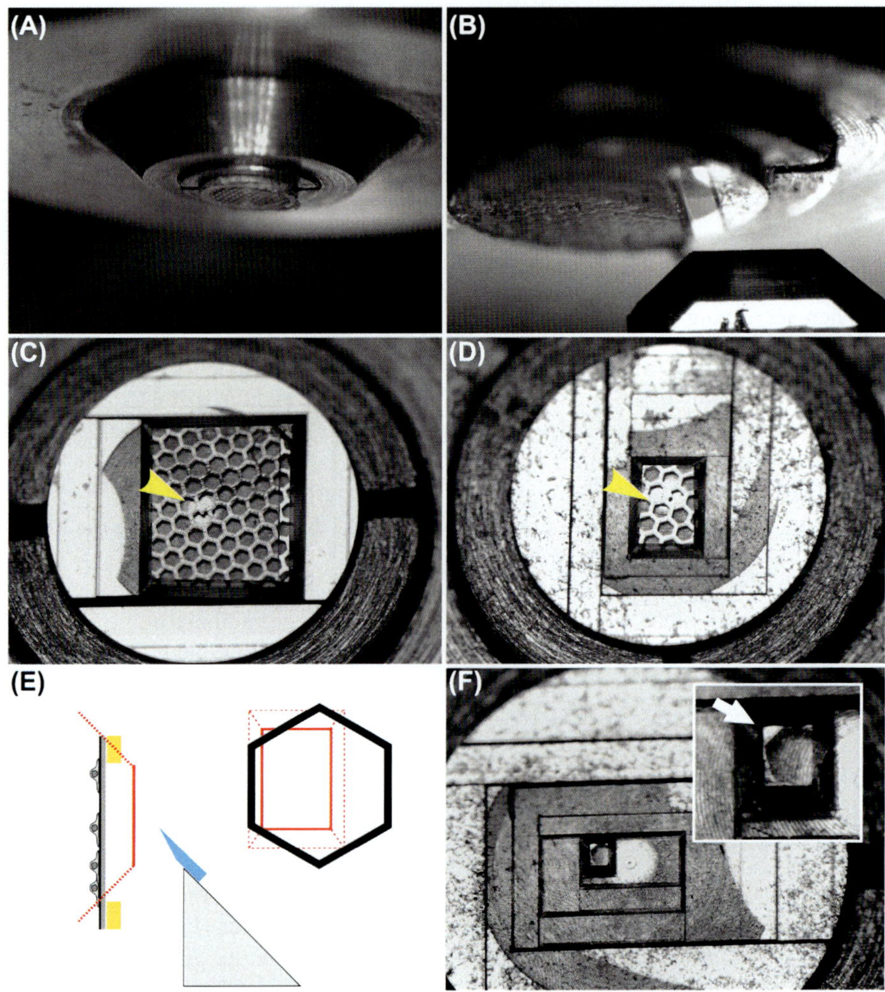

FIGURE 6

Trimming method for sectioning parallel to the cell substrate. (A) Top view of the carrier inserted into the chuck with the grid exposed to the front. (B) The beginning of the trimming process starting from the right side of the carrier. (C) Initial, rough trimming of the surface, targeting closer to the region of interest (ROI). (D) Further trimming of the surface closer to the ROI. (E) Schematic diagram showing the approach of the diamond knife to the carrier with the cells (left), and a schematic diagram showing how the pyramid will be trimmed with reference to the grid bars (right). (F) The block face shows the many trimming steps leading to the final pyramid. The inset shows a magnified view of the pyramid, with the leftover portion of the grid at the top left of the block face (*white arrow*). *Yellow arrowheads* are pointing to the ROI.

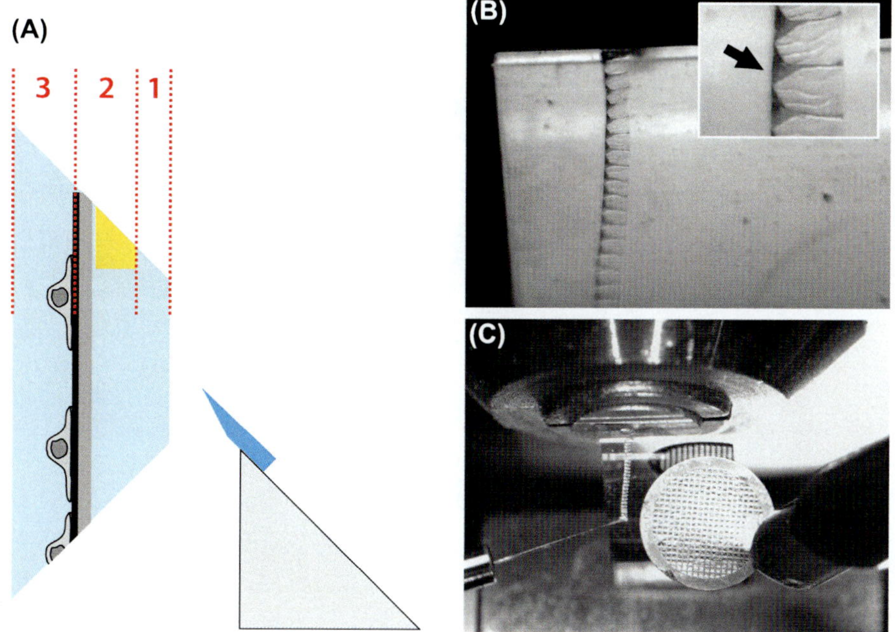

FIGURE 7

Ultrathin cryo-sections. (A) A schematic diagram showing the different layers or zones within the sample pyramid. Zone 1: This area contains a thin vitrified film of cryoprotectant, which is above the cells and the grid bars. Zone 2: This area contains the grid bar. The bar is around 18 μm. Zone 3: This area is below the grid bar and contains the cells. (B) An image of a ribbon of sections coming off of the diamond knife when trimming through Zone 2. The inset shows a magnified view of the ribbon. The *black arrow* points to the portion of the gold grid embedded into the sections. (C) Low-magnified image of the sections to be retrieved onto the grid. With use of the micromanipulator the hair on the left guides the ribbon, while the forceps on the right holds the grid in place for retrieval.

4. Collect two ribbons per grid. Each ribbon should contain around 20 sections each, covering about 2 μm of the total cell height. Once eight ribbons have been collected the complete volume of the monolayer has been sectioned.

2.9 CORRELATIVE LIGHT AND ELECTRON MICROSCOPY OPTION D: CRYO-CORRELATIVE LIGHT AND ELECTRON MICROSCOPY

Depending on the type of correlation, the EM grids would be transferred either directly to a cryo-TEM or be inspected first by cryo-light microscopy. In toto, CLEM as performed with options A, B, and C aims at selecting one cell among a heterogeneous population. The targeting strategy, as described here, is thus sufficiently precise to discern the exact same cells in the EM. Ultrastructural analysis of these cells

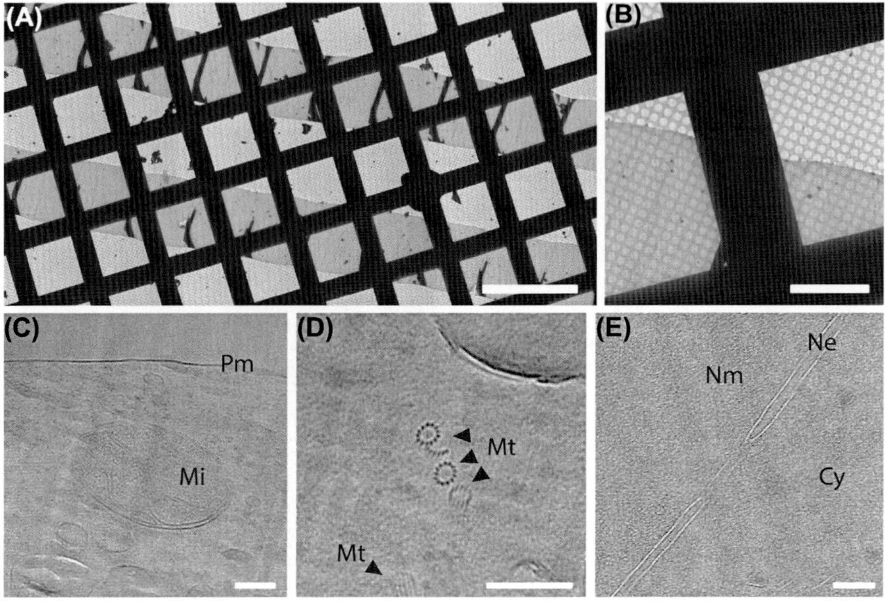

FIGURE 8

Cryo-electron microscopy. (A and B) Low-magnification view of two ribbons collected on a carbon film (Scale bar 50 μm in A and 10 μm in B). (C–E) Higher magnification images showing organelles in HEK293 cells. *Cy*, cytoplasm; *Mi*, mitochondria; *Mt*, microtubules; *Ne*, nuclear envelope; *Nm*, nucleoplasm; *Pm*, plasma membrane (Scale bars 100 nm).

is then directly performed in the TEM (Fig. 8). The consecutive steps of a high-precision CLEM workflow (option D), on the cryo-section, have been described elsewhere (Nolin et al., 2012; Schorb et al., 2017) and are not detailed further here. Handling cryo-sections between the different pieces of hardware is very tedious and can lead to ice contamination or even devitrification. Specific care should thus be given to strictly control the temperature of the sample, below the devitrification point, and to protect them from ambient air. When transferred to the cryo-light microscope, grid maps can be performed to localize precisely the position of the regions of interest that will be further imaged by cryo-EM [see chapter: Matrix MAPS—An Intuitive Software to Acquire, Analyze, and Annotate Light Microscopy Data for CLEM by Schorb & Sieckmann, 2017 for a detailed description of the cryo-CLEM software for light microscopy (LM) to EM grid registration].

3. INSTRUMENTATION, MATERIALS, AND REAGENTS

3.1 INSTRUMENTS

- HPF machine HPM010 (Abra Fluid, Switzerland)
- ACE600 carbon coating machine (Leica Microsystems, Vienna)

- UC6/FC6 Ultramicrotome (Leica Microsystems, Vienna)
- Micromanipulator (Diatome, Switzerland)
- EM Crion—Ionizer (Leica Microsystems, Vienna)
- Cryo-AFM holder (Leica Microsystems, 16702447)
- Chuck (M. Wohlwend GmbH, 742-1)
- Confocal or epifluorescent microscope
- Cryo-light microscope—cryo-CLEM (Leica Microsystems, Vienna)
- Fluorescence module GFL LP (Leica Microsystems, 10446143) for the cryo-ultramicrotome (Leica Microsystems, Vienna)
- Cryo-TEM: Tecnai Spirit (FEI, Eindhoven) operated at 120 kV
- Cryo-TEM Holder: model #626 (Gatan, Pleasanton)

3.2 MATERIALS AND REAGENTS

- Gold finder grids, 135 mesh, diameter 3.05 mm hexagonal (Plano, 8GG135)
- Formvar (Electron Microscopy Sciences, 15800)
- Chloroform (Merck, 1.02445.1000)
- Glass slides for making the Formvar film (Thermo Scientific, AAAA000001##12E)
- Aclar, thickness 51 μm, Fluoropolymer-Film (Science Services, A50426-10)
- Carbon thread (Leica, 16771511116)
- Matrigel, Basement Membrane Matrix Growth Factor reduced (Corning, 356230)
- Cell culture medium—low glucose DMEM 1× (Gibco, 31885-023)
- Fetal Bovine Serum (Life Technologies, 10270-106)
- Penicillin/Streptomycin (Sigma, P0781-100 ML)
- L-Glutamine 200 mM (Sigma, G7513)
- Dextran from *Leuconostoc* spp. Mr ~ 40.000 (Sigma, 31389-25G)
- HEPES p.A. *N*-2-Hydroxyethylpiperazine-*N′*-2-ethane sulfonic acid (Biomol, 05288.100)
- Holey carbon grids, C-flat 200 mesh, copper, hole diameter 4 μm, hole spacing 1 μm (Protochips, CF-4/1−2C)
- B carriers, (aluminum), recess 0.3 mm, 0.5 mm thickness (M. Wohlwend GmbH, cat. no. 242)
- Gold-coated copper platelet, recesses 0.1/0.2 mm, thickness 0.5 mm, (M. Wohlwend GmbH, art. no. 662)
- Petri dish, diameter 6 cm (Thermo Scientific, Nunclon Delta Surface, cat. no. 150288)
- 1-Hexadecane (Merck, 8.22064.0500)
- Acetone (Merck, 1.00014.2500)
- 45 degree cryo-trim diamond knife (Diatome, Switzerland)
- 35 degree cryo-immuno diamond knife (Diatome, Switzerland)

4. RESULTS AND DISCUSSION

Here we describe a new way to make thin sections through a vitrified monolayer of cells, while sectioning parallel to the culture substrate. As a result, large portions of cells can be investigated to study fine ultrastructural details of well-preserved samples (Fig. 8).

To efficiently target cells of interest for sectioning, we have integrated a workflow where the cells are grown on gold grids and high-pressure frozen. Cell selection by CLEM happens at various steps in the workflow. Remarkably, the viewing angle in both the LM and the EM modalities is similar. This is especially advantageous when rare events are sought or when focusing on particular subcellular compartments, since the overview images performed at the LM can be directly overlaid to the vitrified samples (whole cells or sections) for precise targeting in the EM. Such approach is extremely interesting when studying genetically modified cell lines or transfected cells, specifically when the transfection efficiency is very low. Other applications include, but are not limited to, studies of cell—cell contacts, cell cycle—specific events, or membrane trafficking. Furthermore, one could take advantage of performing high-accuracy CLEM directly on the section to precisely target macromolecular complexes of interest.

The workflow presented here is to be performed by advanced microscopists with a solid background in sample preparation techniques for cryo-EM. Even if we now routinely produce sections (parallel to the grid) with a success rate of about 80%, we think that the approach would benefit from further improvement. To vitrify the cells, this workflow involves HPF that is performed with specimen carriers having a smaller diameter to the cell-bearing grids. As a result, the rim of the grid is often bent, making the sample much more difficult to process. With a matching diameter, fewer samples would be lost. Trimming the finder grid is also challenging when mounted parallel to the cutting stroke of the microtome because it often leads to further bending of the grid that would come off as small chips, hindering the precise targeting of the ROI. More problematic is the potential consecutive breaking of the cell-containing vitreous ice. Another difficulty, inherent to any cryo-sectioning protocol is the fact that cryo-sections are never perfectly flat on the support grid. Since cryo-EM imaging is preferably performed on those regions of the sections that fall within a hole of the thin carbon film, the probability that the structure of interest (fluorescent spot) locates on the carbon film and in a hole is therefore very low. Inspecting the cryo-sections by fluorescence microscopy helps selecting for these events; however, low throughput is still expected. Additionally, we have found cryo-sections to be more prone to devitrification than vitrified liquid films (obtained by plunge freezing), and it is something that has to be examined more carefully. The fact that the sections only touch the cold supporting material with a, respectively, small surface area as compared to embedded ice could be a reason. Specific care should thus be taken keeping the temperature of the grids below the devitrification point, especially during the fluorescence microscopy steps.

The recent breakthrough in cryo-EM opens new avenues for collecting high-resolution structural information of protein complexes. Mostly used on purified protein preparations, the challenge is now to perform cryo-EM in the context of the cell, which involves more complex sample preparation steps. Among them is the thinning of the cells, which enables cryo-EM of deep portions of the cell (e.g., perinuclear region, nucleoplasm). One very promising solution is FIB milling under cryo-conditions (Mahamid et al., 2016), which can also be used in a correlative pipeline to select specific cells or subcellular regions within a cell (Arnold et al., 2016). Nevertheless, such approach is destructive in a way that only one thin lamella within the cell of interest is preserved. CEMOVIS on the other hand will produce a series of thin sections that can be analyzed sequentially to address a larger volume within cells. It comes with sectioning artifacts that could hamper fine structural analysis (Alamoudi, Studer, & Dubochet, 2005; Chang, McDowall, & Lepault, 1983; Han, Zuber, & Dubochet, 2008), but as of today, it remains a technique that is accessible to more laboratories and can offer a higher throughput.

Considering the importance of visualizing functional units of the cells in their subcellular context and as close to their native state as possible, we think correlative light and cryo-EM is the method of choice. When relying on thinning methods, such approaches are still difficult to implement. With technical improvements such as the one presented here, we believe cryo-sectioning vitrified monolayers of cells will have numerous applications in the field of cellular structural biology, and that its implementation in core facilities will open access to cryo-EM for a large community.

ACKNOWLEDGMENTS

We would like to thank Ralf Schubert (Leica Microsystems) for his support with the fluorescent module that we mounted on the cryo-ultramicrotome. From EMBL, we warmly thank Paolo Ronchi and Wim Hagen for fruitful discussions about sample preparation and imaging. We thank Claudio Bussi and Pablo Iribarren (CIBICI-CONICET Argentina) for providing cells for our multiple trials. Important feedback on cryo-ultramicrotomy and advices on diamond knives handling were provided by Helmut Gnägi (Diatome). Finally, we would like to thank Rob Mesman for the inspiration and advice on this method, and Wanda Kukulski for precious feedback on the cryo-CLEM workflows.

REFERENCES

Al-Amoudi, A., Chang, J.-J., Leforestier, A., McDowall, A., Salamin, L. M., Norlén, L. P. O., … Dubochet, J. (2004). Cryo-electron microscopy of vitreous sections. *The EMBO Journal, 23*(18), 3583–3588.

Alamoudi, A., Studer, D., & Dubochet, J. (2005). Cutting artefacts and cutting process in vitreous sections for cryo-electron microscopy. *Journal of Structural Biology, 150*(1), 109–121.

Anon. (2015). Method of the year 2015. *Nature Methods, 13*(1), 1.

Arnold, J., Mahamid, J., Lucic, V., de Marco, A., Fernandez, J. J., Laugks, T., ... Plitzko, J. M. (2016). Site-specific cryo-focused ion beam sample preparation guided by 3D correlative microscopy. *Biophysical Journal*, 1–10.

Callaway, E. (2015). The revolution will not be crystallized: A new method sweeps through structural biology. *Nature, 525*(7568), 172–174.

Chang, J. J., McDowall, A. W., & Lepault, J. (1983). Freezing, sectioning and observation artefacts of frozen hydrated sections for electron microscopy. *Journal of Microscopy, 132*(1), 109–123.

Dubochet, J. (2011). Cryo-EM-the first thirty years. *Journal of Microscopy, 245*(3), 221–224.

Dubochet, J., Adrian, M., Chang, J. J., Homo, J. C., Lepault, J., McDowall, A. W., & Schultz, P. (1988). Cryo-electron microscopy of vitrified specimens. *Quarterly Reviews of Biophysics, 21*(2), 129–228.

Fernandez-Moran, H. (1953). A diamond knife for ultrathin sectioning. *Experimental Cell Research, 5*(1), 255–256.

Griffiths, G. (1993). *Fine structure immunocytochemistry.* Berlin, Heidelberg: Springer Berlin Heidelberg.

Han, H.-M., Zuber, B., & Dubochet, J. (2008). Compression and crevasses in vitreous sections under different cutting conditions. *Journal of Microscopy, 230*(Pt 2), 167–171.

Jiménez, N., Van Donselaar, E. G., De Winter, D. A., Vocking, K., Verkleij, A. J., & Post, J. A. (2010). Gridded Aclar: Preparation methods and use for correlative light and electron microscopy of cell monolayers, by TEM and FIB-SEM. *Journal of Microscopy, 237*(2), 208–220.

Leforestier, A., Levitz, P., Preat, T., Guttmann, P., Michot, L. J., & Tchénio, P. (2014). Imaging *Drosophila* brain by combining cryo-soft X-ray microscopy of thick vitreous sections and cryo-electron microscopy of ultrathin vitreous sections. *Journal of Structural Biology, 188*(2), 1–6.

Mahamid, J., Pfeffer, S., Schaffer, M., Villa, E., Danev, R., Cuellar, L. K., ... Baumeister, W. (2016). Visualizing the molecular sociology at the HeLa cell nuclear periphery. *Science, 351*(6276), 969–972.

Marko, M., Hsieh, C., Schalek, R., Frank, J., & Mannella, C. (2007). Focused-ion-beam thinning of frozen-hydrated biological specimens for cryo-electron microscopy. *Nature Methods, 4*(3), 215–217.

McDowall, A. W., Chang, J. J., Freeman, R., Lepault, J., Walter, C. A., & Dubochet, J. (1983). Electron microscopy of frozen hydrated sections of vitreous ice and vitrified biological samples. *Journal of Microscopy, 131*(Pt 1), 1–9.

Medalia, O., Weber, I., Frangakis, A. S., Nicastro, D., Gerisch, G., & Baumeister, W. (2002). Macromolecular architecture in eukaryotic cells visualized by cryoelectron tomography. *Science, 298*(5596), 1209–1213.

Mesman, R. J. (2013). A novel method for high-pressure freezing of adherent cells for frozen hydrated sectioning and CEMOVIS. *Journal of Structural Biology, 183*(3), 527–530.

Nolin, F., Ploton, D., Wortham, L., Tchelidze, P., Bobichon, H., Banchet, V., ... Michel, J. (2012). Targeted nano analysis of water and ions using cryocorrelative light and scanning transmission electron microscopy. *Journal of Structural Biology, 180*(2), 352–361.

Peters, P. J., & Pierson, J. (2008). Chapter 8 Immunogold labeling of thawed cryosections. In : Introduction to electron microscopy for biologists, *Methods in cell biology*, (pp. 131–149). Elsevier.

Porter, K. R., Claude, A., & Fullam, E. F. (1945). A study of tissue culture cells by electron microscopy methods and preliminary observations. *The Journal of Experimental Medicine, 81*(3), 233–246.

Resch, G. P., Brandstetter, M., Wonesch, V. I., & Urban, E. (2011). Immersion freezing of cell monolayers for cryo-electron tomography. *Cold Spring Harbor Protocols, 2011*(7). http://dx.doi.org/10.1101/pdb.prot5643.

van Rijnsoever, C., Oorschot, V., & Klumperman, J. (2008). Correlative light-electron microscopy (CLEM) combining live-cell imaging and immunolabeling of ultrathin cryosections. *Nature Methods, 5*(11), 973–980. Available at http://www.nature.com/nmeth/journal/v5/n11/full/nmeth.1263.html.

Sartori, A., Gatz, R., Beck, F., Rigort, A., Baumeister, W., & Plitzko, J. M. (2007). Correlative microscopy: Bridging the gap between fluorescence light microscopy and cryo-electron tomography. *Journal of Structural Biology, 160*(2), 135–145.

Schorb, M., Gaechter, L., Avinoam, O., Sieckmann, F., Clarke, M., Bebeacua, C., … Briggs, J. A. (2017). New hardware and workflows for semi-automated correlative cryo-fluorescence and cryo-electron microscopy/tomography. *Journal of Structural Biology, 197*(2), 83–93.

Schorb, M., & Sieckmann, F. (2017). Matrix MAPS—an intuitive software to acquire, analyze, and annotate light microscopy data for CLEM. In T. Mueller-Reichert, & P. Verkade (Eds.), *Methods in Cell Biology* (Vol. 140, pp. 321–334).

Steinbrecht, R. A., & Zierold, K. (1987). *Cryotechniques in biological electron microscopy.* Springer Berlin Heidelberg.

Studer, D., Klein, A., Iacovache, I., Gnaegi, H., & Zuber, B. (2014). A new tool based on two micromanipulators facilitates the handling of ultrathin cryosection ribbons. *Journal of Structural Biology, 185*(1), 125–128.

CHAPTER

Correlative light and electron microscopic detection of GFP-labeled proteins using modular APEX

Nicholas Ariotti[a], Thomas E. Hall, Robert G. Parton[1]

The University of Queensland, Brisbane, QLD, Australia

[1]Corresponding author: E-mail: r.parton@imb.uq.edu.au

CHAPTER OUTLINE

[a]Current address: University of New South Wales, Sydney, NSW, Australia.

Methods in Cell Biology, Volume 140, ISSN 0091-679X, http://dx.doi.org/10.1016/bs.mcb.2017.03.002

Abstract

The use of green fluorescent protein (GFP) and related proteins has revolutionized light microscopy. Here we describe a rapid and simple method to localize GFP-tagged proteins in cells and in tissues by electron microscopy (EM) using a modular approach involving a small GFP-binding peptide (GBP) fused to the ascorbate peroxidase–derived APEX2 tag. We provide a method for visualizing GFP-tagged proteins by light and EM in cultured cells and in the zebrafish using modular APEX-GBP. Furthermore, we describe in detail the benefits of this technique over many of the currently available correlative light and electron microscopy approaches and demonstrate APEX-GBP is readily applicable to modern three-dimensional techniques.

High-resolution analyses are crucial to understand the cellular functions of proteins and for understanding the dysfunction of proteins in disease. The most frequently used electron microscopy (EM)-based detection method involves immunogold labeling of thin frozen sections using methods developed by Tokuyasu (1986). These methods are technically demanding, reliant upon high-quality antibodies for the detection of antigens and require specialized cryo-ultramicrotomy equipment. A further disadvantage is that immunogold labeling is predominantly restricted to the surface of an ultrathin section making the technique less useful for modern three-dimensional EM methods such as electron tomography and serial block face scanning EM. A recently described genetic tag that allows for "green fluorescent protein (GFP)"-like localization at the EM level throughout the cell represents a major step forward in biology (Martell et al., 2012).

The APEX2 tag, derived from soybean ascorbate peroxidase (Lam et al., 2015), fulfills many of the criteria required for such a genetic marker. APEX is an approximately 28 kDa protein that converts 3,3′-diaminobenzamidine (DAB) into an insoluble osmiophillic polymer at the site of the tag in the presence of the cofactor hydrogen peroxide. Expression of fusion proteins between APEX and the protein of interest has been shown to be a powerful method for electron microscopic detection of the proteins of interest. However, this requires the generation and characterization of new fusion proteins, with each protein of interest conjugated to APEX. For studies in animal systems a considerable investment would be required to generate animals expressing the new fusion proteins with no guarantee that the fusion proteins would be functional.

As an alternative to this approach, we have developed a system that relies on the recruitment of APEX2 to GFP-labeled proteins (Ariotti et al., 2015). This method involves the sequestration of APEX to the GFP-tagged protein of interest through fusion of APEX to a GFP-binding nanobody (Kirchhofer et al., 2010). Modular APEX represents a rapid, simple, and robust technique for correlative light and electron microscopy (CLEM) whereby the GFP-tagged protein of interest can be tracked using fluorescence microscopy, and after EM-processing, the electron density generated by APEX at the site of the GFP-tag can be resolved.

1. VECTORS FOR CORRELATIVE LIGHT AND ELECTRON MICROSCOPY IN MAMMALIAN CELL CULTURE AND WHOLE ZEBRAFISH

We have developed multiple constructs that are compatible with CLEM-based analyses of subcellular protein distributions. Each of these constructs is used for a unique set of applications:

1. APEX2-GBP (GFP-binding peptide) in a pCSDEST2 vector (APEX-GBP, Addgene #67651)
   We routinely use this standard APEX-GBP construct as our initial screening method for broad analyses of proteins with multiple cellular localizations. It is essential to screen for changes (if any) to protein distribution by fluorescence microscopy before high-resolution transmission EM can be performed and subcellular localization inferred. Moreover, we utilize this construct when performing cotransfections of constructs with two (or more) fluorescent tags as APEX-GBP lacks a fluorescent reporter.
2. mKate2-P2A-APEX2-GBP in a pCSDEST2 vector [APEX-GBP (mKate), Addgene #67650]
   The APEX-GBP (mKate) construct is our specific CLEM vector and has been extensively used here. The P2A sequence from porcine teschovirus-1 2A facilitates self-cleavage (Szymczak-Workman, Vignali, & Vignali, 2012) of the fluorescent mKate reporter upstream from the APEX-GBP domain. This cleavage allows for the simultaneous detection of cells expressing both the GFP-tagged protein of interest and cells expressing the APEX-GBP by mKate detection and fluorescence microscopy (Fig. 1). The P2A self-cleavage sequence was utilized to minimize the size of the complex directly linked to the protein of interest.
3. bact2-APEX2-GBP in a pDEST-Tol2-pA2 vector (*zf*APEX-GBP, Addgene #67668)
   The *zf*APEX-GBP construct is under the control of the constitutive beta-actin2 promoter, which induces APEX-GBP expression in all cells of the zebrafish. The pDEST-Tol2-pA2 vector backbone possesses a mCherry sequence under the control of the alpha-crystallin promoter that results in red fluorescence in the eye of all transgenic zebrafish. The red fluorescent eye allows for easy determination of transgenic animals (Hall, Ariotti, Ferguson, Xiong, & Parton, 2016).

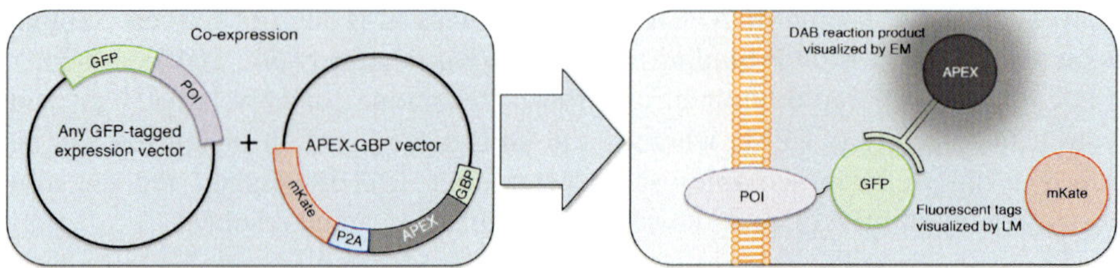

FIGURE 1

Schematic of modular APEX-GBP—based detection of GFP-tagged POI by correlative light and electron microscopy. *GBP*, GFP-binding peptide; *GFP*, green fluorescent protein; *POI*, protein of interest.

4. HSP701-APEX2-GBP in a pDEST-Tol2-pA2 vector (*inducible zf*APEX-GBP, Addgene #71282)
   The *inducible zf*APEX-GBP construct is under the control of the hsp701 promoter, which induces protein expression in response to a short heat shock. Zebrafish are subjected to a minor heat treatment (37°C) for 2 h and subsequently returned to standard tank temperature for one to two days before fluorescence imaging and subsequent processing for EM. The inducible control of APEX-GBP expression avoids potential effects on protein distribution during zebrafish development. This construct also possesses the red lens system for ease of identification of transgenic embryos (Hall et al., 2016).

2. CORRELATIVE LIGHT AND ELECTRON MICROSCOPY IN CELL CULTURE

2.1 METHOD

2.1.1 Cell culture and transient transfections

For all cell culture CLEM experiments MatTek 35 mm No. 1.5 gridded coverslip (14 mm glass diameter) dishes were used. Each grid square of the coverslip base possesses a unique alphanumeric code to allow for simple localization of cells of interest by bright field microscopy (Fig. 2A—C).

The following transfection conditions were optimized in HeLa cells although this protocol has also been successfully employed to localize proteins to high-resolution in baby hamster kidney cells, A431 cells, MDA-MB-231 cells, LNCaP cells, PC3 cells, MDCK cells, and Caco-2 cells.

2.1.2 Transfections

1. HeLa cells were grown in Dulbecco's modified eagle medium (DMEM; Gibco/Invitrogen GmbH, Germany) supplemented with 10% Fetal bovine serum (FBS; Serana, France) and 2 mM L-glutamine (Sigma—Aldrich, St. Louis, MO). Cells

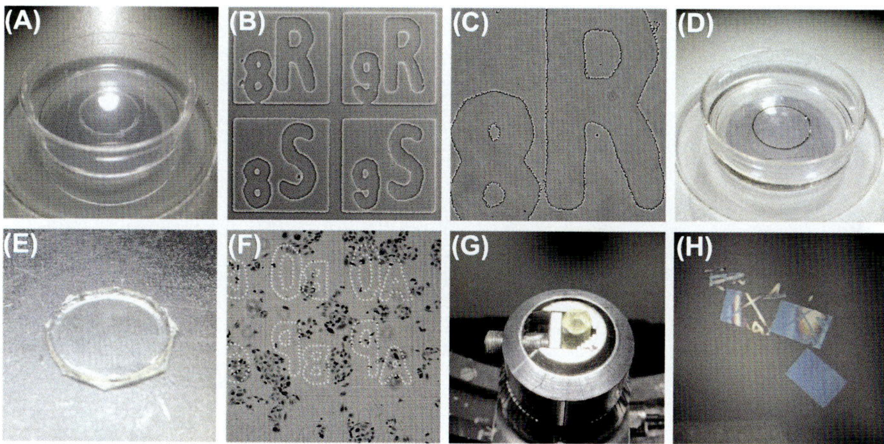

FIGURE 2

(A) MatTek 35 mm tissue culture dish with 14 mm gridded coverslip. (B) Bright field image of the gridded coverslip demonstrating the alphanumeric coded engineered into the glass coverslip. (C) Higher-magnification image of (B). (D) Cells embedded in LX112 resin after infiltration and polymerization. (E) Central disc of resin after the removal of the excess resin and the glass coverslip. Note the inverted imprinted grid pattern on the resin. (F) HeLa cells expressing nls-GFP and APEX-GBP demonstrate significant density in the nuclei of transfected cells. The grid pattern is inverted on the block face and the coordinates surrounding the region of interest are highlighted. (G) Remounted region of interest adhered to a blank LX112 resin stub. (H) The very first sections cut of a region of interest demonstrating the grid pattern in the sections.

were passaged at 37°C with 5% CO_2 and 98% humidity. Cells were seeded onto MatTek dishes 24 h prior to transfection (Fig. 2A–C).

2. Transfections were performed using Lipofectamine 3000 (Life Technologies, Carlsbad, CA) as per the manufacturer's instruction. nls-GFP (nuclear localization signal) was cotransfected with APEX-GBP (mKate) in a 1:1 ratio. A reduction in the ratio of APEX-GBP (mKate) to nls-GFP DNA would result in reduced total signal, which can be difficult to observe in the transmission electron microscope. Concurrently, an increase in the DNA ratio of APEX-GBP (mKate) would increase the proportion of unbound APEX-GBP in the cytoplasm, which in turn would reduce the specific signal to noise at the site of the nls-GFP (Ariotti et al., 2015).
3. Cell culture medium was replaced 3 h after transfection and cells were left for 24 h before imaging for light microscopy.
4. Heme is an essential cofactor for the generation of the insoluble precipitate after the DAB reaction. HeLa cells possess sufficient free heme such that exogenous addition into the cell culture medium is not required. However, in certain cell types (for example, Caco-2 cells) low levels of free heme can result in poor

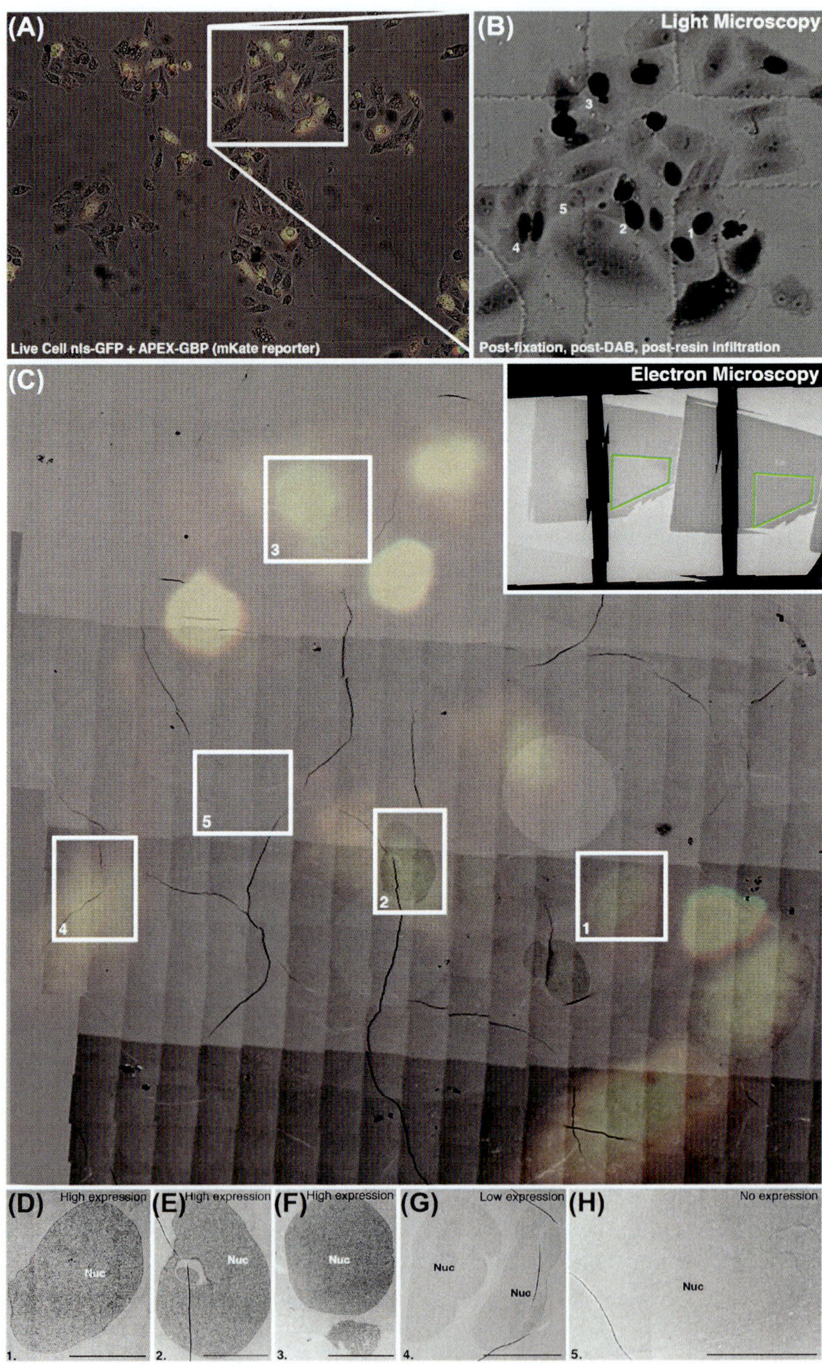
(A)
Live Cell nls-GFP + APEX-GBP (mKate reporter)
(B)
Light Microscopy
Post-fixation, post-DAB, post-resin infiltration
(C)
Electron Microscopy
(D)
High expression
Nuc
(E)
High expression
Nuc
(F)
High expression
Nuc
(G)
Low expression
Nuc
Nuc
(H)
No expression
Nuc

formation of DAB reaction product (Martell et al., 2012). When exogenous heme addition is required we utilize the protocol described by the Ting group. A 483 μM stock solution of bovine Hemin (Sigma–Aldrich) was dissolved in 0.01 M NaOH with vigorous vortexing. This solution was diluted to a final concentration of 7 μM in normal cell culture medium at 37°C and added to the cells of interest for 16 h. Cells were washed in normal nonheme containing media at 37°C prior to imaging for light microscopy.

2.1.3 Light and fluorescence microscopy

The application and use of live-cell screening prior to EM-based imaging has revolutionized the field. Fluorescence microscopy allows for real-time analysis of the redistribution of any fluorescent tag under certain experimental conditions and allows for the imaging of highly dynamic and rare cellular events. The selective imaging of cells of interest reduces the often time-intensive screening required to find these rare events in the TEM.

5. Dishes were screened on an $EVOS_{FL}$ Cell Imaging System for cells transfected with both nls-GFP and APEX-GBP (mKate) by fluorescence microscopy.
6. Transfected HeLa cells were imaged at 4× and 10× magnification. Bright field images were acquired to record the grid coordinates. Fluorescent images were acquired with standard 488 and 587 nm filter settings at the same magnifications (Fig. 3A).
7. It is critical to avoid any delay between light microscopy and the initial fixation step. Immediately after fluorescence imaging, HeLa cells were washed 3 times with phosphate buffered saline (2.7 mM KCl, 10 mM Na_2HPO_4, 1.8 mM KH_2PO_4, and 137 mM NaCl; pH 7.4) for 2 min each. Cells were fixed in 2.5% glutaraldehyde (Electron Microscopy Sciences, Hatfield, PA) in 0.1 M sodium

FIGURE 3

(A) A composite of bright field and fluorescent images demonstrating HeLa cells expressing nls-GFP and APEX-GBP (mKate reporter) highlighting the region of interest. (B) A bright field image after EM processing demonstrating significant DAB reaction product (contrasted by OsO_4) in the nuclei of the same cells from the region of interest highlighted in (A). Five specific regions of interest (1–5) have been selected. (C) A montaged series of electron micrographs taken at 4400× magnification reconstructing the same area of interest in (B) overlayed with the composite image from (A) (by increasing the transparency) demonstrating the cells with fluorescence correspond to the same cells with increased electron density. Inset: The 130× magnification initial "sloppy" map of the whole grid generated in the navigator program in SerialEM. The regions of interest are designated by the *green polygons*. (D–H) Higher-magnification images showing the nuclei of transfected cells corresponding to regions 1–4 and an untransfected cell (region 5), respectively. Scale bars = 10 μm. *GBP*, GFP-binding peptide; *GFP*, green fluorescent protein.

cacodylate buffer (pH 7.4; Sigma–Aldrich) for 1 h at room temperature. The fixative solution was made from a 25% glutaraldehyde stock solution added to 0.2 M sodium cacodylate buffer stock solution (Sigma–Aldrich) and diluted to 0.1 M with double distilled water.

2.1.4 Sample processing for transmission electron microscopy

8. The DAB reaction should be performed immediately after fixation and so the DAB reaction mixture can be prepared during the fixation period. This step is critical for the generation of specific reaction product at the site of the APEX-GBP modular tag. 3,3′-diaminobenzidine tetrahydrochloride (DAB; 10 mg tablets; Sigma–Aldrich) was dissolved in double distilled H_2O to a concentration of 2 mg/mL with vigorous vortexing for 5 min. Next, 0.2 M sodium cacodylate buffer (pH 7.4) was added to the solution to a final concentration of 0.1 M sodium cacodylate and 1 mg/mL of DAB. The solution was subjected to vigorous vortexing for an additional 5 min and any undissolved precipitate was removed with syringe filtration using a 0.2 μm filter (Millipore). The solution was then divided into two separate aliquots: (1) a wash mixture (described earlier) and (2) a final DAB reaction mixture. Hydrogen peroxide (H_2O_2; Sigma–Aldrich) was added to a final concentration of 5.88 mM and mixed vigorously to generate the final DAB reaction mixture.
9. After fixation, HeLa cells were washed three times for 5 min in 0.1 M sodium cacodylate buffer. These wash steps were performed to remove residual glutaraldehyde.
10. Cells were washed in 1 mL of the (1) 1 mg/mL DAB/cacodylate wash mixture for 2 min. This represents a sufficient volume to cover a 35 mm MatTek dish without the cells drying out.
11. The wash mixture was then removed and 1 mL of the (2) DAB/cacodylate final reaction mixture was added to the dish and incubated for 30 min at room temperature. The addition of H_2O_2 results in the oxidation of DAB into a DAB precipitate and this insoluble reaction product, generated at the site of APEX-GBP, can be contrasted by postfixation with osmium tetroxide (OsO_4). It is important to note that heme is also a critical cofactor for the generation of the insoluble DAB precipitate; the addition of heme (if required) is described in Step 4.
12. The cells were washed three times for 5 min with 0.1 M sodium cacodylate buffer to remove all free DAB.
13. HeLa cells were subsequently postfixed with 1% OsO_4 (EMS) in 0.1 M sodium cacodylate buffer for 2 min to convert the DAB reaction product into a discernable electron dense stain surrounding the modular APEX-GBP at the site of the nls-GFP. It should be noted that this osmication time has been optimized for cell culture. Longer osmication times demonstrated increased background electron density but were required for even contrasting of thicker tissue samples.

14. Cells then were washed three times for 2 min in 0.1 M cacodylate buffer to remove the remaining OsO_4. Cells were washed for additional three times in double distilled H_2O.
15. Serial dehydration was then performed with increasing concentrations of ethanol in a PELCO BioWave microwave at 250 W fitted with a PELCO ColdSpot Pro system for temperature control. Cells were initially subjected to a 40 s incubation in 30% ethanol (in double distilled H_2O; vol/vol) in the BioWave then to subsequent incubations in 50%, 70%, 90%, and 100% ethanol twice.
16. Cells were then serially infiltrated with increasing concentrations (25%, 50%, 75%) of LX112 resin (Ladd, Williston, VT) in ethanol (vol/vol) in the BioWave at 250 W under vacuum for 3 min per step. Cells were then infiltrated twice with 100% LX112 resin under the same BioWave conditions. LX112 resin is preferred for cells grown in tissue culture, as this mixture (unlike standard Epon) does not interact with the plastic on the MatTek dish.
17. Cells were flat embedded. LX112 resin was polymerized to hardness at 60°C for 16–24 h. Samples were removed from the oven and allowed to cool to RT (Fig. 2D).
18. Resin dishes were trimmed such that the central coverslip with flat embedded cells was removed from the remaining dish (Fig. 2E). The coverslip remains associated with this central disc and must be removed to visualize the MatTek grid pattern imprinted on the resin. To remove the glass without damaging the sample, the resin disc was cooled in liquid nitrogen and the glass coverslip was lifted off with forceps. Once removed, the alphanumeric code (now inverted) was visualized on the block face (Fig. 2E) using a dissecting microscope (Leica EZ4, Leica microsystems).
19. The disc with embedded HeLa cells was imaged by bright field microscopy on an $EVOS_{FL}$ Cell Imaging System at 4× and 10× magnification to find the region of interest (Figs. 2F and 3B). The DAB reaction product was visible under these imaging conditions (Fig. 3B; regions of interest 1–3).
20. The region of interest was trimmed and super glued onto a blank resin stub and allowed to dry to hardness (Fig. 2G).
21. Sections were cut on an ultramicrotome (Leica EM UC6, Leica Microsystems). The initial sections demonstrated the imprinted grid pattern (Fig. 2H). 60 nm ultrathin sections were then cut using a 45 degrees diamond knife (Diatome) and placed on a formvar and carbon coated 2 bar slot grid (ProSciTech, Australia). Grids were not poststained.

2.1.5 Transmission electron microscopy

22. Grids were imaged on a Philips T12 transmission electron microscope at 120 kV. Digital micrographs were collected using a Direct Electron LC1100 camera under the control of the Navigator program in SerialEM (Boulder, Colorado). Navigator allows for the automated collection of high-resolution

images of entire grids and retains these coordinates in the header information of each image. An initial map of the whole grid was generated at 130× magnification (Fig. 3C; inset). A polygon was then plotted on the map containing the region of interest (Fig. 3C; inset), which was subsequently imaged at 4400× magnification with two-fold binning (Fig. 3C–H). The images were exported as a single .mrc file. Acquiring large montaged data sets improve the confidence and efficiency for unambiguous determination of regions of interest as multiple reference points can be correlated over large areas.

2.1.6 Postimage processing

23. A composite image combining the fluorescent and bright field images (Fig. 3A) acquired in Step 5 was generated using ImageJ (National Institutes of Health, USA).
24. The high-resolution transmission electron micrograph output file was montaged using IMOD (Boulder, Colorado) to generate a single aligned image (Fig. 3C). This program has been the most successful for consistently generating well-aligned montaged images. The piece list file was extracted from the header information of the .mrc file using the program "edmont." "Blendmont" was then utilized to generate a final blended image. Alternatively, images can be manually aligned using Adobe Photoshop CS6 (Adobe Inc., USA) or in an automated process using the Photomerge program also in Adobe Photoshop CS6.
25. The montaged image was then imported into Adobe Photoshop CS6 and manually aligned with the composite image generated in Step 23 (Fig. 3C).
26. The regions of interest (1–4) highlighted in the post-DAB, postresin infiltration step (Fig. 3B) were correlated with the same areas in the montaged electron micrograph overlayed with the fluorescent images of nls-GFP and APEX-GBP (mKate) (Fig. 3C). Correlated high-resolution images demonstrated that regions 1–3 with the highest expression of APEX-GBP and nls-GFP (by fluorescence intensity) also demonstrated the greatest electron density in their nuclei by EM. Cells lacking expression of nls-GFP and APEX-GBP did not possess any electron density within the nucleus (Fig. 3G; region 5).

2.2 MATERIALS AND INSTRUMENTATION

2.2.1 Cell culture

Instrumentation: BH-EN Class II Biological Safety cabinet (Gelaire, QLD, Australia), MCO-18AC CO_2 Incubator (SANYO Electric Co. Ltd, Japan),
Materials: T75 Cell culture flask (Nunc EasYFlask 75 cm^2 Nunclon Δ Surface, Thermoscientific, Denmark), 35 mm Dish, No. 1.5 Gridded coverslip, 14 mm glass diameter (MatTek, Ashland, MA).
Reagents: HeLa cervical epithelial cells (ATCC, CCL-2), DMEM (Gibco), Fetal bovine serum (Serana), L-Glutamine (Sigma–Aldrich), Opti-MEM (Gibco), Lipofectamine 3000 (Life Technologies).

2.2.2 Light microscopy

Instrumentation: $EVOS_{FL}$ epifluorescence Cell Imaging System fitted with DAPI, GFP, and RFP Light cubes (Advanced Microscopy Group, Bothell, WA). The microscope was fitted with 4× (NA 0.13), 10× (NA 0.25), 20× (NA 0.4) and 40× (NA 0.65) lenses.
Software: ImageJ (National Institutes of Health, USA).

2.2.3 Electron microscopy

Instrumentation: EMS 150T E carbon coater (Quorum Technologies Ltd, United Kingdom), T12 Transmission Electron Microscope (120 kV; Philips), 4k × 4k LC-1100 lens coupled CCD camera (Direct Electron, USA), PELCO BioWave fitted with a SteadyTemp Thermocube and vacuum chamber (Ted Pella, Inc, Redding, CA), Leica EZ4 dissecting microscope (Leica Microsystems, Australia), Oven (Scientific Equipment Manufacturers, Australia), Leica EM UC6 ultramicrotome (Leica Microsystems), Frontier FM Floor Mounted fume hood (Esco Micro Pte. Ltd., Singapore).
Materials: Dumont No. 5 tweezers (ProSciTech), Ultra 45 diamond knife (Diatome), copper slot grids (ProSciTech).
Reagents: Sodium cacodylate (Sigma–Aldrich), 25% glutaraldehyde (EMS), 4% aqueous osmium tetroxide (EMS), 3,3′-diaminobenzidine tetrahydrochloride (10 mg tablets; Sigma–Aldrich), LX112 resin kit (Ladd), Formvar (Merck, White House Station, NJ), UHU Super Glue (GmbH & Co KG).
Software: SerialEM (Mastronarde, 2005).

2.2.4 Image processing

Instrumentation: 3.4 GHz Intel Core i7 iMac with 32 GB memory fitted with an NVIDIA GeForce GTX 680MX graphics card.
Software: IMOD 4.7.15 (Kremer, Mastronarde, & McIntosh, 1996), Adobe Photoshop CS6 (Adobe Inc.), ImageJ (NIH).

3. SUBCELLULAR PROTEIN DISTRIBUTION ANALYSIS OF TRANSGENIC ZEBRAFISH

3.1 METHODS

Two different zebrafish lines were generated and are described in detail in Section 1. All zebrafish embryos were harvested for imaging and processing three days postfertilization.

3.1.1 Zebrafish crossing

1. Transgenic carriers were crossed and the offspring sorted on a fluorescent dissecting microscope (Nikon SMZ1500) for presence of the GFP transgene and the APEX2-GBP cassette (using the red lens reporter).

3.1.2 Mounting fish samples for confocal microscopy

2. Dual transgenic zebrafish expressing GFP-CAAX and *inducible zf*APEX-GBP were subjected to a minor heat treatment (39°C) for 2 h and subsequently returned to standard tank temperature for 2 days.
3. Fish for live imaging were anaesthetized in 0.001% tricaine (Ethyl 3-aminobenzoate methanesulfonate, Sigma–Aldrich)
4. Zebrafish were mounted in 1% low melting point agarose on a microscope slide, under a 22 mm × 22 mm coverslip (Menzel).
5. Transgenic zebrafish were imaged on a Zeiss LSM 710 Meta with 40× objective. Fluorescent images were acquired with standard 488 and 405 nm excitation settings (Fig. 4A).

3.1.3 Sample processing for transmission electron microscopy

Sample processing was performed as described earlier; however, differences between cell culture processing and whole organism zebrafish processing are highlighted further.

6. *Step 7*—Zebrafish were fixed in 2.5% glutaraldehyde in E3 (5 mM NaCl, 0.17 mM KCl, 0.33 mM $CaCl_2$, 0.33 mM $MgSO_4$) containing 0.001% tricaine at 80 W for 6 min with 2-minutes-on-2-minutes-off-2-minutes-on cycling under vacuum in a PELCO BioWave. Zebrafish were washed five times for 5 min in E3 media and the head (including the yolk) and tail were removed. The remaining trunk muscle was refixed in 2.5% glutaraldehyde in 0.1 M sodium cacodylate buffer (pH 7.4) for 3 min under vacuum at 80 W in the PELCO BioWave.
7. *Step 13*—DAB and wash steps were preformed exactly as described earlier; however, we have observed that longer osmication times are required for even contrasting of three days postfertilization zebrafish embryos. Therefore, embryos were postfixed with 1% OsO_4 (EMS) in 0.1 M sodium cacodylate buffer for 30 min then subjected to three washes of 0.1 M cacodylate buffer for 2 min each.
8. *Step 16*—Dehydration was performed exactly as described earlier; however, infiltration was performed using EMbed 812 resin (EMS) rather than LX112 resin. Embryos were serially infiltrated with increasing concentrations (25%, 50%, 75%) of EMbed 812 resin in ethanol (vol/vol) in the BioWave at 250 W under vacuum for 3 min per step. Embryos were then infiltrated three times with 100% EMbed 812 resin under the same BioWave conditions.
9. *Step 17*—Embryos were transferred to rubber molds containing EMbed 812 resin and polymerized to hardness at 60°C for 48 h.
10. *Step 21*—Thin (60 nm) and thick (180 nm) sections were cut using a 45 degrees diamond knife (Diatome) and placed on a formvar and carbon coated 2 bar slot grid (ProSciTech, Australia).

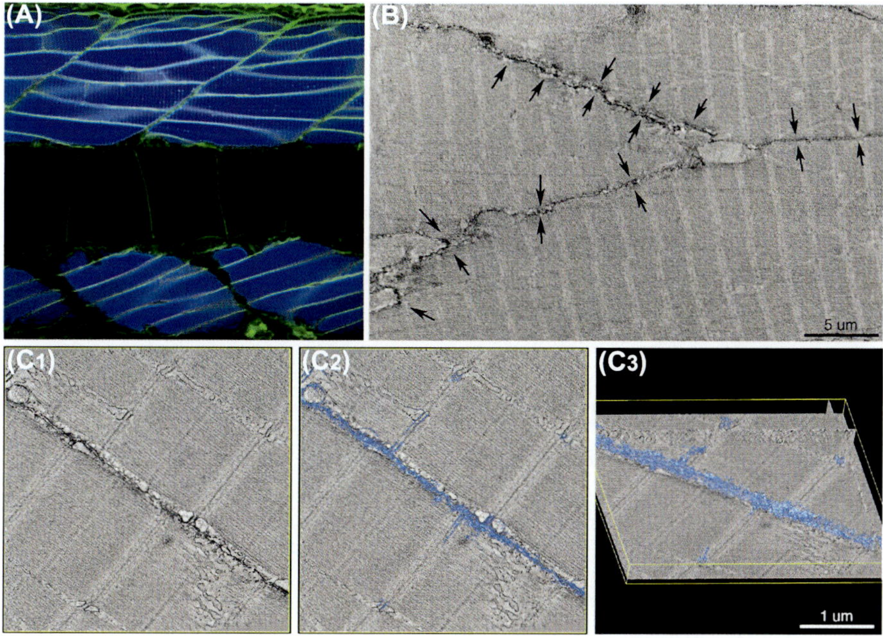

FIGURE 4

(A) Confocal slice of a transgenic zebrafish expressing a GFP-CAAX (membrane anchor), BFP (under the control of a muscle cell–specific promoter) to denote muscle cells and APEX-GBP. (B) A transmission electron micrograph of GFP-CAAX and *inducible zf*APEX-GBP double transgenic zebrafish demonstrating significant electron density at the plasma membrane of zebrafish muscle cells; *arrows* denote electron dense areas. ($C_{1 \text{ and } 2}$) Optical slices from a reconstructed electron tomogram two adjacent zebrafish muscle cells from a double transgenic (GFP-CAAX and *inducible zf*APEX-GBP) zebrafish. ($C_{2 \text{ and } 3}$) Density thresholding was applied to the whole reconstructed volume of the electron tomogram to generate an unbiased render of the areas with the greatest electron density (blue). GFP-CAAX is detected at the cell surface and also within the transverse tubule network. Yellow = bounding box.

3.1.4 Transmission electron microscopy

11. *Step 22*—Thick sections were placed on a droplet of 10 nm colloidal gold for 5 min as fiducial markers and washed three times for 5 min in double distilled water; this process was repeated for each side of the grid. Grids were lightly carbon coated each side and imaged on a Philips T12 transmission electron microscope at 120 kV. Digital micrographs were collected using a Direct Electron LC1100 camera under the control of SerialEM (Boulder, Colorado). Dual axis tilt series were acquired from −60 to +60 degrees at 1 degree

increments without binning. Thin sections were imaged on a Philips T12 transmission electron microscope at 120 kV under the control of SerialEM (Fig. 4B).

3.1.5 Postimage processing

12. *Step 24*—Dual-axis tilt series were reconstructed using weighted back-projection in eTomo in the IMOD software suite (Boulder, Colorado). The isosurface render program in IMOD was used to generate the full three-dimensional segmentation of areas with the greatest electron density. The entire tomographic volume was subjected to equivalent density-based thresholding (Fig. $4C_1$–C_3).

3.2 MATERIALS AND INSTRUMENTATION

3.2.1 Subcellular protein distribution analysis of transgenic zebrafish

3.2.1.1 Zebrafish

Animals: Compound transgenic animals from an incross of GFP-expressing protein of interest line (e.g., bact2-GFPcaax^{pc10}) with APEX2-GBP expressing line (HSP70l-APEX2-GBP^{uq4rp} or bact2-APEX2-GBP^{uq3rp}).

3.2.1.2 Confocal microscopy

Instrumentation: Zeiss LSM 710 Meta with ×40 objective (1.3 NA).
Materials: Microscope slides (Sigma–Aldrich) and coverslips (Menzel, 22 mm × 22 mm).
Reagents: 1% low-melting point agarose (Sigma–Aldrich) in E3 media (5 mM NaCl, 0.17 mM KCl, 0.33 mM $CaCl_2$, 0.33 mM $MgSO_4$), tricaine (Sigma–Aldrich).

3.2.1.3 Electron microscopy

Instrumentation: EMS 150T E carbon coater (Quorum Technologies Ltd, United Kingdom), T12 Transmission Electron Microscope (120 kV; Philips), 4k × 4k LC-1100 lens coupled CCD camera (Direct Electron, USA), PELCO BioWave fitted with a SteadyTemp Thermocube and vacuum chamber (Ted Pella, Inc, Redding, CA), Leica EZ4 dissecting microscope (Leica Microsystems, Australia), Oven (Scientific Equipment Manufacturers, Australia), Leica EM UC6 ultramicrotome (Leica Microsystems), Frontier FM Floor Mounted fume hood (Esco Micro Pte. Ltd., Singapore).
Materials: Dumont No. 5 tweezers (ProSciTech), Ultra 45 diamond knife (Diatome), copper slot grids (ProSciTech).
Reagents: Sodium cacodylate (Sigma-Aldrich), 25% glutaraldehyde (EMS), 4% aqueous osmium tetroxide (EMS), 3,3′-diaminobenzidine tetrahydrochloride (10 mg tablets; Sigma-Aldrich), EMbed 812 resin kit (EMS), Formvar (Merck, White House Station, NJ).
Software: SerialEM (Mastronarde, 2005).

3.2.1.4 Image processing

Instrumentation: 3.4 GHz Intel Core i7 iMac with 32 GB memory fitted with an NVIDIA GeForce GTX 680MX graphics card.
Software: IMOD 4.7.15 (Kremer et al., 1996), Adobe Photoshop CS6 (Adobe Inc.), ImageJ (NIH).

3.3 DISCUSSION

The APEX-GBP method is a simple and robust tool for CLEM. It allows for fast and reliable detection of the subcellular distribution of any GFP-tagged protein of interest by fluorescence microscopy and, with the cotransfection of the APEX-GBP vector, the high-resolution detection of the corresponding region by EM.

We have routinely utilized confocal microscopy to confirm that GFP-tagged proteins of interest are not disrupted by the expression of the modular APEX construct. The initial APEX-GBP vector (Ariotti et al., 2015) lacked the coexpression of a fluorescent reporter. To determine if the expression of this construct was disruptive to the protein of interest, it was necessary to image large numbers of GFP-expressing cells with and without cotransfection of APEX-GBP. This proved cumbersome for CLEM studies as we lacked a direct readout to determine if a cell of interest was expressing the APEX-GBP construct; we reasoned a system with a fluorescent reporter would improve the technique's applicability for CLEM-based analyses. Moreover, as APEX-GBP is ~64 kDa in size when bound to GFP (and ~41 kDa alone), the direct addition of another fluorescent protein into the complex would likely have resulted in mislocalization of any GFP-tagged protein. Therefore, we chose to pursue a nonconjugated fluorescent reporter system.

The APEX-GBP (mKate) construct used in this study has been optimized for CLEM-based analysis. It has been engineered to express a nonconjugated mKate fluorescent reporter with a porcine teschovirus 2A sequence between the mKate and the APEX-GBP domains. The addition of the nonconjugated reporter allows for the unambiguous and simultaneous demarcation of doubly transfected cells by fluorescence microscopy. Moreover, the P2A site results in self-cleavage of the mKate from the APEX-GBP during protein translation (Szymczak-Workman et al., 2012), which results in an ~1:1 expression ratio between the mKate and the modular APEX protein. This ratio means that the fluorescence intensity of the mKate in any given cell correlates directly with the level of APEX-GBP expression. This is not necessarily the case with other vectors that also express fluorescent reporters (i.e., the pIRES vector) (Szymczak-Workman et al., 2012). The direct correlation between the nonconjugated mKate fluorescence intensity and APEX-GBP expression means that potential changes to the subcellular distribution of the GFP-tagged protein of interest can be directly assessed by fluorescence microscopy over a wide range of APEX-GBP (mKate) expression levels.

The use of APEX-GBP is not restricted to cell culture methods and transient expression of fluorescently tagged proteins of interest. We have developed two different APEX-GBP expressing transgenic zebrafish lines for the analysis of the protein distributions in whole organisms (Hall et al., 2016). By crossing either the *zf*APEX-GBP line (which constitutively expresses APEX-GBP in all tissues) or the *inducible zf*APEX-GBP line (which is under the control of a heat shock promoter) with any established GFP/YFP transgenic zebrafish line, it is possible to analyze protein distribution in all tissues (Ariotti et al., 2015). Unlike traditional immuno-EM techniques, where antigens can only be detected on the surface of an ultrathin section, APEX-mediated detection of protein distribution is compatible with volume-EM methods including electron tomography (Fig. 4C) and serial block face sectioning scanning EM (Ariotti et al., 2015). An enzymatic approach with the APEX tag and the DAB reaction results in three-dimensional electron density at the site of the protein of interest, which is clearly demonstrated in Fig. 4C using electron tomography. By using APEX-GBP, it is possible to analyze the subcellular localization of any given GFP-tagged protein in whole cells and potentially whole organisms, which provides a more dynamic view of the cellular distributions of proteins of interest.

APEX-GBP has many significant advantages over current CLEM technologies, but this method is not without some caveats. It is critical that controls are used for each experiment and that all GFP-tagged constructs are screened for potential subcellular mislocalization. Additionally, certain considerations must be taken into account before attempting CLEM using this method. (1) The expression level of APEX-GBP. Given that APEX-GBP is a soluble cytosolic protein when expressed in the absence of GFP (Ariotti et al., 2015), careful optimization of expression conditions are required prior to high-resolution analyses. If the expression of APEX-GBP far exceeds the expression of the GFP-tagged protein of interest there will be increased cytoplasmic electron density, which could potentially obscure the distribution of the protein of interest. We have optimized transfection conditions such that we routinely employ a 1:1 ratio of GFP DNA to APEX-GBP DNA. Furthermore, the use of APEX-GBP (mKate) vector provides a cell-to-cell indication of the level of APEX-GBP expression. A recent study demonstrated that the GBP could be mutated into a conditionally stable state such that when bound to GFP the protein is stable but when unassociated the protein is targeted for proteosomal degradation (Tang et al., 2016). The adaption of a conditionally stable form of APEX-GBP could potentially overcome any potential saturation effects caused by the overexpression of APEX-GBP. (2) The localization of cytosolic proteins and proteins with multiple subcellular distributions, including soluble pools, need to be carefully assessed as APEX-GBP is also cytosolic marker. (3) The DAB reaction product can diffuse away from the site of the GFP-bound APEX tag; this is not the case with immunogold labeling. While the potential diffusion of this product could result in reduced resolution, our studies have demonstrated a quantifiable reduction in electron density by line scan analysis below the resolution limit of many immunogold labeling techniques (Ariotti et al., 2015) so this remains only a minor concern.

The modular expression of the APEX2 tag engineered to the camelid-derived GFP-binding peptide relies on the most commonly used fluorescent tag in cell biology today, GFP. The tracking of the GFP-tag by fluorescence microscopy and the subsequent detection of GFP by an enzymatic-tag for EM offers a simple and easily applied alternative for many of the complex CLEM techniques currently available.

REFERENCES

Ariotti, N., Hall, T. E., Rae, J., Ferguson, C., McMahon, K. A., Martel, N., … Parton, R. G. (2015). Modular detection of GFP-labeled proteins for rapid screening by electron microscopy in cells and organisms. *Developmental Cell, 35*(4), 513–525. http://dx.doi.org/10.1016/j.devcel.2015.10.016.

Hall, T. E., Ariotti, N., Ferguson, C., Xiong, Z., & Parton, R. G. (2016). New transgenic lines for localization of GFP-tagged proteins by electron microscopy. *Zebrafish, 13*(3), 232–233. http://dx.doi.org/10.1089/zeb.2016.29002.hal.

Kirchhofer, A., Helma, J., Schmidthals, K., Frauer, C., Cui, S., Karcher, A., … Rothbauer, U. (2010). Modulation of protein properties in living cells using nanobodies. *Nature Structural & Molecular Biology, 17*(1), 133–138. http://dx.doi.org/10.1038/nsmb.1727.

Kremer, J. R., Mastronarde, D. N., & McIntosh, J. R. (1996). Computer visualization of three-dimensional image data using IMOD. *Journal of Structural Biology, 116*(1), 71–76. http://dx.doi.org/10.1006/jsbi.1996.0013.

Lam, S. S., Martell, J. D., Kamer, K. J., Deerinck, T. J., Ellisman, M. H., Mootha, V. K., & Ting, A. Y. (2015). Directed evolution of APEX2 for electron microscopy and proximity labeling. *Nature Methods, 12*(1), 51–54. http://dx.doi.org/10.1038/nmeth.3179.

Martell, J. D., Deerinck, T. J., Sancak, Y., Poulos, T. L., Mootha, V. K., Sosinsky, G. E., … Ting, A. Y. (2012). Engineered ascorbate peroxidase as a genetically encoded reporter for electron microscopy. *Nature Biotechnology, 30*(11), 1143–1148. http://dx.doi.org/10.1038/nbt.2375.

Mastronarde, D. N. (2005). Automated electron microscope tomography using robust prediction of specimen movements. *Journal of Structural Biology, 152*(1), 36–51. http://dx.doi.org/10.1016/j.jsb.2005.07.007. pii: S1047-8477(05)00152-8.

Szymczak-Workman, A. L., Vignali, K. M., & Vignali, D. A. (2012). Design and construction of 2A peptide-linked multicistronic vectors. *Cold Spring Harbor Protocols, 2012*(2), 199–204. http://dx.doi.org/10.1101/pdb.ip067876.

Tang, J. C., Drokhlyansky, E., Etemad, B., Rudolph, S., Guo, B., Wang, S., … Cepko, C. L. (2016). Detection and manipulation of live antigen-expressing cells using conditionally stable nanobodies. *eLife, 5*. http://dx.doi.org/10.7554/eLife.15312.

Tokuyasu, K. T. (1986). Application of cryoultramicrotomy to immunocytochemistry. *Journal of Microscopy, 143*(Pt 2), 139–149.

CHAPTER

Correlation of live-cell imaging with volume scanning electron microscopy

Miriam S. Lucas*,[1], Maja Günthert*, Anne Greet Bittermann*, Alex de Marco[§], Roger Wepf[¶]

**ETH Zurich, Zurich, Switzerland*

[§]Monash University, Clayton, VIC, Australia

[¶]The University of Queensland, Brisbane, QL, Australia

[1]Corresponding author: E-mail: miriam.lucas@scopem.ethz.ch

CHAPTER OUTLINE

Methods in Cell Biology, Volume 140, ISSN 0091-679X, http://dx.doi.org/10.1016/bs.mcb.2017.03.001

Abstract

Live-cell imaging is one of the most widely applied methods in live science. Here we describe two setups for live-cell imaging, which can easily be combined with volume SEM for correlative studies. The first procedure applies cell culture dishes with a gridded glass support, which can be used for any light microscopy modality. The second approach is a flow-chamber setup based on Ibidi μ-slides. Both live-cell imaging strategies can be followed up with serial blockface- or focused ion beam-scanning electron microscopy. Two types of resin embedding after heavy metal staining and dehydration are presented making best use of the particular advantages of each imaging modality: classical en-bloc embedding and thin-layer plastification. The latter can be used only for focused ion beam-scanning electron microscopy, but is advantageous for studying cell-interactions with specific substrates, or when the substrate cannot be removed. En-bloc embedding has diverse applications and can be applied for both described volume scanning electron microscopy techniques. Finally, strategies for relocating the cell of interest are discussed for both embedding approaches and in respect to the applied light and scanning electron microscopy methods.

INTRODUCTION

Correlative light and electron microscopy (CLEM), not being one single technique but rather a family of techniques, offers a wide range of applications and combinations of various light and electron microscopy modalities (de Boer, Hoogenboom, & Giepmans, 2015; Jahn et al., 2012; Karreman, Hyenne, Schwab, & Goetz, 2016; Loussert Fonta & Humbel, 2015; Lucas, Günthert, Gasser, Lucas, & Wepf, 2012). This chapter describes options for correlating live-cell imaging with volume scanning electron microscopy (SEM).

In order to achieve optimum conditions for live-cell imaging controlled environmental conditions are required in terms of temperature and gas concentrations, plus a constant supply of fresh cell culture medium. For the control of the environmental conditions, various solutions are available, including flow-chambers, incubator boxes, or heated microscope tables. While the latter two are primarily used to control temperature and gas concentration, flow-chambers have the advantage of controlled liquid handling for media changes, including onset of perfusion with effector media, e.g., to apply trigger molecules and signals, or finally chemical fixatives (Droste et al., 2005). Additionally, the imaging conditions need to be chosen in order to prevent the incident light dose from being harmful to the cells. This can e.g., be achieved by applying fast imaging and detection techniques such as conventional wide field light microscopy (LM), reducing the photon dosage by light sheet fluorescence microscopy, two photon laser scanning or spinning disk confocal imaging techniques, or by reducing the interaction volume of the laser beam with the cells, as with total internal reflection fluorescence (TIRF) microscopy (Ettinger & Wittmann, 2014). With the exception of two photon laser scanning microscopy, all of these techniques allow monitoring and capturing of fast processes, which is one of the major attractions of live-cell imaging (Gibson, Vorkel, Meissner, & Verbavatz, 2014; Karreman et al., 2016; Spiegelhalter et al., 2010).

Theoretically, live-cell imaging can be correlated with any available electron microscopy (EM) technique. However, not all biological questions require the highest achievable imaging resolution in EM, but on the other hand benefit greatly from three-dimensional portrayal of the structure of interest in its natural context and at the ultrastructural level. Therefore, volume SEM, in particular focused ion beam-SEM (FIB-SEM), or serial blockface-SEM (SBF-SEM) is well suited as follow-up technique (Denk & Horstmann, 2004; Peddie & Collinson, 2014; Russell et al., 2017). Both techniques require heavy metal staining to highlight e.g., membranes, the main compartmentalization structure of biology and most important descriptor of intracellular structures, followed by dehydration and resin-embedding (T. Deerinck, Bushong, Lev-Ram, Shu, Tsien, & Ellisman, 2010; Knott, Marchman, Wall, & Lich, 2008). For FIB-SEM, two approaches to resin embedding are available: classical en-bloc embedding and thin-layer plastification (TLP). The former encases the specimen in a volume of resin, whereby the structure of interest, i.e., the cell monolayer, needs to be positioned at the edge of the block in order to be accessible for FIB-SEM. Here, the cell substrate needs to be removed prior to FIB-SEM and the samples are mounted and imaged upside-down, with the cell's basal region exposed on the topside of the resin block. By contrast, TLP just barely covers the specimen with a thin film of resin, and thus renders shape and position of the respective specimen visible, which is particularly useful for the investigation of cells on support types that cannot be removed, or when the contact between cells and substrate are to be investigated. In this case, the cells remain attached to their substrate, with the interface included in the field of view of the FIB-SEM images, i.e., the samples are mounted upright, with the cell substrate at the bottom (Bittermann, Schaer, Mitsi, Vogel, & Wepf, 2012; Kizilyaprak, Bittermann, Daraspe, & Humbel, 2014).

One of the major challenges in CLEM is the relocation of the region of interest (ROI) identified by fluorescence LM, in the electron microscope (Karreman et al., 2014). It is therefore of utmost importance to document the ROI, its surroundings, and any structures that could potentially be used as fiducial markers at each point in time of the workflow, i.e., not only during live-cell imaging, but also during sample preparation and subsequent EM imaging. Commercial solutions are available from the major microscope suppliers for documentation and relocation, and establishing a connectivity between LM and EM (e.g., Shuttle & Find, or Atlas by Carl Zeiss Microscopy, FEI MAPS by Thermo Fischer Scientific, MirrorCLEM by Hitachi High-Tec and Riken, or MiXcroscopy by Jeol and Nikon). These software and hardware solutions can support and enhance CLEM workflows, making them not only more comfortable, but also more reproducible and reliable.

1. RATIONALE

Two types of setups for live-cell imaging are described in this chapter. First, a fluidic setup fitted to the CorrSight LM platform (FEI Company), the "Live Module." The core of this module is a microscope slide-shaped flow-chamber with six cell culture wells, each containing an embossed finder grid. This is attached to a fluid handling system designed for controlled, reproducible perfusion in a closed system. This module is used in a controlled environment, including a heated microscope table and optional gas control in the imaging chamber of the CorrSight. The second setup employs standard cell culture dishes with gridded coverslip glass bottom fitted into the CorrSight imaging chamber.

Both described live-cell CLEM workflows are based on chemical fixation, followed by standard room temperature dehydration and resin embedding for volume SEM. Both, en-bloc embedding and TLP were used for FIB-SEM, while for SBF-SEM, only en-bloc embedding can be applied.

Connectivity between LM and EM and relocation of the ROI were established either manually (for SBF-SEM) or software assisted (for FIB-SEM), using FEI MAPS or the Atlas software (Zeiss). Cell substrates with finder grids were used in both approaches, in order to simplify the relocation of the cells of interest.

Both approaches have advantages and disadvantages. Among these are the applicability of TIRF or spinning disk imaging, speed of liquid exchanges, or orientation of the imaging plane in EM.

2. METHODS

Here we describe two exemplary workflows for correlating live-cell imaging with volume SEM: one employing TIRF imaging to capture fast events near the cell substrate interface, and a second approach making use of benefits of the "Live Module" fitted to the FEI CorrSight. Both approaches can be followed up by similar sample preparation for EM and can be combined with multiple EM imaging modalities (Fig. 1). A suitable workflow needs to be chosen and adapted according to the aim of the respective study.

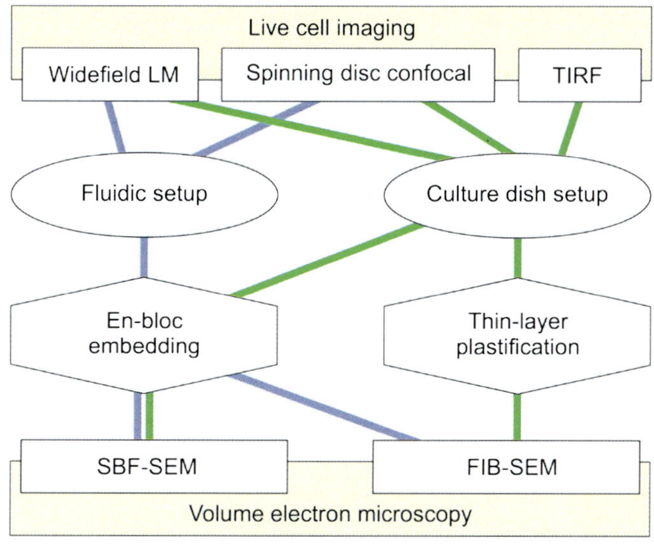

FIGURE 1

Schematic representation of the described workflows for live-cell CLEM. Both live-cell imaging approaches can be combined with all described LM modalities, with the exception of TIRF imaging, which cannot be used with the fluidic setup based on the Ibidi μ-slides. Due to geometric properties of the Ibidi μ-slides, this approach is better suited for en-bloc embedding, while culture dishes can be used for both embedding techniques. Thin-layer plastification can be used only for FIB-SEM. But en-bloc embedding is suitable for both volume SEM methods.

2.1 CHOICE OF CELL CULTURE SUBSTRATES

2.1.1 Flow-chamber setup

The approach employing the "Live Module" attached to the CorrSight is based on the Ibidi culture slides "μ-Slide CorrSight Live." These are small flow-chambers, molded into a microscope slide-shaped carrier, with a polymer bottom (Fig. 2). These slides contain three parallel channels with each two wells in a row, which can be attached to a fluid-handling system for perfusion. Each well contains a finder grid, which can be visualized by brightfield or fluorescence LM. Cells are seeded in the open wells and cultured until appropriate for imaging. The wells are then sealed with an adhesive foil to create the flow-chamber. Although this provides ideal conditions for live-cell imaging, the polymer-bottom with the etched finder grid is not compatible with TIRF imaging, due to the heterogeneity of the material. However, it is perfectly suited for wide field LM or confocal imaging, and compatible with any type of resin embedding.

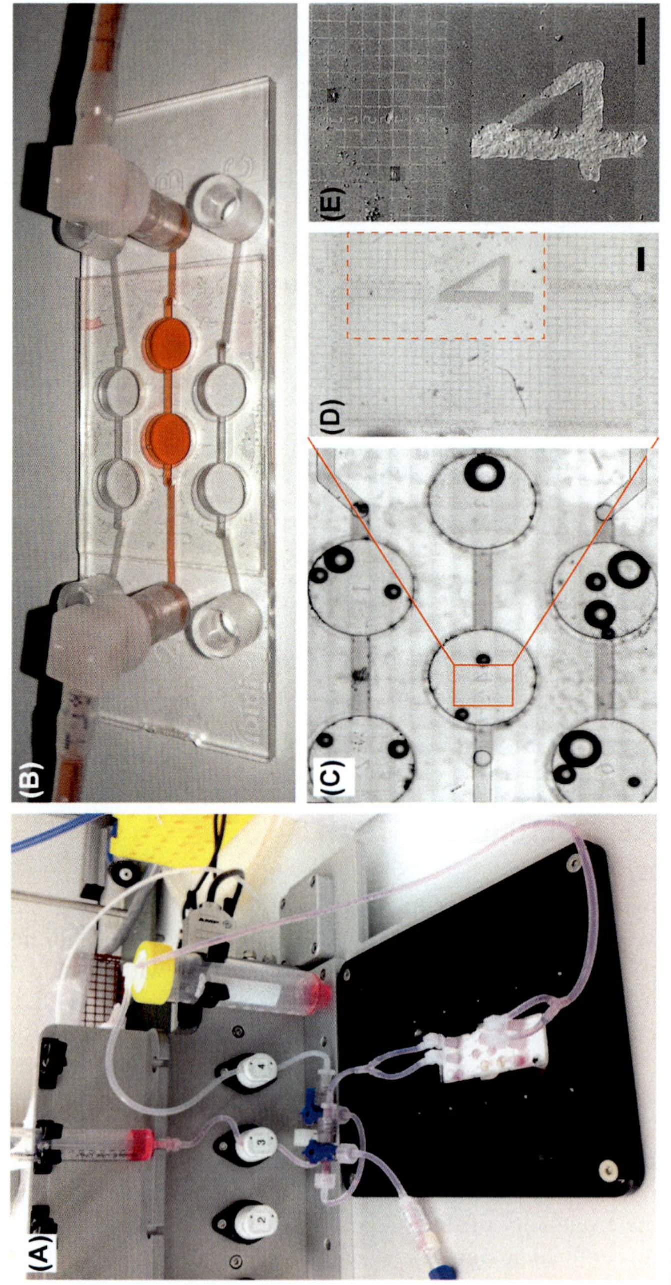

FIGURE 2

Fluidic setup. (A) The tray of the "Live Module," holding the Ibidi μ-slide attached to the reservoir and waste container for perfusion. (B and C) Close-up and LM overview of the Ibidi μ-slide CorrSight Live, with one of the three parallel channels connected to the pump for perfusion. (D) A finder grid is etched in the polymer coverslip bottom, which is well visible in LM. (E) It is also well perceivable in SEM on the resin surface after removing the polymer foil after en-bloc embedding. Scale bars: 300 μm.

2.1.2 Cell culture dishes with gridded glass substrate

The second approach employs glass-bottom cell culture dishes. The major benefit of these is the applicability for TIRF imaging. A finder grid either in the same focal plane as the cells or in a different focal plane, i.e., on the bottom of the glass, is beneficial, although not implicitly needed when using the connectivity offered by the above mentioned correlative software and hardware solutions. These culture dishes are available in different sizes and glass thickness from several suppliers. We used 35-mm diameter dishes from MatTek and Ibidi. Both have advantages for different applications and it is recommended to choose the cell substrate according to the intended workflow for CLEM.

The Ibidi culture dishes offer a very large glass bottom (approx. 2.5 cm in diameter) and thus a large sample area for imaging. However, these dishes are not optimal for en-bloc embedding, as detaching the large glass from the resin block is difficult. This dish is therefore best suited for TLP approaches. The MatTek dishes on the other hand offer a smaller glass inlay (approx. 1 cm^2). Despite the smaller sample area, these dishes are better suited for en-bloc embedding, because the glass can be easily detached from the cured resin. The location of the finder grid, either in the same or a different focal plane as the cells should also be considered with respect to the embedding technique. For en-bloc embedding it is beneficial if the finder grid is in the same focal plane as the cells. As finder grids are etched into the glass, the grid will be molded onto the resin surface and thus remain visible for LM and SEM after detaching the glass. As for TLP, the glass is not detached from the embedded sample, it is recommended to choose a dish with the finder grid in a different focal plane. That way it will remain visible at least in LM, while a grid on the topside of the glass would become invisible by covering it with resin; however this renders relocation of the ROI in the SEM more challenging.

2.2 CELL CULTURE

An U2OS-derived stable cell line co-expressing a mitochondrial and an endoplasmatic reticulum marker (mtBFP and sec61a-GFP, respectively), called KERMIT (Kanfer et al., 2015), was used to demonstrate both approaches. These cells were grown in the above described culture dishes or slides and kept under culture conditions until imaging. Live-cell imaging was carried out at 37°C, but without CO_2 incubation, as the imaging period did not exceed 15 min and the cells did not show any negative effects due to lack of CO_2.

When using the flow-chamber setup, the cells were seeded in the open wells of the Ibidi μ-slides; after allowing the cells to settle and adhere for 2 h the fluidic channels were filled with medium. The cells were then cultured overnight until the desired confluency was reached. Prior to each experiment, the fresh culture medium was added and the wells were closed with the adhesive foil. Then the channels were filled by carefully sucking the medium through the chambers and channels using a syringe connected to the outlet channel via a 15-cm long piece of tube, while slightly tilting the slide to allow air bubbles to be washed out. When disconnecting the slides from the tubing and reconnecting it to the pump of the fluidic device, one has to pay attention not to cause any pressure or suction in the flow-chambers to prevent new air bubble formation. In the fluidic setup the slide is connected in such a way that the medium is pressed through the flow-chambers at flow-rates of 1−10 mL/min, depending on the viscosity of the respective solution. The outlet can be directly and safely collected in a waste container (Fig. 2).

2.3 LIVE-CELL IMAGING, FIXATION, AND 3D IMAGING

For both approaches, the imaging procedure was the same, starting with a low-magnification (5× objective) mosaic covering the entire field of view of the respective culture dish. This was recorded in transmission mode to visualize the finder grid, as well as in fluorescence mode showing the cells. This map was used to define an ROI, showing the desired confluency, viable cells, and the desired fluorescence signal. The ROI was then imaged in a smaller mosaic using the 40× objective, including the neighboring cells and the finder grid coordinates to enable easy relocation of the cell of interest in subsequent imaging steps. In case of the finder grid being in a different focal plane, a second tile-set was collected with the respective focus settings.

Once the cell of interest was chosen, time-lapse image series were recorded using TIRF. Directly after the acquisition of the time-lapse images the cells were fixed. In the dish setup, fixative was pipetted directly into the culture dish, adding an equal amount of double concentrated fixative to the existing cell culture medium. In order to prevent the culture dishes from shifting position when removing the lid to add the fixative buffer, the dishes were fixed on the microscope table with Blu-Tack, a reusable putty-like adhesive. When using the fluidic device, the flow was changed from culture medium to standard strength fixative solution.

Once the cells were fixed additional images were acquired to capture the final, i.e., fixed state and position of the cells previously recorded in the time-lapse series. After the final image acquisition, the culture dishes were removed from the microscope, the fixative-medium mixture was replaced by fresh, standard strength fixative buffer and stored on ice until further processing. In case of the fluidic setup it would be possible to continue with fixation and subsequent washing, staining, and dehydration steps directly on the microscope. However, considering the toxicity of certain staining solutions applied during sample preparation for EM, it is recommended to move the entire tray holding the fluidic setup and the pump into a fume hood. Alternatively, the Ibidi μ-slides can be removed from the tray, opened up by detaching the adhesive foil to perform further preparation steps on the open wells.

2.4 SAMPLE PREPARATION FOR ELECTRON MICROSCOPY

2.4.1 Staining and dehydration

For both approaches the samples were prepared for EM following a shortened version of the protocol for SBF-SEM as described by T. J. Deerinck, Bushong,

Thor, and Ellisman (2010). Briefly, cells were rinsed with cacodylate buffer and post-fixed in potassium ferrocyanide-reduced osmium tetroxide in the same buffer, containing calcium chloride. This is followed by incubation in thiocarbohydrazide, osmium tetroxide, an overnight uranyl acetate step, and en-bloc Walton's lead aspartate staining. Then, cells were dehydrated in an ascending ethanol series, embedded in epoxy resin, and cured at 60°C for three days.

All described staining, dehydration, and embedding steps were performed in the respective culture vessels. Therefore, the use of acetone should be omitted due to its plastic dissolving properties. Theoretically, all steps using nonviscous solutions, i.e., until the start of the epoxy-infiltration, could be performed using the fluid handling system of the "Live Module." However, it has proven more practical to disconnect the μ-slides from the flow setup to free the microscope and fluid handling system for further image acquisition. For manual processing, it is more convenient to remove the foil covering the wells and pipet the respective solutions directly into the sample containers. This has the additional economic benefit of reducing the volumes of the respective solutions used for each embedding step.

2.4.2 En-bloc embedding

Depending on which volume SEM technique will be applied, the cell monolayers can either be en-bloc embedded or thin-layer plastified. For en-bloc embedding, the indentation of the culture dish holding the gridded cover glass is filled with an approx. 1-mm thick layer of resin, while making sure that the resin is not smeared over the rim of this notch. After polymerization, the cover glass can be removed by carefully separating it from the still warm resin and plastic dish with a fine razor blade, and the resin disk can be pressed out of the plastic dish (Fig. 3C). This works best with the MatTek dishes. When using the Ibidi μ-slides, the wells can be filled with resin to create a disk-shaped specimen analogous to the above described approach. After curing the resin, the polymer foil at the bottom is peeled off and the resin tablet pressed out of the slide. Alternatively, a resin-filled BEEM capsule can be placed upside down onto each well, which after polymerization can be used as a handle to remove the embedded samples from the μ-slides (Fig. 3A).

The resin disks can easily be trimmed into small pieces using a razor blade. As the finder grid is imprinted on the surface of the resin, the respective square of the finder grid holding the ROI can easily be relocated and extracted. This type of sample preparation is well suited for FIB-SEM and SBF-SEM, with the cells embedded upside down beneath the surface of the resin bloc. Although trimming is imperative for SBF-SEM, the resin disk can be mounted as a whole for FIB-SEM.

2.4.3 Thin-layer plastification

TLP is achieved by positioning the cell substrate upright after the final epoxy infiltration steps, to allow the resin to drain off before polymerization, leaving only a minimum layer covering the cells. Still in upright position, the dishes are placed in the cold polymerization oven, which is then slowly heated to 40°C. After 30 min at 40°C, the temperature is raised to 60°C for polymerization. After resin

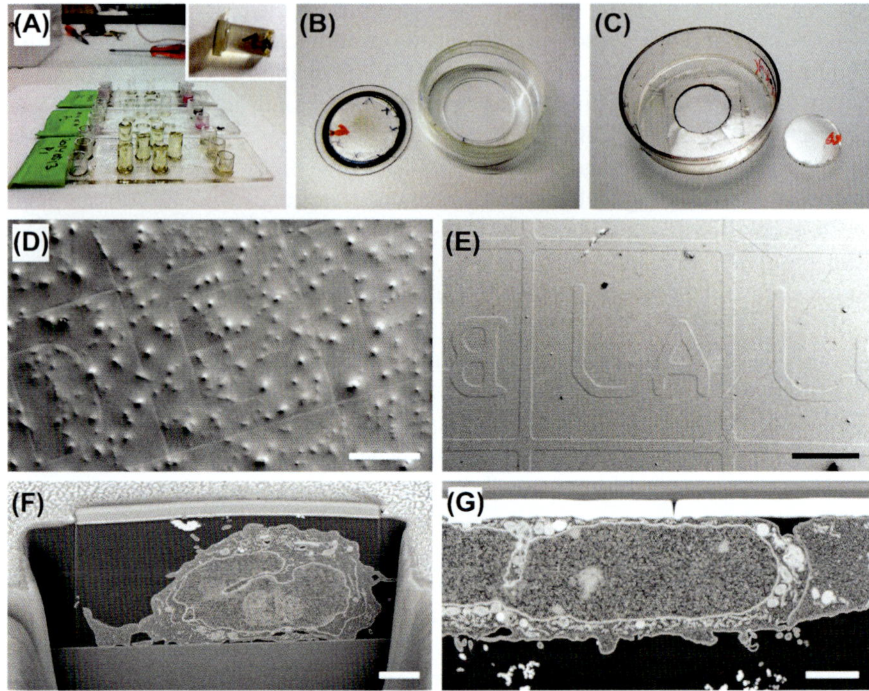

FIGURE 3

Thin-layer embedding versus en-bloc embedding. (A) Ibidi μ-slides after dehydration and resin embedding. The wells can either be filled with resin to form a flat tablet-shaped sample, or resin-filled BEEM capsules can be placed upside down onto each well to form a larger resin block (inset). (B) Ibidi μ-dish after TLP and removal of the glass inlay holding the embedded cells. (C) MatTek culture dish after removal of the gridded cover glass and pressing out the resin disk. (D) Thin-layer plastified cell monolayer on an Ibidi μ-dish 500. Cells and finder grid are well perceivable in SEM. (E) SEM image of the surface of an en-bloc embedded sample showing the imprint of the MatTek finder grid. (F) FIB-SEM cross-section of a thin-layer plastified cell, still attached to the glass substrate of the Ibidi μ-dish. The cell is imaged upright, with the FIB milling into the sample from the top. (G) After en-bloc embedding, the gridded cover glass has to be removed to become accessible for FIB-SEM. The cells are imaged upside down, "hanging" from the resin surface. The sample was coated with a platinum deposition (white) as a protection layer to prevent beam damage during ion milling, which is topped with a carbon deposition (dark) to highlight the registration marks in the platinum layer. Scale bars: (D and E) 200 μm; (F and G) 2 μm.

curing, the whole glass substrate holding the embedded cells has to be carefully detached from the plastic culture dish (Fig. 3B). This preparation approach is well suited for FIB-SEM and prevents any unwanted effects caused by detaching the glass substrate. It thus enables investigation of contact sites and interaction of cells with their substrate or as here described studying structures in the proximity of the cover glass previously imaged by TIRF.

In principle, this approach can also be applied to the flow-chamber slides, although the small diameter of the wells does not allow the resin to drain off completely, causing a slope towards the downward part of the well.

2.4.4 Specimen mounting for volume SEM

For FIB-SEM entire resin disks or the glass substrates prepared by TLP are glued onto SEM stubs, using conductive epoxy glue. For SBF-SEM, the piece of the resin disk containing the ROI needs to be extracted and trimmed to a size of maximum $1 \times 1\ mm^2$. This resin piece is then glued onto the sample pin for SBF-SEM, preferably perpendicular or at a shallow angle towards the diamond knife of the SBF-SEM. The conductive epoxy glue can be cured at room temperature overnight. However, in order to achieve optimum conductivity and adhesion, we bake the mounted samples at 100°C for 10 min before allowing the epoxy glue to harden overnight. The specimen is then sputter-coated with a 5–10 nm layer of gold or platinum to render them conductive for SEM.

2.5 RELOCATION OF CELLS OF INTEREST AND VOLUME SEM IMAGING

Prior to trimming and mounting the embedded specimen for EM, the resin disks were imaged again by LM. The intense staining with heavy metal salts quenches the fluorescence, which renders the cells dark brown to black, making them easy to discern in transmission bright field LM. This serves as a control ensuring the cells of interest have not detached during sample preparation. Tile-scans including the ROI and the finder grid, or any noticeable features that could be helpful when relocating the ROI in EM, were performed again using the MAPS acquisition software and stored in the same project as the live-cell imaging data. In case the finder grid is not perceivable in transmission mode, the reflection signal of the sample surface can be used and superimposed on the images showing the position of the cells. These post-embedding images were aligned with respect to the live-cell data using the multipoint alignment function of MAPS.

2.5.1 Focused ion beam-scanning electron microscopy

The correlative workflow implemented with the MAPS or Atlas software facilitates the relocation of an ROI. MAPS projects containing the previously recorded LM data can be directly loaded on FEI FIB-SEMs. However, the LM images can also be used with the Atlas software. In both cases, a multipoint alignment can again be used to align the top view of the specimen recorded in SEM with the existing LM images. For this purpose, the finder grid imprinted on the surface of en-bloc–embedded samples (Figs. 3E and 4A) or any other well recognizable feature on the sample surface can be used.

With TLP specimen, the cell shapes are molded into the very thin layer of resin covering them, so the positions and cell shapes can be well discerned, helping with

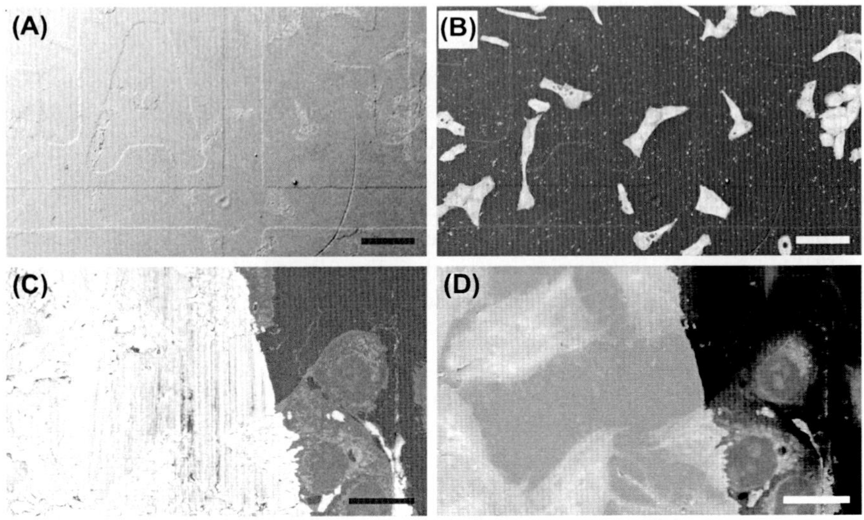

FIGURE 4

Relocating the ROI in volume SEM. (A) FIB-SEM low-voltage SE2 image (2 kV) of an en-bloc–embedded sample, showing the imprint of the finder grid. (B) By increasing the acceleration voltage to 15 kV and thus increasing the interaction volume of the electron beam with the sample, BSE and SE2 can be detected from within the resin block, revealing the cells stained with heavy metal salts for precise positioning of the trench-milling. (C) 2 kV BSE image of the sample surface in SBF-SEM. The right side of the sample was already cut with the diamond knife, exposing the cells. The left part of the sample is not yet cut. Here the conductive gold coating (white) is still intact. (D) Imaging the same region imaged at 15 kV allows determination of the position of the cells beneath the resin surface. This can be used to choose a region of interest and position the imaging window correctly before starting a stack acquisition. Scale bars: (A and B) 100 μm; (C and D) 20 μm.

relocation of a specific ROI. But the finder grid positioned in the same focal plane as the cells remains visible only on the Ibidi dishes, while on the MatTek dishes, with the grid being a very shallow etching in the glass, it is filled with resin and cannot be detected anymore by LM (Fig. 3D). In this case it can be beneficial to add some additional marks, e.g., by touching the sample surface lightly with a razor blade and creating random lines and crossings along the edges of the specimen, thus ensuring not to damage the ROI. These marks can be perceived well by both LM and SEM. After aligning the LM and SEM images, any of the existing image data can be used to navigate the FIB-SEM and relocate the cell of interest. However, considering the mismatch in resolution between LM and SEM images, the precision of the alignment strongly depends on how well the alignment markers are perceivable.

In order to be able to accurately target the cell of interest for FIB-SEM volume acquisition, fine-tuning the alignment is essential. This can be achieved by increasing the acceleration voltage and detecting secondary (SE2) and/or backscattered electron (BSE) signals from beneath the resin surface (Fig. 4). The cells strongly impregnated with heavy metal salts will become clearly visible, enabling determination of their exact position by correlating these images with surface features of the resin block. About 15–20 kV has proven to be a good compromise between penetration depth of electrons, i.e., size of interaction volume of the electron-beam with the sample, and potential beam damage to the resin. With TLP specimen, locating the cells is facilitated due to their shapes protruding from the covering resin layer. If necessary, additional markers facilitating this fine-tuning of the alignment can easily be created by ion or electron beam-induced deposition in the FIB-SEM.

Once the cell of interest has been located, the ROI for the 3D stack is marked with a 0.5–1 μm thick platinum deposition. The location of this deposition can again be correlated with fluorescence LM data to help positioning the imaging window for FIB-SEM stack acquisition. The entire area for the 3D stack is additionally covered by a carbon deposition (300–500 nm thick) for additional protection. The protective layer prevents ion beam damage and enables smooth milling. Trenches of minimum 10 μm depth are milled perpendicular to the sample surface in front of the ROI and, optionally, also at the sides flanking the cell. Milling conditions vary depending on the applied FIB-SEM system. Generally, 30 kV acceleration voltage and milling currents between 6.5 and 13 nA for trench milling and 30 kV and 1.5–2.5 nA for polishing the cross-section before stack acquisition are useful settings when working with resin-embedded specimen. Trenches and imaging planes for the 3D stacks are milled perpendicular to the sample surface.

3D volumes were automatically acquired by sequentially ion milling and SEM-imaging, using the automatic routines of the respective FIB-SEM system (Slice & View on FEI Helios, or Atlas 5 on Zeiss NVision 40). SEM-images of cross-sections were recorded at 2 kV, using either the through-the-lens detector (TLD) in BSE mode with a dwell time of 30 μs (FEI Helios) or the EsB detector and a dwell time 2×5.0 μs (NVision 40). Dynamic focus and tilt correction are applied. Volumes are recorded with isotropic voxels, i.e., the nominal pixel size (i.e., x/y-resolution) for imaging was chosen to match the average slice thickness (=z-resolution). In general, the voxel size is chosen for each sample according to the respective aim of the study, usually ranging from 4 to 20 nm. For this application, we chose 8 nm voxels. These milling and imaging conditions should be considered as guidelines. It is strongly recommended to adapt all parameters for different specimen or embedding resins, but implicitly for other FIB-SEM systems.

2.5.2 Serial blockface-scanning electron microscopy

For SBF-SEM, en-bloc embedded specimen of cell monolayers are mounted upside down, i.e., with the substrate-facing side of the cells on top. As the disk-shaped specimens are sufficiently flat to be mounted on the aluminum specimen pins used for SBF-SEM, it is usually not necessary to correct them for tilt. On the contrary, a slight tilt of the surface allows controlled approach cuts, exposing the vicinity of the cells of interest first, thus enabling to choose appropriate imaging and cutting parameters prior to stack acquisition without destroying the actual ROI (Fig. 4). It is advantageous to extract the piece of resin holding the ROI and trim it in such a fashion that the ROI is in the center of the block. That way, starting the approach cuts from either corner or edge of the block will not endanger the ROI. Approaching the sample towards the knife and ensuring that the cutting process has started is usually done with open SEM chamber to allow visual control via binoculars. Once this is achieved, the SEM chamber is pumped to the desired vacuum. As cell monolayers are rarely completely confluent and the empty regions between cells tend to charge up heavily when scanning under high-vacuum conditions, low-vacuum conditions are preferable. Vacuum settings of 20–30 Pa using water vapor have proven to be suitable for this application.

As correlative software packages are not available in conjunction with the employed Gatan 3View system, a different approach had to be applied to identify the cells of interest. For this purpose, and to ensure accurate approach-cuts without damaging the ROI, the acceleration voltage is again increased to 15–20 kV in order to visualize the position of the selected cell under the sputter-coated resin surface. The resulting image can then be used as a map to correlate surface features also visible in the images acquired with the low-voltage settings used for stack acquisition (Fig. 4). The cutting process is then continued, tightly controlled by high-voltage images, just until the cell of interest is reached. Image acquisition is commenced right before starting to expose it. Imaging parameters are optimized in a field of view adjacent to the ROI. Typically, section thickness is chosen between 40 and 50 nm, imaging at acceleration voltages between 1.8 and 2.2 kV, according to the signal quality acquired from the respective specimen. We usually record images with 5–10 nm pixel size with dwell times of maximum 2–4 μs, hereby balancing field of view, scan time, and potential beam damage to the resin samples. Again, these image parameters need to be adapted for each sample type and more importantly for different microscope platforms.

2.6 VISUALIZING THE CORRELATION

Visualization and merging of correlative light and electron microscopic data is challenging due to the intrinsic mismatch in resolution and preparation inherent distortions. For most FIB-SEM data, the different orientation of the primary imaging planes, with the FIB-SEM image plane being perpendicular to that of the LM data, adds complexity to the task. Therefore, recording FIB-SEM volume data with isotropic voxels, i.e., equally sized pixel in all directions, is auxiliary as it offers the option of virtually reslicing the data to match the imaging orientation of the LM data and thus facilitating the correlation (Fig. 5C and D). SBF-SEM data, on the other hand, cannot be reasonably recorded with isotropic voxels, because of

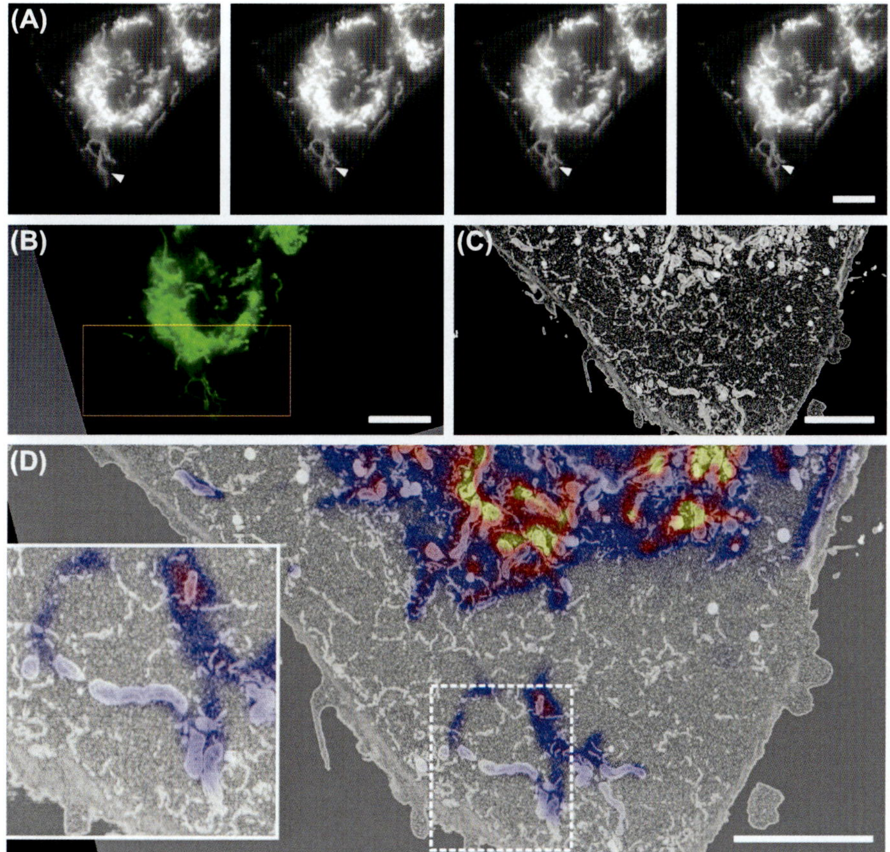

FIGURE 5

Correlation of live-cell TIRF imaging with FIB-SEM. (A) A time-lapse series of TIRF images was acquired capturing the movement of mitochondria (arrowhead) over 100 s. The four images show the images number 0, 30, 60, and 100 (from left to right). (B) TIFR image of the same ROI after chemical fixation. The orange rectangle marks the position of the corresponding FIB-SEM stack. (C) The image shows a virtual slice of the FIB-SEM volume matching the primary imaging orientation of the LM to facilitate the correlation. (D) Overlay of images B and C produced in Amira. The LM overlay is depicted using a lookup table ranging from blue to yellow for better visibility. The FIB-SEM dataset has a much finer depth-resolution (8 nm isotropic voxels) compared to the TIRF image, therefore fluorescence signal recorded in a single imaging plane is generated from a larger volume (z-resolution ~200 μm), explaining the fluorescence signal in areas where no mitochondria are depicted in a single virtual slice of the FIB-SEM stack (inset). Scale bars: (A and B) 10 μm; (C and D) 5 μm.

the sectioning thickness being dictated by the physical cutting process. However, for correlation purposes this issue is overcome by the fact that the primary imaging plane of both LM and SBF-SEM is practically the same (Fig. 6). Any preprocessing of the datasets, such as registration to compensate e.g., for drift, shift, or distortions

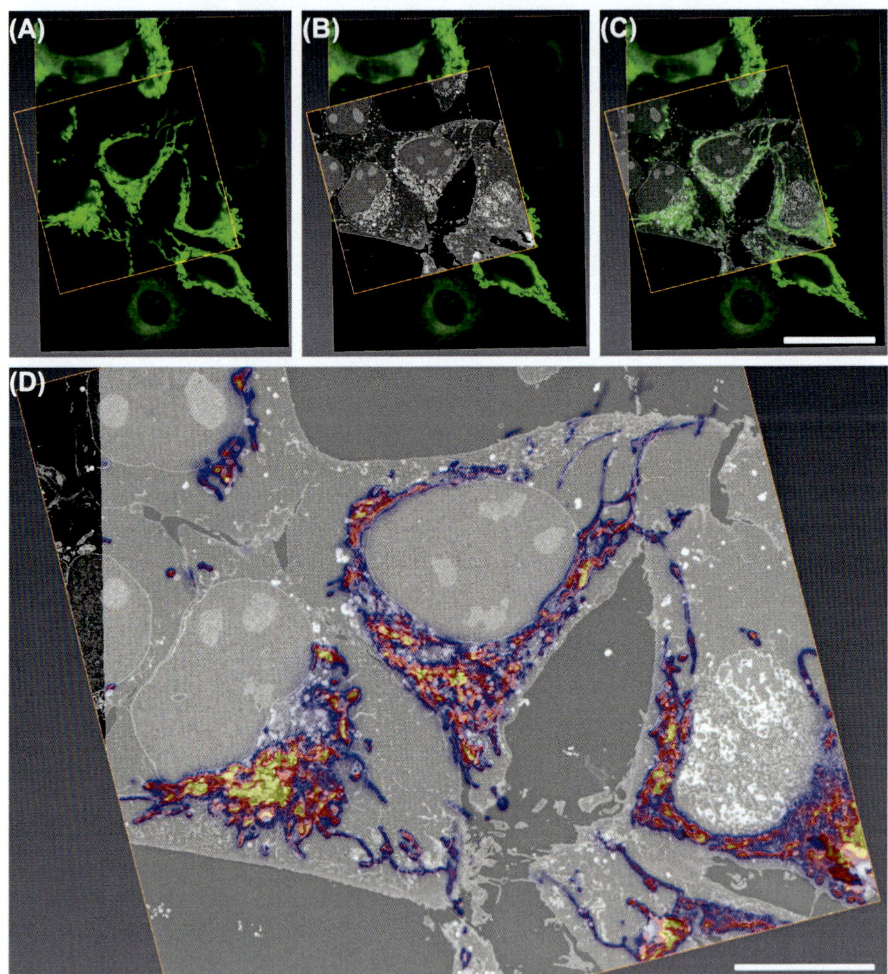

FIGURE 6

Correlation of live-cell TIRF imaging with SBF-SEM. (A) TIRF image acquired after chemical fixation. (B) An SBF-SEM stack was acquired of the region marked with the orange square. The SEM image shows one plane of the stack acquired close to the sample surface, just after complete removal of the covering gold layer. (C) The SEM image was rendered transparent to visualize the exact positioning of the overlay. (D) Correlation of LM and SBF-SEM, using a lookup table ranging from blue to yellow to overlay the fluorescence signal onto the SEM image. Scale bars: (A–C) 50 μm; (D) 20 μm.

induced by charging of the sample during volume recording, may additionally reduce the precision of the correlation.

Fully automated merging of these two data types is to date impossible and requires manual interaction. But software tools, such as Amira, which enable simultaneous visualization and processing of two datasets, offer algorithms for automatic alignment after manual prealignment.

All image post-processing steps, visualization, and merging of correlative datasets were done using the Amira 3D visualization and analysis software. Volume SEM data from both applied modalities were registered using the align-function, excluding rotational adaptations, to compensate for image drift during stack acquisition. FIB-SEM data were additionally sheared to correct the tilt of the surface intrinsic to FIB-SEM. Correlation of LM and volume SEM data was performed using the "Multiplanar View" and associated tools for registering datasets in Amira.

3. MATERIALS

3.1 EQUIPMENT

- Cell culture equipment: Cell culture hood, incubator (37°C, 5% CO_2), water bath, refrigerator, freezer (−20°C), inverted microscope, and autoclave
- Light microscope: FEI CorrSight (Thermo Fischer Scientific, Eindhoven, The Netherlands), equipped with a Hamamatsu ORCA-Flash 4.0 v2 camera for wide field and TIRF imaging; 5× 0.16 NA Plan-Neofluar and 40× 0.9 NA Plan-Neofluar objectives for overview imaging and a 63× 1.46 NA alpha Plan-Apochromat Korr M27, TIRF (all objectives: Carl Zeiss Microscopy, Oberkochen, Germany). Software: FEI LiveAcquisition and FEI MAPS 2.1 for correlative workflows. The CorrSight "Live Module" is attached for live-cell imaging using the fluidic setup
- FIB-SEM: (1) FEI Helios 600i (Thermo Fischer Scientific), with Elstar Schottky field-emission SEM technology and Tomahawk FIB with low kV operation, TLD, which can be operated in SE- or BSE mode, and Everhart-Thornley SE detector, AutoSlice and View G3 1.51 software for automated stack acquisition, FEI MAPS 2.1 for correlative and large area scan workflows, and a gas injection (GIS) system for local deposition of carbon or platinum (2) CrossBeam workstation NVision 40 (Carl Zeiss Microscopy), equipped with a GIS for Pt and C-deposition, SE in-lens and EsB (BSE) detectors, and an Atlas 5 system to facilitate correlative workflows, acquisition of 3D stacks, and large-area scans
- Serial blockface-SEM: FEI Quanta 250 FEG variable pressure SEM (Thermo Fischer Scientific), equipped with a Gatan 3View 2XP system and dedicated backscatter detector (Gatan, Inc., Pleasanton CA, USA)
- Sputter coating unit to render resin blocks conductive for (FIB-)SEM: CCU-010 high vacuum coater (Safematic GmbH, Bad Ragaz, Switzerland), fitted with Au- or Pt-Pd-target
- Amira 3D visualization and analysis software (FEI Company), version 6.2

3.2 CONSUMABLES AND CHEMICALS

- Cell culture slides for fluidic setup: μ-Slide CorrSight Live (Ibidi, Martinsried, Germany)
- Cell culture dishes: Ibidi μ-Dish 35 mm with glass bottom imprinted with a 500 μm relocation grid (Ibidi), or MatTek 35 mm glass-bottom dishes, No. 1.5 with gridded coverslip, either with grid in the same focal plane as the cells, or with grid and cells in different focal planes (MatTek Corporation, Ashland MA, USA)
- Cell culture consumables: aspiration vacuum pump, cell culture flasks, glass pipettes, waste containers, medium (DMEM), antibiotics (penicillin/streptomycin), media supplement L-glutamine (1%), FCS (10%)
- Cell line: KERMIT: U2OS-derived stable cell line co-expressing a mitochondrial (mtBFP) and an ER marker (sec61a-GFP) (Kanfer et al., 2015)
- Fine-tipped forceps
- Gilson pipettes (20 and 200 mL) and tips and disposable Pasteur pipettes
- Fixative: 2.5% glutaraldehyde (EM grade), 2% paraformaldehyde in 0.15 M cacodylate buffer, pH 7.4, supplemented with 2 mM calcium chloride
- Contrasting agents: 2% osmium tetroxide and 1.5% potassium ferrocyanide in 0.15 M cacodylate buffer, pH 7.4, with 2 mM calcium chloride; 1% thiocarbohydrazide in double distilled water; 2% osmium tetroxide solution, 1% uranyl acetate solution, Walton's lead aspartate solution
- Double distilled water
- Ethanol (100% and anhydrous)
- Epoxy embedding resin (Fluka epoxy embedding kit #45359, Sigma-Aldrich Chemie GmbH, Buchs, Switzerland), using the standard mixture recommended by Fluka
- Reusable adhesive putty for fixing of culture dishes on the microscope holder: Blu-Tack (Bostik GmbH, Borgholzhausen, Germany)
- Razor blades for manual trimming
- SEM sample stubs (Plano GmbH, Germany)
- Aluminum specimen pins for SBF-SEM (Gatan, Pleasanton, CA)
- Conductive epoxy (CircuitWorks; Chemtronics, Hoofddorp, The Netherlands) for mounting resin specimen to sample stubs and SBF-SEM sample pins

4. DISCUSSION

CLEM comprises a large family of microscopy techniques, offering manifold combinations of LM and EM applications. Here we have described two approaches to correlative live-cell imaging and volume SEM.

4.1 LIVE-CELL IMAGING SETUP

The use of culture dishes is extremely simple and effective. It offers control of temperature and gas environment as well as direct accessibility, and allows TIRF imaging in addition to other transmission, reflection, and fluorescence imaging modes. Although this approach does not benefit from the advantages of controlled fluid handling, it is more flexible. Fixation can be evoked at a desired moment, by simply adding the fixative buffer into the culture medium, preferably adding double concentrated fixative to an equal amount of culture medium. Special care has to be taken to avoid any stress to the cells prior to fixation, which could be caused by liquid flow during mixing of the solutions. However, by using this approach, unwanted ingredients of the culture medium may be attached and fixed onto the cells, thus contaminating the sample with debris, an effect that could be avoided when using the flow-chamber setup. Nevertheless, this approach can easily be applied in any LM platform, preferably equipped with an environmental chamber, and is not tied to a specific hardware or acquisition software to facilitate the relocation in EM.

A flow-chamber setup on the other hand, coupled with a closed system for fluid handling, such as the described "Live Module" for the CorrSight platform has its own advantages. It offers a constant supply of fresh media to the cell culture at a desired flux rate, and controlled onset of perfusion when changing perfusion media, e.g., for fixation or to study effects of a particular treatment, as e.g., injection of activator or drug molecules. The fluidic handling system allows automated and controlled onset of perfusion and delivery of media for "triggered" experiments. This provides full control of the experiments and increases reproducibility. With the system being closed and waste fluids being collected in a proper way, the flow-chamber setup additionally increases ease of handling and working safety. For CLEM studies this would e.g., allow in situ observation and documentation of the cells during fixation, and potentially even follow-up steps of the sample preparation for EM. This approach can e.g., be of use to document the loss of fluorescence signal during dehydration and/or staining with heavy metal salts. Theoretically, the EM sample preparation can be performed directly on the microscope until during the resin infiltration the resin-alcohol mixture becomes too viscous to be pumped through the channels of the μ-slides. Additionally, the current version of μ-slides is equipped with standard plug connectors to attach the tubing for the fluid handling system. These connectors can easily become loose or detach completely and thus pose the risk of leakage, which may contaminate the equipment, or even endanger the health of the operator. Therefore, it seems adequate to perform the more critical steps of perfusion in a fume hood. For this, the entire tray of the Live-Module holding the flow-chamber slide, tubing, liquid storage, and waste container can be detached from the CorrSight and transferred into the fume hood to proceed using the fluidic system. Alternatively, the μ-slide can be removed to manually perform the following media changes. In the latter case, we prefer to additionally remove the cover foil, so that fluids can be pipetted directly into the cell culture wells.

Replacing the standard plug connectors by Luer lock connectors would undoubtedly improve the liquid handling, not only during perfusion, but also when setting up the system and initially filling the chambers and supply channels with media. In addition, liquid handling could be improved and leakages prevented, by sucking the media through the chamber, instead of pumping it (Droste et al., 2005). However, for the current study the major drawback of the fluidic setup is the incompatibility with TIRF microscopy, due to the polymer foil bottom of the Ibidi μ-slides.

4.2 THE INFLUENCE OF SAMPLE PREPARATION

Correlation of volume SEM data with LM data acquired prior to fixation, dehydration, and resin embedding, poses a challenge. Precisely merging the two data types may even be impossible due to change of shape or size of the cells, caused by chemical fixation, dehydration, and resin embedding (Karreman et al., 2014; Kushida, 1962). Additionally, chemical fixation, although being relatively easy to apply, is not fast enough to arrest cellular processes immediately when adding the fixative buffer (Droste et al., 2005). Complete fixation of a single cell may take several seconds up to a couple of minutes and depends on the speed of penetration and thus also on the thickness and composition of the respective cell. The fastest available technique for fixation is high-pressure freezing, arresting cellular processes within milliseconds (Dahl & Staehelin, 1989; Mohr, 1973; Reipert, Fischer, & Wiche, 2004; Riehle & Hoechli, 1973). Freezing cells at a desired point in time after monitoring them by LM has been demonstrated by Verkade (2008) and Heiligenstein et al. (2014). With these fast transfer systems from the LM to high-pressure freezer the time gap between LM imaging and the actual fixation of the cells can be reduced to 2–4 s.

Chemical fixation is still the most commonly used technique and does not require expensive equipment such as a high-pressure freezer. However, one has to consider the time lag between observation of an interesting cellular event and the effective fixation of the cell. Documenting the entire process by e.g., recording time lapse image series throughout the entire experiment, including chemical fixation, or at least capturing a reference image during or after fixation, will allow a good correlation.

The type of embedding, en-bloc embedding or TLP, does not have a direct influence on the correlation with LM data. In both cases an FIB-SEM image series needs to be registered and sheared, which works equally well for both types of samples. Merging these results with the LM data is thus influenced by both the same potential imprecision and the mismatch due to post-processing of the raw data. However, the relocation of an ROI needs to be addressed differently, depending on the type of embedding.

In TLP samples, the individual cells are readily detectable by SEM. Culture dishes with the finder grid in the same focal plane as the cells require different procedures, depending on the type and supplier. While the etching of the finder grids in Ibidi dishes is directly visible, the more shallow lines of the MatTek dishes are fully masked by resin and not available to assist the relocation. In this case additional preparation and imaging steps may become necessary. These may include the addition of landmarks e.g., by randomly touching the resin surface with a razor blade and thereby creating arbitrary line crossings, or ion-beam induced deposition of landmarks. Additional imaging of these landmarks in LM and SEM and correlating these images with the preembedding LM images can be of help. In case the finder grid is positioned in a different focal plane than the cells, an additional step of imaging the resin embedded samples in LM for correlation with live-cell imaging data is required. Here correlative software packages can be very helpful.

By contrast, the finder grid imprinted on the surface of en-bloc embedded specimen makes the relocation rather straightforward. The glass itself is removed, but the transfer of the finder grid works for all described cell substrates with the finder grid in the same focal plane as the cells. The advantages and disadvantages of the applied cell culture substrates are summarized in Table 1.

4.3 RELOCATION OF THE STRUCTURE OF INTEREST AND CORRELATION OF LM AND EM DATA

Relocation of the ROI in the SEM can be simplified by the use of software connectivity, such as MAPS, Shuttle & Find, Atlas, MirrorCLEM, or MiXcroscopy. However, it is not indispensable. Finder grid supports for cell culture alone facilitate the relocation of a cell of interest very well, provided the finder grid is detectable in both imaging modalities (LM and SEM). SEM imaging at higher acceleration voltage (15–30 kV) and thus detecting SE2 and/or BSE from beneath the resin surface can be applied to locate the heavy metal stained, embedded cells. The use of relocator software tools can greatly simplify this relocation. This is particularly important for FIB-SEM, as only relatively small volumes can be acquired and precise milling spares valuable instrument time (Peddie & Collinson, 2014), and additionally saves the ROI from fatal damage. Overlays of LM and SEM images are mandatory for high-precision relocation. These can be prepared using any image processing software. All acquisition software tools designated for correlation include functions to import LM image data, overlays, and alignment with current SEM images, and finally direct navigation of the SEM stage based on these images. However, some of these software tools limit the connectivity to LM data of certain microscope brands.

In SBF-SEM, the entire sample is ablated from top to bottom. Mounting the specimen with the surface at a slight tilt angle is beneficial, because it allows a controlled approach to the cell of interest. By first exposing the vicinity of the cell one can choose appropriate imaging and cutting parameters prior to stack acquisition without destroying the actual ROI. To date, the employed SFB-SEM does not allow the use of any correlative software connectivity in conjunction with 3View operation. Therefore, a manual approach to relocating the ROI is imperative.

For some applications, a full 3D correlation of the two datasets may not be necessary to answer a biological question. Especially for cell monolayers, it may be sufficient to locate the cell of interest correctly for further EM imaging, without the need to precisely correlate the fluorescence signal with the volume EM data in 3D. However, if this exact correlation is needed to identify a structure of interest

Table 1 Advantages and Disadvantages of the Described Workflows for Live-Cell CLEM

	Flow-Chamber Setup	Dish Setup		
	μ-Slide CorrSight Live (Ibidi) Polymer Bottom With Fluorescent Finder Grid in the Same Focal Plane as the Cells	**μ-Dish 35 mm, 500 (Ibidi) Glass Bottom Culture Dish With Finder Grid and Cells *in the Same* Focal Plane**	**35-mm Glass-Bottom Dish, No. 1.5 (MatTek), With Finder Grid and Cells *in the Same* Focal Plane**	**35-mm Glass Bottom Dish, No. 1.5 (MatTek), With Finder Grid and Cells *in Different* Focal Planes**
Advantages	• Optimum conditions for long-period live-cell imaging, due to constant supply with fresh culture medium • Controlled onset of perfusion with fixative or effector media • Closed system for handling potentially harmful liquids • Fluorescent finder grid is easy to visualize and identify • En-bloc embedding works well, finder grid well detectable on resin surface	• Large field of view • Fast imaging of cells and grid in one focal plane • All LM modes applicable • Finder grid is directly visible and easily detectable in SEM after TLP • Easy removal of glass inlay after TLP	• Large field of view • Fast imaging of cells and grid in one focal plane • All LM modes applicable • Easy handling for en-bloc embedding	• Large field of view • All LM modes applicable • Finder grid well detectable in LM after TLP
Disadvantages	• No TIRF microscopy • Relatively small well, therefore TLP difficult • Air bubbles may interfere with imaging and/or flow • Tubing and connectors become leaky when using osmium-solutions	• Fixation by adding fixative buffer into the cell culture medium may cause debris from media ingredients on the cells • Not ideal for en-bloc embedding	• Fixation by adding fixative buffer into the cell culture medium may cause debris from media ingredients on the cells • TLP masks the finder grid: relocation of ROI requires partial removal of the cured resin layer • Removal of glass inlay after TLP very delicate	• Finder grid not directly detectable in SEM after TLP, relocation of ROI requires landmark- or software-assisted workflow and additional LM-imaging steps • Not compatible with en-bloc embedding, due to lack of finder grid imprint on resin surface • Removal of glass inlay after TLP very delicate
Recommended application	• Long-term live-cell imaging, shear force experiments, 3D image acquisition • FIB-SEM and SBF-SEM	• 3D imaging, live-cell imaging, TIRF • FIB-SEM	• 3D imaging, live-cell imaging, TIRF • FIB-SEM and SBF-SEM	• 3D imaging, live-cell imaging, TIRF • FIB-SEM and SBF-SEM

within a cell, the correlation needs to be as exact as possible. To achieve this, not only the relocation of the ROI is important, but also the modality for volume EM. Both imaging modalities presented here, FIB-SEM and SBF-SEM, can achieve comparable imaging resolution in the primary image plane, i.e., in x- and y-direction of the original image (Kremer et al., 2015; Villinger et al., 2012). With standard FIB-SEM applications, this imaging plane is situated perpendicular to the surface of the sample, i.e., also perpendicular to the imaging plane of LM. However, this drawback can be overcome by recording isotropic voxel, i.e., equally sized pixel in x-, y-, and z-orientation. This enables virtual reslicing of the volume dataset in silico and thus facilitates matching with the LM data (Armer et al., 2009). SBF-SEM on the other hand is limited in the z-resolution by the process of physically cutting the resin samples with a diamond knife (Denk & Horstmann, 2004). Reproducibly cutting sections thinner than 30–40 nm is hardly possible for resin-embedded specimen, resulting in a significant mismatch in pixel size between x-/y- and z-dimension. However, the geometrical positioning of the image plane in SBF-SEM is parallel to the sample surface (and perpendicular to that of FIB-SEM). The individual images of the 3D stack are therefore acquired in the same imaging plane as the LM images, which simplifies the correlation. Theoretically it would be possible to mount samples for FIB-SEM at a 90 degrees angle to allow milling parallel to the cell substrate. However, the acquisition of a 3D stack would become very challenging, because the recording of images would have to start prior to effectively milling into the cell of interest, i.e., optimization of the imaging parameters can only be done when already milling into the ROI. Milling at a shallow angle and slowly approaching the cell of interest as done in SBF-SEM could overcome this problem, but on the other hand, the field of view in FIB-SEM may not be sufficient to achieve this. In both cases, SBF-SEM or FIB-SEM, any larger tilt angles in the primary image plane complicate correlation with the LM data.

CONCLUSIONS

Live-cell imaging correlated with volume SEM combines two powerful and maybe most widely applied imaging techniques in biological research. The two described approaches to live-cell imaging, using either a simple, but effective culture dish setup, or the rather sophisticated one involving a flow-chamber, both yield excellent and comparable results. While the culture dish setup is very flexible and can be combined with every LM imaging mode and any available microscope platform, the flow-chamber setup, although not compatible with TIRF imaging, allows a better control of cell culture conditions and precise perfusion with different media and enables longer live-cell experiments. Both approaches offer gridded cell substrates to facilitate relocation of an ROI in the follow-up volume SEM technique. Again, the dish-based approach offers more flexibility, as these samples can be prepared using either TLP for FIB-SEM or en-bloc embedded for both FIB-SEM and SBF-SEM. The flow-chamber setup on the other hand, is better suited for en-bloc embedding. Although en-bloc embedding at first sight may appear to be the more versatile method, TLP has advantages for the investigation of cells on support types, which cannot be removed as e.g., medical implants, or when the contact between cells and substrate are of interest. For the investigation of larger areas or volumes en-bloc embedded samples offer the additional option of (serial) sectioning for large area SEM imaging either in 2D or 3D (array-tomography), enabling the collection of statistical data on the ultrastructural level (Lucas et al., 2012; Oberti, Kirschmann, & Hahnloser, 2010). The applicability of gridded cell culture substrates for live-cell imaging to different volume SEM technique completes the picture of a versatile and comprehensive tool for site-specific correlative ultrastructure investigations in live science.

ACKNOWLEDGMENTS

We thank Prof. Dr. Benoît Kornmann (Institute of Biochemistry, ETH Zurich) for providing the KERMIT cell line and Dr. Simona Rodighiero and Dr. Tobias Schwarz (ScopeM, ETH Zürich) for their help with culturing the cells. We also thank Dr. Liesbeth Hekking for her help with setting up the "Live Module" and Dr. Kristian Wadel (both from Thermo Fisher Scientific, formerly FEI) for assisting with TIRF imaging. Dr. Alexandra Graff and Dr. Christel Genoud are gratefully acknowledged for helping out with recording of SBF-SEM data at the electron microscopy facility of the Friedrich Miescher Institute for Biomedical Research.

REFERENCES

Armer, H. E. J., Mariggi, G., Png, K. M. Y., Genoud, C., Monteith, A. G., Bushby, A. J., … Collinson, L. M. (2009). Imaging transient blood vessel fusion events in zebrafish by correlative volume electron microscopy. *PLoS One, 4*(11), e7716.

Bittermann, A. G., Schaer, D. J., Mitsi, M., Vogel, V., & Wepf, R. (2012). Thin layer plastification vs. block embedding: Two alternative preparation strategies for 3D-imaging of cultured cells and biofilms by FIB/SEM. In *Paper presented at the European microscopy congress EMC2012, Manchester, United Kingdom.*

de Boer, P., Hoogenboom, J. P., & Giepmans, B. N. G. (2015). Correlated light and electron microscopy: Ultrastructure lights up! *Nature Methods, 12*(6), 503–513.

Dahl, R., & Staehelin, L. A. (1989). High-pressure freezing for the preservation of biological structure: Theory and practice. *Journal of Electron Microscopy Technique, 13*(3), 165–174.

Deerinck, T., Bushong, E., Lev-Ram, V., Shu, X., Tsien, R., & Ellisman, M. (2010). Enhancing serial block-face scanning electron microscopy to enable high resolution 3-d nanohistology of cells and tissues. *Microscopy and Microanalysis, 16*(Suppl. S2), 1138–1139.

Deerinck, T. J., Bushong, E. A., Thor, A., & Ellisman, M. H. (2010). *NCMIR methods for 3D EM: A new protocol for preparation of biological specimens for serial block-face SEM.* Retrieved from https://ncmir.ucsd.edu/sbem-protocol.

Denk, W., & Horstmann, H. (November 2004). Serial block-face scanning electron microscopy to reconstruct three-dimensional tissue nanostructure. *PLoS Biology, 2*(11), e329.

Droste, M. S., Biel, S. S., Terstegen, L., Wittern, K. P., Wenck, H., & Wepf, R. (2005). Noninvasive measurement of cell volume changes by negative staining. *Journal of Biomedical Optics, 10*(6), 064017.

Ettinger, A., & Wittmann, T. (2014). Fluorescence live cell imaging. In C. W. Jennifer, & W. Torsten (Eds.), *Methods in cell biology* (Vol. 123, pp. 77–94). Academic Press (Chapter 5).

Gibson, K. H., Vorkel, D., Meissner, J., & Verbavatz, J. M. (2014). Fluorescing the electron: Strategies in correlative experimental Design. In T. MullerReichert, & P. Verkade (Eds.), *Methods in cell biology: Correlative light and electron microscopy II* (Vol. 124, pp. 23–54). San Diego: Elsevier Academic Press Inc.

Heiligenstein, X., Hurbain, I., Delevoye, C., Salamero, J., Antony, C., & Raposo, G. (2014). Step by step manipulation of the CryoCapsule with HPM high pressure freezers. In T. MullerReichert, & P. Verkade (Eds.), *Correlative light and electron microscopy II* (Vol. 124, pp. 259–274). San Diego: Elsevier Academic Press Inc.

Jahn, K. A., Barton, D. A., Kobayashi, K., Ratinac, K. R., Overall, R. L., & Braet, F. (2012). Correlative microscopy: Providing new understanding in the biomedical and plant sciences. *Micron, 43*(5), 565–582.

Kanfer, G., Courtheoux, T., Peterka, M., Meier, S., Soste, M., Melnik, A., … Kornmann, B. (2015). Mitotic redistribution of the mitochondrial network by Miro and Cenp-F. *Nature Communications, 6*, 8015.

Karreman, M. A., Hyenne, V., Schwab, Y., & Goetz, J. G. (2016). Intravital correlative microscopy: Imaging life at the nanoscale. *Trends in Cell Biology, 26*(11), 848–863.

Karreman, M. A., Mercier, L., Schieber, N. L., Shibue, T., Schwab, Y., & Goetz, J. G. (2014). Correlating intravital multi-photon microscopy to 3D electron microscopy of invading tumor cells using anatomical reference points. *PLoS One, 9*(12), e114448.

Kizilyaprak, C., Bittermann, A. G., Daraspe, J., & Humbel, B. M. (2014). FIB-SEM tomography in biology. In J. Kuo (Ed.), *Methods in molecular biology: Electron microscopy* (3rd ed., Vol. 1117, pp. 541–558). New York: Humana Press.

Knott, G., Marchman, H., Wall, D., & Lich, B. (2008). Serial section scanning electron microscopy of adult brain tissue using focused ion beam milling. *Journal of Neuroscience, 28*(12), 2959–2964.

Kremer, A., Lippens, S., Bartunkova, S., Asselbergh, B., Blanpain, C., Fendrych, M., … Guérin, C. J. (2015). Developing 3D SEM in a broad biological context. *Journal of Microscopy.* http://dx.doi.org/10.1111/jmi.12211.

Kushida, H. (1962). A study of cellular swelling and shrinkage during fixation, dehydration and embedding in various standard media. *Journal of Electron Microscopy, 11*(3), 135–138.

Loussert Fonta, C., & Humbel, B. M. (2015). Correlative microscopy. *Archives of Biochemistry and Biophysics, 581*, 98–110.

Lucas, M. S., Günthert, M., Gasser, P., Lucas, F., & Wepf, R. (2012). Bridging microscopes: 3D correlative light and scanning electron microscopy of complex biological structures. In T. Müller-Reichert, & P. Verkade (Eds.), *Methods in cell biology: Correlative light and electron microscopy* (Vol. 111, pp. 325–356). San Diego: Elsevier Academic Press.

Mohr, H. (1973). Cryotechnology for the structural analysis of biological material. In E. L. Benedetti (Ed.), *Freeze-etching: Techniques and applications* (pp. 11–19). Paris, France: Societe Francaise de Microscopie Electronique.

Oberti, D., Kirschmann, M. A., & Hahnloser, R. H. (2010). Correlative microscopy of densely labeled projection neurons using neural tracers. *Frontiers in Neuroanatomy, 4*, 24.

Peddie, C. J., & Collinson, L. M. (June 2014). Exploring the third dimension: Volume electron microscopy comes of age. *Micron, 61*, 9–19. http://dx.doi.org/10.1016/j.micron.2014.01.009.

Reipert, S., Fischer, I., & Wiche, G. (2004). High-pressure freezing of epithelial cells on sapphire coverslips. *Journal of Microscopy, 213*(1), 81–85.

Riehle, U., & Hoechli, M. (1973). The theory and technique of high pressure freezing. In E. L. Benedetti, & P. Favard (Eds.), *Freeze-etching techniques and applications* (pp. 11–19). Paris: Societe Francaise de Microscopic Electronique.

Russell, M. R. G., Lerner, T. R., Burden, J. J., Nkwe, D. O., Pelchen-Matthews, A., Domart, M.-C., … Laporte, J. (2017). 3D correlative light and electron microscopy of cultured cells using serial blockface scanning electron microscopy. *Journal of Cell Science, 130*(1), 278–291.

Spiegelhalter, C., Tosch, V., Hentsch, D., Koch, M., Kessler, P., Schwab, Y., … Laporte, J. (2010). From Dynamic live cell imaging to 3D ultrastructure: Novel integrated methods for high pressure freezing and correlative light-electron microscopy. *PLoS One, 5*(2).

Verkade, P. (2008). Moving EM: The rapid transfer system as a new tool for correlative light and electron microscopy and high throughput for high-pressure freezing. *Journal of Microscopy, 230*(2), 317–328.

Villinger, C., Gregorius, H., Kranz, C., Hohn, K., Munzberg, C., von Wichert, G., … Walther, P. (2012). FIB/SEM tomography with TEM-like resolution for 3D imaging of high-pressure frozen cells. *Histochemistry and Cell Biology, 138*(4), 549–556.

CHAPTER

A fully integrated, three-dimensional fluorescence to electron microscopy correlative workflow

Claudia S. López*[,1], Cedric Bouchet-Marquis[§], Christopher P. Arthur[§,¶], Jessica L. Riesterer[§], Gregor Heiss[§], Guillaume Thibault*, Lee Pullan[§], Sunjong Kwon*, Joe W. Gray*[,1]

**Oregon Health and Sciences University, Portland, OR, United States*

[§]Thermo Fisher Scientific, Hillsboro, OR, United States

[¶]Genentech, San Francisco, CA, United States

[1]Corresponding authors: E-mail: lopezcl@ohsu.edu; grayjo@ohsu.edu

CHAPTER OUTLINE

Abstract

While fluorescence microscopy provides tools for highly specific labeling and sensitive detection, its resolution limit and lack of general contrast has hindered studies of cellular structure and protein localization. Recent advances in correlative light and electron

Methods in Cell Biology, Volume 140, ISSN 0091-679X, http://dx.doi.org/10.1016/bs.mcb.2017.03.008

microscopy (CLEM), including the fully integrated CLEM workflow instrument, the FEI CorrSight with MAPS, have allowed for a more reliable, reproducible, and quicker approach to correlate three-dimensional time-lapse confocal fluorescence data, with three-dimensional focused ion beam—scanning electron microscopy data. Here we demonstrate the entire integrated CLEM workflow using fluorescently tagged MCF7 breast cancer cells.

INTRODUCTION

Historically, light microscopy (LM) and fluorescence microscopy (FM) have produced stunning images of cells and cellular structures that have led to a number of integral discoveries in the field of cellular biology (Booth et al., 2016; van Driel, Valentijn, Valentijn, Koning, & Koster, 2009; Keene et al., 2014; Keene, Tufa, Lunstrum, Holden, & Horton, 2008; Kolotuev, Schwab, & Labouesse, 2009; Kong & Loncarek, 2015; Kuipers et al., 2015; Lucas, Günthert, Gasser, Lucas, & Wepf, 2012; Moore, Cheng, Shami, & Murphy, 2016; Murphy et al., 2011; Padman, Bach, & Ramm, 2014; van Rijnsoever, Oorschot, & Klumperman, 2008; Russell et al., 2016; Wang et al., 2016). FM has allowed researchers to label very specific cellular components of interest and track those components over both space and time. There are, however, many limitations to imaging using visible light. Only cellular components labeled for FM imaging are visible, and the resolution of conventional light microscopy lies in the submicrometer range, whereas the molecules to be imaged are in the 0.1—10 nm range. To overcome some of these limitations, researchers have turned to electron microscopy (EM) (Sosinsky, Giepmans, Deerinck, Gaietta, & Ellisman, 2007). Multiscale biological imaging allows researchers to survey broad areas of interest to pinpoint exact locations of specific molecular interactions. Correlative light and electron microscopy (CLEM) is rapidly becoming a mainstream biological research technique (de Boer, Hoogenboom, & Giepmans, 2015; Gibson, Vorkel, Meissner, & Verbavatz, 2014). With increased interest in CLEM approaches, comes a greater need for workflows and instrumentation which support data and sample transfer across hardware platforms as well as correlation between them.

There are many fields of research, such as cancer biology and neurosciences (Cazemier, Clascá, & Tiesinga, 2016; Kempen et al., 2015; Knott, Holtmaat, Trachtenberg, Svoboda, & Welker, 2009; Maco et al., 2013; Revach et al., 2015), which could potentially benefit from a CLEM workflow, and a recent collaboration between Oregon Health and Science University (OHSU) and Thermo Fisher Scientific has highlighted the utility of such a workflow in the field of breast cancer research. Here we present a correlative workflow involving a fluorescently tagged membrane protein HER2, which is overexpressed in 20% of human breast cancers and associated with drug resistance (Mukohara, 2011). The workflow combines spinning disk confocal imaging using the FEI CorrSight, along with specially designed microfluidic incubation slides (μ-Slide CorrSight Live, ibidi), which allow for live cell imaging and immediate flowing of reagents of interest into the microfluidic slide. The integrated ibidi well μ-Slide system enables the user to process the sample from fluorescence imaging to electron microscopy sample preparation all in the same slide. The FEI MAPS software allows for cross-platform image acquisition of multiple areas of interest at freely defined areas (through tiling and stitching). The acquired images can be correlated across imaging modalities on-the-fly and used to direct image acquisition along the workflow.

1. MATERIALS AND METHODS

MOLECULAR BIOLOGY

The human AKT2 cDNA cloned into pcDNA3 vector, with an HA-tag at the N-terminus of the protein, was generously provided by Dr. Gordon Mills (MD Anderson Cancer Center). To construct the AKT2-mCherry plasmid, the HA-tagged AKT2 protein mentioned before was fused to the N-terminus of mCherry fluorescent protein (Shaner et al., 2008). For this construct cDNA encoding HA-tagged human AKT2 protein was PCR amplified using forward primer (5′TCCGCTCGAGCGCCACC**ATGTACCCATACGATGTTCCAG**3′; XhoI restriction site is underlined and in bold is the sequence encoding for the N-terminal amino acids of HA tag) and reverse primer (5′CGGGGTAC**CTCGCG GATGCTGGCCGAGTAGG**3′; KpnI restriction site is underlined and in bold is the sequence encoding for the C-terminal amino acids of AKY2 protein). The PCR product was then inserted using XhoI/KpnI restriction sites in-frame with N-terminus of pmCherry-N1 (Clontech). The subcloning step created an 11-amino acid spacer sequence (VPRARDPPVAT) between AKT2 and the mCherry fusion protein. To construct the AKT2-tagRFP expression plasmid used in this work, cDNA encoding tagRFP (Shaner et al., 2008) was kindly obtained from Dr. Xiaolin Nan (OHSU), and was PCR-amplified using forward primer (5′CCGGGGTACCGCGGGCCCGGGATCCACCGGTCGCCACC**ATGTCTGA GCTGATTAAGGAG**3′; KpnI restriction site is undefined and in bold is the sequence encoding for N-terminal amino acids of tagRTF) and reverse primer (5′ CTAGTCTAGAGTCGCGGCCGC**TTTAATTAAGTTTGTGCCCCAGTTT GC**; XbaI restriction site is underlined and in bold is the sequence encoding for the C-terminal amino acids of tagRFP). This PCR product was used to replace the mCherry ORF in the AKT2-mCherry plasmid via subcloning using KpnI/XbaI restriction sites. This subcloning step created an 11-amino acid spacer (VPRARDPPVAT) between AKT2 and tagRFP fusion proteins. pHER2-eGFP was a gift from Dr. Martin Offterdinger (Addgene plasmid # 39321) (Offterdinger & Bastiaens, 2008). All three fusion genes of AKT2-mCherry, AKT2-tagRFP, and HER2-eGFP, were located downstream of cytomegalovirus promoter.

Day 1: Cell culture and transfection

MCF7 breast cancer cells (American Type Culture Collection) were grown in Dulbecco's modified Eagle's medium (DMEM) supplemented with 10% fetal bovine serum (FBS) in 35-mm culture dish. Cells were transiently cotransfected with 500 ng of DNA of HER2-eGFP and AKT2-tagRFP expression plasmids each using X-tremeGENE HP DNA Transfection Reagent (Roche) for 24 h.

Day 2: μ-Slide cell seeding

After the transfection step, the cells were trypsin treated and then transferred into the μ-Slide CorrSight Live chambers previously coated with poly-L lysine (Fig. 1A). The μ−Slide CorrSight Live is an array of six microwells (well diameter, 5.5 mm) where cells can be cultured and, subsequently, investigated with the CorrSight. This μ−Slide has high optical quality for high end microscopy. Pairs of wells are connected to form a total of three fluidic channels. Each channel is then connected to the ibidi microfluidics pump to perfuse cells. Each of the wells contains a Grid−100 structure on its bottom for relocating events, which is clearly visible in phase contrast and bright field microscopy (Fig. 1B). It provides 100

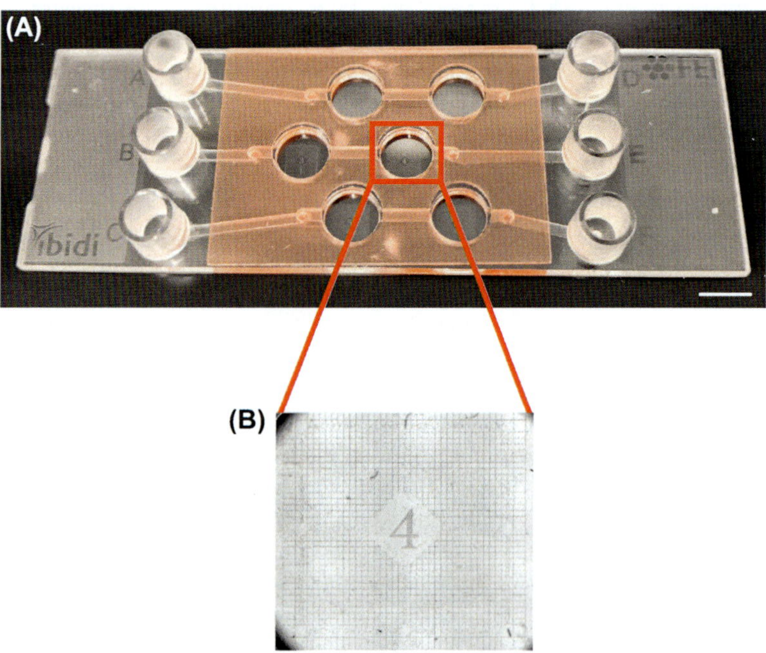

FIGURE 1

(A) μ-Slide CorrSight Live ibidi. This slide consists of six wells (well diameter, 5.5 mm), each containing an etched grid pattern with a 100 μm repeat distance. (B) Pairs of wells are connected via microfluidic channels allowing delivery of multiple reagents during imaging. Scale bar: 5 mm.

distinguishable observation squares of 100 μm edge length. Moreover, this grid is also visible by EM. For more details on this product please go to: http://ibidi.com/xtproducts/en/ibidi-Labware/Correlative-Light-and-Electron-Microscopy-CLEM/m-Slide-CorrSightTM-Live.

The μ-Slide's wells were treated with 80 μL of 0.01% poly-L lysine solution (Sigma–Aldrich) for 5 min at room temperature. This solution was then removed by aspiration, and the wells were rinsed with sterile tissue culture grade water before seeding the cells. To seal the wells, the enclosed polymer coverslip needs to be removed from its protective foil and then attached to the coverslip using the adhesive side of it. For the activation of AKT signaling pathway (Baxi, Tan, Murphy, Smeal, & Yin, 2012), cotransfected MCF7 cells incubated in the μ-Slide CorrSight Live chambers were serum-starved for 15 h and then using the ibidi Pump system (Fig. 2A), insulin (final concentration 10 μg/mL) (Sigma-Aldrich) was added to media for 5 min while temperature of the stage was maintained at 37°C (Fig. 2B).

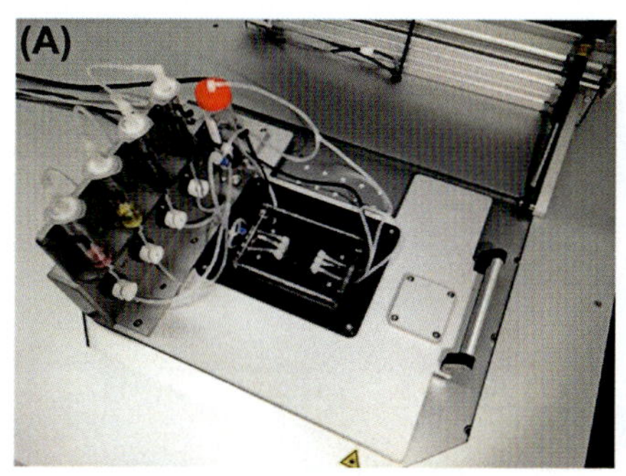

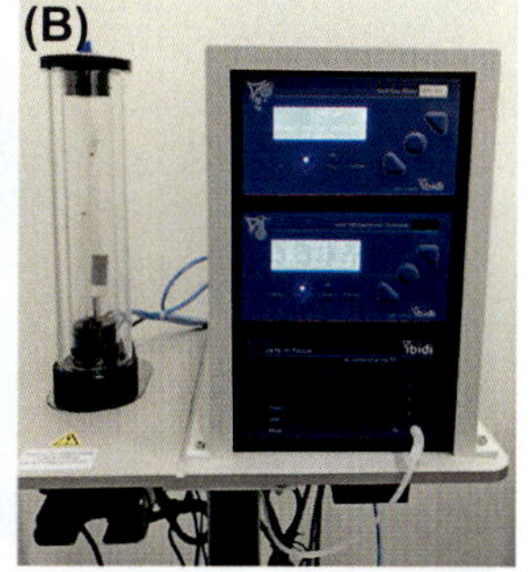

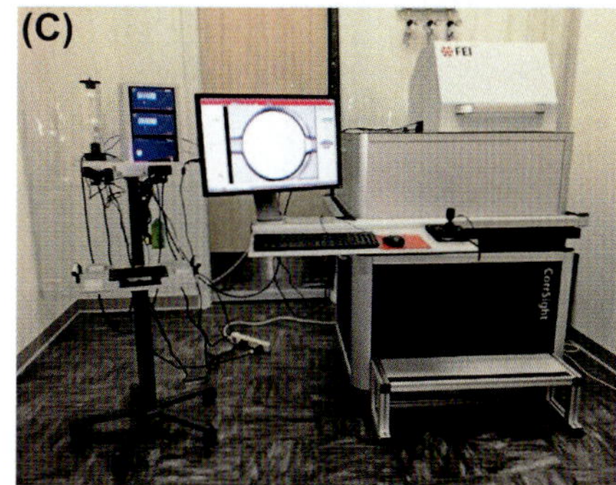

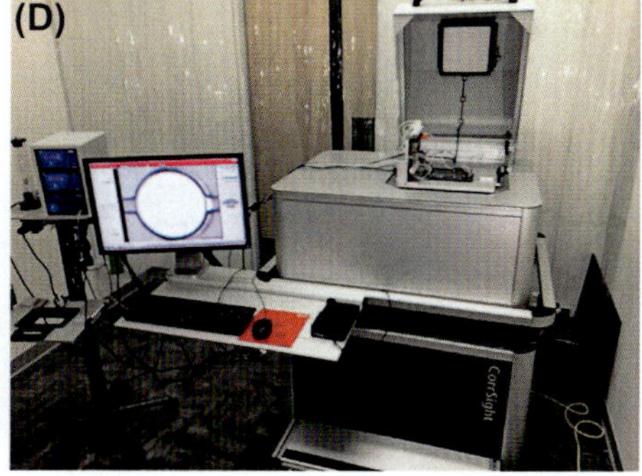

FIGURE 2

(A) CorrSight Live imaging module showing the microfluidics stage module.
(B) ibidi microfluidics control module: stage temperature, pump, and CO_2 incubation controller. (C) CorrSight showing all the hardware components and with the microscope stage cover closed. (D) CorrSight Live imaging module, stage cover is open showing the mounted microfluidics stage.

Day 3: FEI CorrSight Live cell imaging

The FEI CorrSight LM used to develop this workflow is equipped for wide field and spinning disk imaging. The microscope is fully motorized and all functions can be controlled by the MAPS software. The microscope is further equipped with an incubation system (CorrSight Live module) able to control temperature, humidity, and CO_2 levels. The incubation system also includes a pump pressurizing four individual tubes with liquid for perfusion of the μ-Slides. The pressure, and in turn the flow rate, is software controlled as well as the timing of the perfusion of each liquid individually. The μ-Slide CorrSight Live chamber described above was loaded onto the CorrSight Live module for live cell imaging (Fig. 2C and D). The FEI MAPS software in conjunction with Live Acquisition software (LA, FEI) was used to capture still and time-lapse image acquisition experiments, respectively.

Using MAPS, tile sets at increasing magnifications were recorded starting at 5× magnification (Zeiss objective lens; NA = 0.16). Based on the 5× overview tile set of whole wells of the fluidic chamber, higher resolved tile set at 20× and 40× (Zeiss objective lenses at NA = 0.80, 0.90, respectively) were acquired, narrowing down the regions of interest to single cells (Fig. 3A). Both brightfield and spinning disk

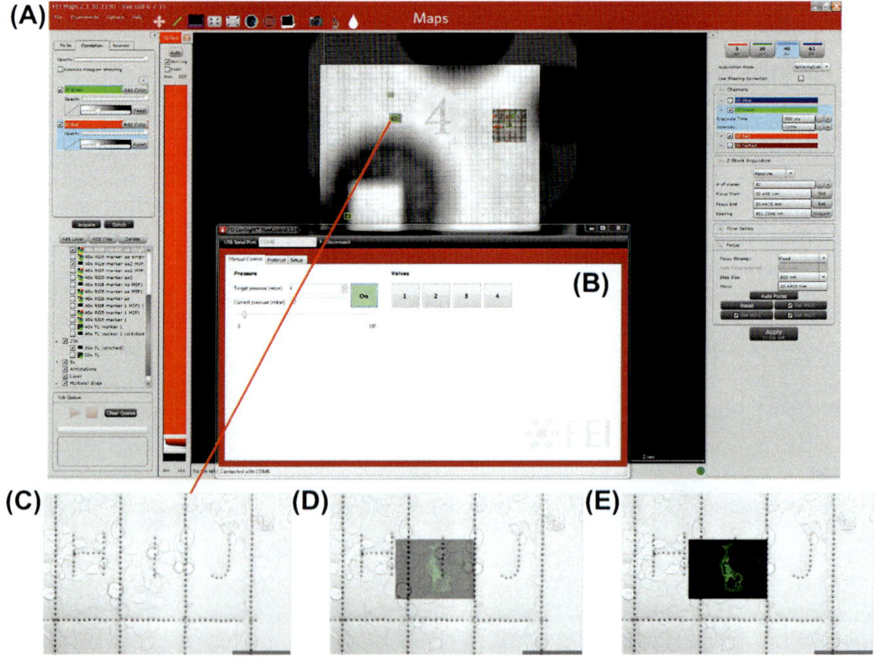

FIGURE 3

(A) MAPS acquisition user interface showing multiple magnifications and transillumination images. (B) FEI Pump controller software. Each of the four valves can be controlled by selecting them on the screen and by adjusting the targeted pressure. (C–E) Automatic overlay of fluorescence spinning disk confocal images over transillumination image.

confocal settings with the appropriate excitation/emission settings were used to perform the fluorescence imaging. In this particular experiment, tiles were acquired setting the excitation to 100% and the exposure time to 50 ms. Each tile is acquired with a 10% overlap of neighboring images to aid stitching. Once the transfected cells of interest have been located using the stitched 40× montage, the ibidi pump system also controlled from MAPS (Fig. 3B) was used to slowly flow (1 mL/min) an insulin solution over the cultured cells stimulating intracellular pathways that promoted fluorescent tag relocalization throughout the cells. Concomitantly, on the six cells selected from the 40× tile sets acquired previously, time-lapse imaging was initiated acquiring *z*-stacks focal series of 15 images at each cell position and looping through all six positions with the imaging parameters set to excitation 100% and 10 ms exposure time. Using these imaging parameters, each cell was imaged every 15 s 100 times resulting in a total imaging time of 24 min (Fig. 3C–E). The use of the LA Software's protocol editor to automate the *z*-stack acquisition at each position also allowed performing an autofocus between each *z*-stack acquisition. After time lapse imaging was completed and fluorescent signal was located at the periphery of the cells of interest, sample preparation for EM started.

Day 4: Electron microscopy sample preparation

Cells grown in μ-Slide CorrSight Live chambers were fixed in cold Karnovsky's (4°C) fixative (2.0% paraformaldehyde, 2.5% glutaraldehyde in 0.1 M sodium cacodylate buffer, pH 7.2) for 30 min using the ibidi pump system with a flow rate of 1 mL/min. After this fixation step the CorrSight Live module was transferred to the fume hood (Fig. 4A). Inside the fume hood, the samples were rinsed using the same flow rate of 1 mL/min with 0.1 M sodium cacodylate buffer and incubated for 5 min. To preserve cellular attachment to the μ-Slide, the samples are treated with a 6% bovine serum albumin dissolved in 0.1 M sodium cacodylate buffer for 1 h at 4°C and then incubated with fixative for 1 h. Samples were treated with 2% (w/v) tannic acid in 0.1 M cacodylate buffer, pH 7.2 for 30 min at room temperature. After this step, the cells were washed in 0.1 M cacodylate buffer and postfixed in 2% reduced osmium tetroxide prepared in 0.1 M sodium cacodylate buffer and 0.8% $K_3Fe(Cn)_6$ for 30 min at room temperature. Cells were washed with dH_2O using the same flow rate and then en bloc stained using a saturated 7% uranyl acetate solution in water for 30 min at room temperature. Following this staining step, the cells were rinsed in dH_2O twice. At this point the μ-Slide CorrSight Live chamber was removed from the microscope stage, the adherent plastic film of the slide was detached from the substrate, and the wells exposed to a graded series of acetone (25%, 50%, 75% and twice 100% for 2 min each at room temperature) (Fig. 4B). Samples were then infiltrated with a 1:1 mix of acetone and Embed 812 resin for 5 h at room temperature (with two exchanges of the 1:1 mix), followed by a 100% Embed 812 resin exchange and overnight incubation at room temperature (Fig. 4C). The next day the resin in the wells is exchanged with 100% freshly prepared Embed 812 resin and the entire slide was moved inside the oven for

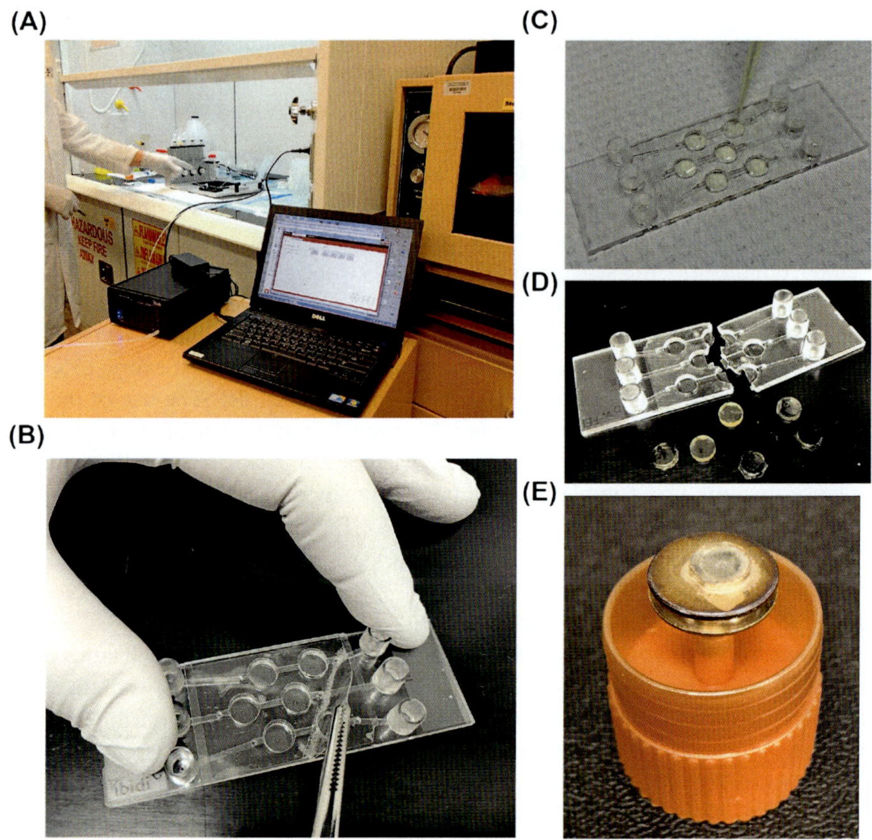

FIGURE 4

(A) ibidi pump system and CorrSight Live imaging module can be moved into the fume hood to start the electron microscopy sample preparation. (B) μ-Slide CorrSight Live is removed from the imaging module, and using standard forceps, the top plastic seal was removed, exposing the six cell culture wells. Samples were dehydrated using an acetone gradient, and finally (C) EPON was added to individual wells and polymerized at 60°C overnight. (D) After polymerization, resin blocks were removed from the slide by simply bending the slide. (E) Resin block was mounted onto standard 12.5-mm flat SEM stub for 3D focused ion beam scanning electron microscopy imaging using silver paint.

polymerization at 60°C for 24 h. The EPON recipe used for this experiment is as follows: Embed (812) 73.5 g, DDSA 45.5 g, NMA 38.5 g, and BDMA 4.2 mL.

Day 5: Mounting and coating blocks for electron microscopy

After complete polymerization, the resulting resin block was recovered from the slide by bending the slide and pushing the polymerized block from underneath (Fig. 4D). Alternatively, a BEEM capsule press can be used with moderate force to recover the polymerized block. The resulting resin block was then polished to remove excess resin using a fine file and mounted on a standard 12.5-mm flat SEM stubs (Ted Pella cat# 16111) using Leitsilber 200 silver paint (Ted Pella cat# 16035) (Fig. 4E). Failure to remove the resin excess from the sample may cause charging effects during SEM imaging obscuring the region of interest (ROI). The silver paint was dried with a heat lamp, and the entire resin sample surface is sputter-coated with a layer of platinum (~20 nm) for grounding purposes.

Relocalization of cell of interest in the DualBeam

The CorrSight LM data acquired previously are directly loaded into the same FEI MAPS software package on the Helios NanoLab 660 DualBeam. The μ-Slide CorrSight Live grid pattern, which was also imprinted into the resin block (Fig. 5A) but as a mirror image, could be clearly imaged using secondary electron mode. FEI MAPS software is again used to locate the ROI by utilizing a one-to-three point alignment procedure and the imprinted fiduciary grid pattern over the regions or features of interest in both LM and SEM images (Fig. 5B). To clearly identify the cells of interest imaged using the CorrSight while still living, backscattered electron (BSE) images are acquired via a dedicated BSE detector and high accelerating voltage to penetrate into empty resin at the block surface and sample the cells.

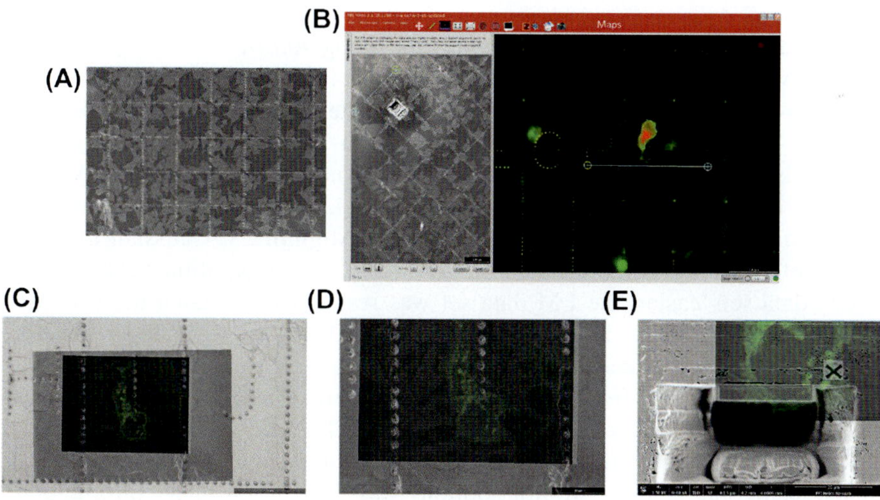

FIGURE 5

(A) μ-Slide CorrSight Live grid pattern can be imaged by secondary electron–SEM imaging on the block face. (B) The grid pattern from the μ-Slide is used in MAPS for the relocalization of the region of interest (ROI) using the one-to-three point MAPS alignment procedure. (C and D) Brightfield, FM, and backscattered electron image overlays with different transparency levels showing the ROI within the sample. (E) Trench generated for focused ion beam scanning electron microscopy image acquisition.

This step is crucial to corroborate that the cells of interest were not lost during the sample preparation step due to poor attachment to the μ-Slide CorrSight Live substrate.

Preparation of the cell of interest for ASV acquisition

Once the ROI was located (Fig. 5C and D), the resin surface was protected from beam damage by coating it with a 2-μm layer of carbon via electron beam–induced deposition (EBID) within the Helios chamber (Fig. 5E). After this step, trenches in front of [45 μm (x) × 40 μm (y) × 20 μm (z)] and along each side (each 10 μm wide) of the identified cells (each 10 μm wide) were created to expose the block face of the ROI (Fig. 5E) using the Gallium-based focused ion beam scanning electron microscope (FIB-SEM). On completion of the site preparation, the ROI was imaged using FEI's Auto Slice & View G3 (ASV) software package for additional high-resolution three-dimensional data collection by automating the serial sectioning and data collection process. The block face was imaged at 45 degrees stage tilt and 2.5 mm working distance, with respect to the electron beam. For each slice, 5 nm of the resin was removed at 52 degrees stage tilt and 4 mm working distance using the Gallium-based FIB column. Images were acquired using 1.5 kV accelerating voltage and beam current of 400 pA in BSE mode using the in-column detector (ICD). Images were acquired at a resolution of 5 nm/pixel using 8-bit gray scale; 2000 slices were obtained for an image stack 10 μm in depth.

Day 9: Image alignment and segmentation using Amira

The last LM z-stack acquired on the cell of interest before fixation from the CorrSight, as well as the EM data sets acquired with the Helios DualBeam FIB-SEM and ASV software, were imported into Amira 6.0.1 (FEI Company, The Netherlands). The EM data were first processed through the DualBeam 3D Wizard to register, crop, and filter the images from the stack. The LM data were intensity normalized across images, aligned, and filtered (AlignSlices; Gaussian filtering, Amira) before being volume rendered. To make the registration between LM and EM data sets easier, the LM data set was resampled to match the EM data set pixel size (x: 5 nm, y: 5 nm, z: 5 nm). LM and EM data were first registered manually in the multiplanar room, which allows loading and manipulating both data sets simultaneously from multiple viewing orientations. During that process, the LM data were loaded as the primary data and the EM data loaded as the overlay data. After a good coarse alignment was found, a refinement of the registration was initiated using the autoregistration options tab (Metric: Mutual Information; Transform: Rigid; Options: Extensive direction). For better results, an optimizer step equivalent to 1–5 voxels should be used. After registration, the volume rendering of the EM data was created, and a portion of the μ-Slide where the cells were grown, the plasma membrane of the cells interacting with each other, two mitochondria, and a nucleus were segmented using the Amira Segmented editor. Segmentation was done using the magic wand segmentation tool and later cleaned up when needed with the brush tool.

The registered, volume rendered, and segmented data sets were finally animated using the Amira Animation Director room (Fig. 6A–F) (Supp. Movie 1).

2. DISCUSSION

The protocol outlined above for a 3D sequential CLEM workflow involves the use of two FEI instruments: the CorrSight fluorescent microscope and the Helios 660 Dual-Beam. One of the advantages of utilizing these two instruments in tandem is that both navigate to, acquire, and align data using FEI's MAPS software. This software facilitates the operator an easy procedure to precisely locate the ROI within the sample in just a few easy steps across imaging modalities; inversion, rotation, and translation alignments at both coarse and fine scales are corrected during the three-point data alignment procedure.

The CorrSight microscope with its "Live module" capability is ideal for live cell imaging experiments requiring the perfusion of specific molecules into the environment of the living cells while continuing image acquisition. This module, equipped with a pump, facilitates the chemical fixation step needed immediately postimage

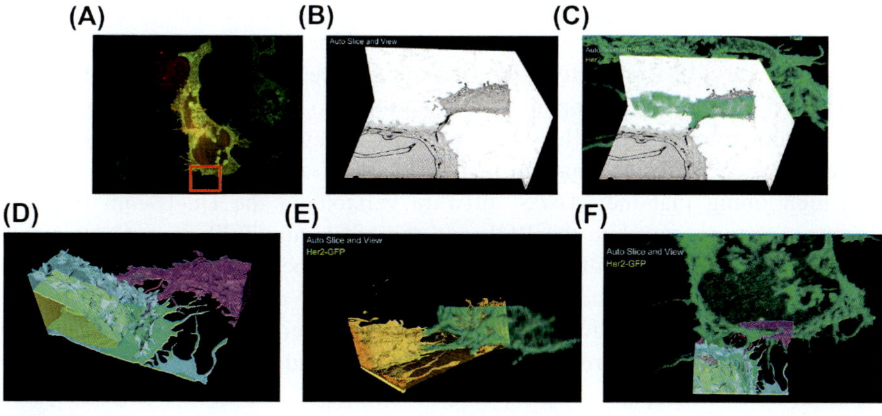

FIGURE 6

(A) Maximum intensity projection of confocal image at zero time point. MCF7 expressing HER2-GFP and AKT2-RFP is shown. The area *boxed in red* corresponds to the region of interest where the Auto Slice & View (ASV) data were acquired. (B) Orthoslices showing ASV data. In the backscatter SEM image, cellular interactions between neighboring cells, membrane protrusions, and nuclear membrane are observed. (C) Overlayed confocal (HER2-GFP signal) and 3D focused ion beam scanning electron microscopy (FIB-SEM) images. (D) 3D volume segmentation performed using Amira software. (E and F) Show the correlation between the fluorescent signal (GFP) and the 3D FIB-SEM data set, from top and side view, respectively.

acquisition in a volume-controlled fashion. This pump system can be utilized also to process the sample for 3D FIB-SEM imaging. Most of the reagents used in this procedure, excluding the resin, can be pumped through this system and into the μ-Slide CorrSight Live substrate making this step easy to the operator. In this chapter we have also introduced the commercially available ibidi μ-Slides for cell cultured CLEM projects. These are versatile slides ideal for locating cells or cell clusters via fiduciary markers that can be used in any fluorescence microscope adaptable with a live cell controlled stage. Its size is compatible with any LM slide stage, and the pump system is also available through ibidi. These μ-Slides have high optical quality similar to that of glass, making them ideal for LM analysis with uncompromised resolution and choice of wavelength.

Many sample preparation methods can be applied to the μ-Slide CorrSight Live, the one described in this chapter is just one example of them. The researcher will have to evaluate the best protocol for their cells under investigation and the features of interest (mitochondria, membranes, cytoskeleton, etc.) to be imaged by 3D FIB-SEM. Moreover, the workflow described here has been designed for adherent cultured cells, so it is important that the researcher uses the correct extracellular matrix on the μ-Slide to ensure cellular attachment. The use of uncoated slides will result in cells growing mostly on the grid's groves and not homogeneously distributed along the chamber well.

The EM imaging can also be adapted to the researcher's needs and instrument access. The imprinted grid pattern on the resulting sample block can be used to reidentify the ROI, the researcher can then trim the block, remount it, and generate thin sections for TEM imaging. Although the groove depth of the μ-Slides grid is less than 5 μm, it gives the operator enough room to approach the sample utilizing an ultramicrotome and collect such sections. Likewise semithick sections can also be collected for SEM imaging in BSE mode if a TEM is unavailable. It is worth mentioning that the MAPS software version on the DualBeam imports only image data to correlate the ROI, and therefore, any TIFF image generated by a light microscope regardless of manufacturer can be imported and aligned to the resulting SEM data. Features in the LM images can then be used as the foundation for this correlation, including simple fiducials such as scratches on the substrate's surface, microparticles, fluorobeads, and of course cellular structures (de Boer et al., 2015).

The resin prepared as described above did not show any major deformation or melting defects during FIB milling. However, if the cells of interest are growing on the grooves of the μ-Slide imprinted grid, potential artifacts during the DualBeam preparation may result in curtaining defects throughout the milling process. To avoid this artifact, the carbon protective pad deposited via EBID and utilized to protect the ROI should be gently deposited to fill these groves.

Finally, the differences that can be observed at the moment of overlaying the LM and the EM images on a sequential CLEM procedure, such as the ones described here, could be due to the movement of cells within the μ-Slide or also deformation of the sample occurring during the dehydration and embedding steps. Both are very common artifacts observed in EM preparation procedures. Moreover, shrinking of the sample is usually nonisotropic, and most of the time, is difficult to correct during the experiment and in data reconstruction. We have also observed that cellular protrusions that were not strongly attached to the substrate were mostly affected by the preparation technique and consequently more difficult to overlay.

3. SUMMARY

Confocal live cell imaging of cultured breast cancer cells, followed by correlated FIB serial sectioning to produce three-dimensional scanning electron microscopy data, reveals localization of specific cellular markers. The workflow described here consists of growing cells in specially designed 6 wells ibidi μ-Slides. These slides are observed in optimal conditions using the FEI CorrSight–controlled environment chamber (Temp/CO_2/Moisture). Thanks to a microfluidic apparatus integrated onto the CorrSight "Live module," cells can be supplemented with fresh media during the experiment and can also be activated and imaged for an extended period of time. The FEI MAPS software installed on this microscope controls the pump of the "Live module." For the data acquisition, large portions of the wells were imaged at regular time intervals and at different magnifications. Once the images were acquired, the cells were chemically fixed, processed for EM observation, and transferred into the FEI Helios 660 DualBeam microscope where a cell of interest was identified using the same FEI MAPS software. A 3D auto slice and view data set was acquired at the interface between the cell of interest and its neighbor cell. Finally, both LM and 3D FIB-SEM data sets were loaded into the FEI Amira Software to allow for easy manipulation of the volumes and correlate information between both imaging modalities.

ACKNOWLEDGMENTS

This work was supported by the National Institutes of Health, the National Cancer Institute grant 5P30CA069533 in support of the Oregon Health & Science University (OHSU) Knight Cancer Institute, as well as the W.M. Keck Foundation and the Prospect Creek Foundation. Electron microscopy was performed at the Multiscale Microscopy Core (MMC) with technical support from the OHSU-FEI Living Lab Collaboration and the OHSU Center for Spatial Systems Biomedicine (OCSSB). This project was also supported by a Pilot Project Grant from the OCSSB to CSL. The content is solely the responsibility of the authors and does not necessarily represent the official views of the National Institutes of Health nor does it reflect the position or the policy of the Government, and no official endorsement should be inferred.

SUPPLEMENTARY DATA

Supplementary data related to this article can be found online at http://dx.doi.org/10.1016/bs.mcb.2017.03.008.

REFERENCES

Baxi, S. M., Tan, W., Murphy, S. T., Smeal, T., & Yin, M.-J. (2012). Targeting 3-phosphoinoside-dependent kinase-1 to inhibit insulin-like growth factor-I induced AKT and p70 S6 kinase activation in breast cancer cells. *PLoS One, 7*(10), e48402. http://dx.doi.org/10.1371/journal.pone.0048402.

de Boer, P., Hoogenboom, J. P., & Giepmans, B. N. (June 2015). Correlated light and electron microscopy: Ultrastructure lights up! *Nature Methods, 12*(6), 503–513. http://dx.doi.org/10.1038/nmeth. 3400 (Review). PMID: 26020503.

Booth, D. G., Beckett, A. J., Molina, O., Samejima, I., Masumoto, H., Kouprina, N., … Earnshaw, W. C. (November 17, 2016). 3D-CLEM reveals that a major portion of mitotic chromosomes is not chromatin. *Molecular Cell, 64*(4), 790–802. http://dx.doi.org/10.1016/j.molcel.2016.10.009. Epub 2016 Nov 10.

Cazemier, J. L., Clascá, F., & Tiesinga, P. H. (November 9, 2016). Connectomic analysis of brain networks: Novel techniques and future directions. *Frontiers in Neuroanatomy, 10*, 110. eCollection 2016.

van Driel, L. F., Valentijn, J. A., Valentijn, K. M., Koning, R. I., & Koster, A. J. (November 2009). Tools for correlative cryo-fluorescence microscopy and cryo-electron tomography applied to whole mitochondria in human endothelial cells. *European Journal of Cell Biology, 88*(11), 669–684. http://dx.doi.org/10.1016/j.ejcb.2009.07.002.

Gibson, K. H., Vorkel, D., Meissner, J., & Verbavatz, J. M. (2014). Fluorescing the electron: Strategies in correlative experimental design. *Methods in Cell Biology, 124*, 23–54. http://dx.doi.org/10.1016/B978-0-12-801075-4.00002-1.

Keene, D. R., Tufa, S. F., Lunstrum, G. P., Holden, P., & Horton, W. A. (August 2008). Confocal/TEM overlay microscopy: A simple method for correlating confocal and electron microscopy of cells expressing GFP/YFP fusion proteins. *Microscopy and Microanalysis, 14*(4), 342–348. http://dx.doi.org/10.1017/S1431927608080306.

Keene, D. R., Tufa, S. F., Wong, M. H., Smith, N. R., Sakai, L. Y., & Horton, W. A. (2014). Correlation of the same fields imaged in the TEM, confocal, LM, and microCT by image registration: From specimen preparation to displaying a final composite image. *Methods in Cell Biology, 124*, 391–417. http://dx.doi.org/10.1016/B978-0-12-801075-4.00018-5.

Kempen, P. J., Kircher, M. F., de la Zerda, A., Zavaleta, C. L., Jokerst, J. V., Mellinghoff, I. K., … Sinclair, R. (January 2015). A correlative optical microscopy and scanning electron microscopy approach to locating nanoparticles in brain tumors. *Micron, 68*, 70–76. http://dx.doi.org/10.1016/j.micron. 2014.09.004.

Knott, G. W., Holtmaat, A., Trachtenberg, J. T., Svoboda, K., & Welker, E. (2009). A protocol for preparing GFP-labeled neurons previously imaged in vivo and in slice preparations for light and electron microscopic analysis. *Nature Protocols, 4*, 1145–1156.

Kolotuev, I., Schwab, Y., & Labouesse, M. (December 4, 2009). A precise and rapid mapping protocol for correlative light and electron microscopy of small invertebrate organisms. *Biology of the Cell, 102*(2), 121–132. http://dx.doi.org/10.1042/BC20090096.

Kong, D., & Loncarek, J. (2015). Correlative light and electron microscopy analysis of the centrosome: A step-by-step protocol. *Methods in Cell Biology, 129*, 1–18. http://dx.doi.org/10.1016/bs.mcb.2015.03.013.

Kuipers, J., van Ham, T. J., Kalicharan, R. D., Veenstra-Algra, A., Sjollema, K. A., Dijk, F., … Giepmans, B. N. (April 2015). FLIPPER, a combinatorial probe for correlated live imaging and electron microscopy, allows identification and quantitative analysis of various cells and organelles. *Cell and Tissue Research, 360*(1), 61–70. http://dx.doi.org/10.1007/s00441-015-2142-7.

Lucas, M. S., Günthert, M., Gasser, P., Lucas, F., & Wepf, R. (2012). Bridging microscopes: 3D correlative light and scanning electron microscopy of complex biological structures. *Methods in Cell Biology, 111*, 325–356. http://dx.doi.org/10.1016/B978-0-12-416026-2.00017-0.

Maco, B., Holtmaat, A., Cantoni, M., Kreshuk, A., Straehle, C. N., Hamprecht, F. A., & Knott, G. W. (2013). Correlative in vivo 2 photon and focused ion beam scanning electron microscopy of cortical neurons. *PLoS One, 8*(2), e57405. http://dx.doi.org/10.1371/journal.pone.0057405. Epub 2013 Feb 28.

Moore, C. L., Cheng, D., Shami, G. J., & Murphy, C. R. (May 2016). Correlated light and electron microscopy observations of the uterine epithelial cell actin cytoskeleton using fluorescently labeled resin-embedded sections. *Micron, 84*, 61–66. http://dx.doi.org/10.1016/j.micron.2016.02.010. Epub 2016 Feb 21.

Mukohara, T. (January 2011). Mechanisms of resistance to anti-human epidermal growth factor receptor 2 agents in breast cancer. *Cancer Science, 102*(1), 1–8. http://dx.doi.org/10.1111/j.1349-7006.2010.01711.x. Epub 2010 Sep 6.

Murphy, G. E., Narayan, K., Lowekamp, B. C., Hartnell, L. M., Heymann, J. A., Fu, J., & Subramaniam, S. (December 2011). Correlative 3D imaging of whole mammalian cells with light and electron microscopy. *Journal of Structural Biology, 176*(3), 268–278. http://dx.doi.org/10.1016/j.jsb.2011.08.013.

Offterdinger, M., & Bastiaens, P. I. (January 2008). Prolonged EGFR signaling by ERBB2-mediated sequestration at the plasma membrane. *Traffic, 9*(1), 147–155. http://dx.doi.org/10.1111/j.1600-0854.2007.00665.x. Epub 2007 Nov 13. PMID: 17956594.

Padman, B. S., Bach, M., & Ramm, G. (April 22, 2014). An improved procedure for subcellular spatial alignment during live-cell CLEM. *PLoS One, 9*(4), e95967. http://dx.doi.org/10.1371/journal.pone.0095967.

Revach, O. Y., Weiner, A., Rechav, K., Sabanay, I., Livne, A., & Geiger, B. (March 30, 2015). Mechanical interplay between invadopodia and the nucleus in cultured cancer cells. *Scientific Reports, 5*, 9466. http://dx.doi.org/10.1038/srep09466.

van Rijnsoever, C., Oorschot, V., & Klumperman, J. (November 2008). Correlative light-electron microscopy (CLEM) combining live-cell imaging and immunolabeling of ultrathin cryosections. *Nature Methods, 5*(11), 973–980. http://dx.doi.org/10.1038/nmeth.1263.

Russell, M. R., Lerner, T. R., Burden, J. J., Nkwe, D. O., Pelchen-Matthews, A., Domart, M. C., … Collinson, L. M. (July 21, 2016). 3D correlative light and electron microscopy of cultured cells using serial blockface scanning electron microscopy. *Journal of Cell Science*. pii: jcs.188433.

Shaner, N. C., Lin, M. Z., McKeown, M. R., Steinbach, P. A., Hazelwood, K. L., Davidson, M. W., & Tsien, R. Y. (June 2008). Improving the photostability of bright monomeric orange and red fluorescent proteins. *Nature Methods, 5*(6), 545–551. http://dx.doi.org/10.1038/nmeth. 1209. Epub 2008 May 4. PMID: 18454154.

Sosinsky, G. E., Giepmans, B. N., Deerinck, T. J., Gaietta, G. M., & Ellisman, M. H. (2007). Markers for correlated light and electron microscopy. *Methods in Cell Biology, 79*, 575–591.

Wang, L., Eng, E. T., Law, K., Gordon, R. E., Rice, W. J., & Chen, B. K. (November 9, 2016). Visualization of HIV T cell virological synapses and virus-containing compartments by 3D correlative light and electron microscopy. *Journal of Virology*. pii: JVI.01605-16.

CHAPTER

CLAFEM: correlative light atomic force electron microscopy

Sébastien Janel, Elisabeth Werkmeister, Antonino Bongiovanni, Frank Lafont[1], Nicolas Barois

Univ. Lille, CNRS UMR 8204, Inserm U1019, CHU Lille, Institut Pasteur de Lille – CIIL – Center for Infection and Immunity of Lille, Lille, France

[1]Corresponding author: E-mail: Frank.lafont@pasteur-lille.fr

CHAPTER OUTLINE

Abstract

Atomic force microscopy (AFM) is becoming increasingly used in the biology field. It can give highly accurate topography and biomechanical quantitative data, such as adhesion, elasticity, and viscosity, on living samples. Nowadays, correlative light electron microscopy is a must-have tool in the biology field that combines different microscopy techniques to spatially and temporally analyze the structure and function of a single sample. Here, we describe the combination of AFM with superresolution light microscopy and electron microscopy. We named this technique correlative light atomic force electron microscopy (CLAFEM) in which AFM can be used on fixed and living cells in association with superresolution light microscopy and further processed for transmission or scanning electron microscopy. We herein illustrate this approach to observe cellular

Methods in Cell Biology, Volume 140, ISSN 0091-679X, http://dx.doi.org/10.1016/bs.mcb.2017.03.010

bacterial infection and cytoskeleton. We show that CLAFEM brings complementary information at the cellular level, from on the one hand protein distribution and topography at the nanometer scale and on the other hand elasticity at the piconewton scales to fine ultrastructural details.

Abbreviations

AFM	Atomic force microscopy
CLAFEM	Correlative light atomic force electron microscopy
CLEM	Correlative light electron microscopy
EM	Electron microscopy
SEM	Scanning electron microscopy
STED	Stimulated emission depletion
TEM	Transmission electron microscopy

INTRODUCTION

Correlative microscopy combines, in a spatially coordinated manner, the use of two or more imaging techniques to analyze the structure and function of a single sample, such as a molecule, a cell, a population of cells, a tissue, or an organism. While the first correlative microscopy has been described more than half a century ago (Bloch, Morgan, Godman, Howe, & Rose, 1957), it has recently undergone a rapid growth with a tremendous number of published articles. Nowadays, it is a must-have tool in numerous fields of biology. We show herein a recently introduced approach by adding three-dimensional topographical and mechanical information given by the atomic force microscope to superresolution microscopy and transmission or scanning electron microscopy (TEM or SEM).

AFM is a rather young technique, based on the breakthrough invention of the scanning tunneling microscope (STM) in 1982 (Binning, Rohrer, Gerber, & Weibel, 1982), and its further developments to work on nonconducting surfaces (Binnig & Quate, 1986). The Nobel Prize in Physics 1986 was awarded to Gerd Binnig and Heinrich Rohrer for their design of the STM. The scanning probe microscopies (SPMs) all consist in raster scanning the sample with the tiniest possible sensor to get the best possible resolution. The movement is usually performed by piezoelectric scanners, either by the sample or the probe, at subnanometer resolution. The measured and feedback signal is different between techniques. In the case of the AFM, the feedback is made on the deflection of a soft cantilever, measured by a laser beam reflected onto a four-quadrant photodiode. At the end of the cantilever, a nanometer-size tip interacts with the sample. The resolution is defined by the convolution of the tip with the surface. It can be as small as an atom, hence the name "atomic force microscope". The spring constant of the cantilever and the feedback parameters govern the forces applied to the sample. When doing imaging, these forces are minimized to scan the sample as gentle as possible, with force in the range of piconewtons (pN). It is also possible to increase these forces and to indent into the sample and by doing so to determine the mechanical properties of the sample. In this context, AFM contributes to the flourishing field of cell mechanics. By retracting the cantilever away from the surface, it is also possible to measure adhesion properties between the tip (and any object grafted onto a cantilever) and the surface. The AFM has the ability to work in air and in cell buffers and at different temperatures. Its first applications on living cells have been performed in the 1990s (for review see Ohnesorge et al., 1997). It was rapidly coupled to an inverted optical microscope to take advantage of the power of bright-field and fluorescence imaging. One of the only downsides of the technique is the relative slowness of the scanning, but recent developments in instrument and cantilever designs [e.g., high-speed AFM (Ando et al., 2001)], as well as scanning modes (e.g., PeakForce by Bruker or QI by JPK), make it more and more suitable for live molecular and cellular events.

In conventional microscopy, the diffraction limit determines the achievable spatial resolution. Thus, two objects closer than a distance $d = \lambda/2n \times \sin(\alpha)$ could not be discriminated with usual microscopy techniques, where λ represents the wavelength, n the refractive index, and $n \times \sin(\alpha)$ the numerical aperture of the objective lens (Abbe, 1873; Rayleigh, 1896). In conventional confocal microscopy, for example, the achieved spatial resolution is hundreds of nanometers [about 200 nm (xy) and 600 nm (z)], which is not sufficient to resolve details about the intracellular structures. Several techniques are now available to overcome the diffraction limit of light and to acquire superresolution fluorescent images. The Nobel Prize in Chemistry 2014 was given jointly to Eric Betzig, Stefan W. Hell, and William E. Moerner for the development of superresolved fluorescence microscopy. Structured illumination microscopy (SIM) enables to double confocal resolution in the three axes (Gustafsson, 1999). Photo-activated localization microscopy (PALM) (Betzig & Trautman, 1992) and stochastic optical reconstruction microscopy (STORM) (Heilemann et al., 2002, 2008) allow to obtain better resolution to date, but are quite time-consuming methods (several minutes to acquire one image). Herein, we are mostly interested in the stimulated emission depletion (STED) technology, on which the first reference was done by Hell and Wichmann (1994). A few years later, applications on biological samples were demonstrated by his team showing a resolution gain of factor 2 in the lateral and 6 in the axial dimension (Klar, Jakobs, Dyba, Egner, & Hell, 2000). The principle of STED microscopy is based on the stimulated emission depletion process. Typically, two different beams are used, one to excite the fluorophore, the second one to deplete the periphery of the excitation area. The second beam has generally a donut shape (zero intensity at the center). All the fluorophores present in the periphery are switched to their ground state, and only the center of the excitation remains fluorescent. To acquire an image, both beams are simultaneously scanning the sample, and the fluorescence signal is collected on an avalanche photodiode, enabling 3-D optical sectioning as for a confocal microscope (Punge et al., 2008).

The most-used correlative microscopy technique, the correlative light electron microscopy (CLEM; reviewed in de Boer, Hoogenboom, & Giepmans, 2015), starts with the light and fluorescence microscopy observation and ends with the electron microscopy observation. Since two decades, correlative microscopy went in several ways of expansions and improvements. The use of fluorescent-tagged proteins and time-lapse microscopy together with TEM rapidly boosted CLEM as it became possible to observe live cellular events that could be further processed to be examined at the ultrastructural level (Polishchuk et al., 2000). Using superresolution, optic microscopy allowed to improve the precision in correlation during cellular events as, for instance, during infection as we have shown using SIM and TEM (Ligeon, Barois, Werkmeister, Bongiovanni, & Lafont, 2015). The photooxidation of 3,3′-diaminobenzidine into electron-dense precipitates, by bleaching of the fluorescence, gives the possibility to see fluorescent-tagged proteins in both light and electron microscopy (Deerinck et al., 1994; Gaietta et al., 2002). Cryopreparation techniques (Verkade, 2008) or cryoobservation with both cryo-light and cryoelectron microscopy (Bos et al., 2014; Schorb & Briggs, 2014) allow observing the cell ultrastructure close to the one present in the living cell. Recently, live-imaging of magnetotactic bacteria in both fluorescence and electron microscopy has been described using a microfluidic chamber (Woehl et al., 2014). Finally, superresolution microscopy (van Engelenburg et al., 2014; Hübner, Cremer, & Neumann, 2013; Kim et al., 2015; Sochacki, Shtengel, van Engelenburg, Hess, & Taraska, 2014; Watanabe et al., 2011) and 3-D electron microscopy by focused ion beam or serial block face with a SEM (Armer et al., 2009; Beckwith et al., 2015; Felts et al., 2010; Shami, Cheng, Henriquez, & Braet, 2014) are now driving up CLEM. Classically, CLEM uses two separate microscopes, one for the fluorescence observation and one for the ultrastructural observation. It then requires a relocation system inside the sample when passing from a microscope to another, such as a coverslip with an alphanumerical photo-etched coordinate system supplied by Bellco Glass (Vineland, NJ, USA) as free coverslip (used in the present article) or MatTek Corporation (Ashland, MA, USA) and ibidi GmbH (Martinsried, Germany) as glass-bottom dish (Polishchuk et al., 2000). Nevertheless, integrated light-electron microscopes are now commercially available, avoiding the relocation question (Agronskaia et al., 2008; Iijima et al., 2014; Jun et al., 2011; Kanemaru et al., 2009; Liv et al., 2013; Zonnevylle et al., 2013).

The combination of light and electron microscopes is not the only correlative microscopy as other microscopy techniques have been associated together with success. A combination of SEM followed by TEM has been used to observe the release of *Flavivirus* from infected cells cultured on microcarrier beads (Burlaud-Gaillard et al., 2014). Microscopic X-ray computed tomography (microCT) or ion beam microprobe analysis (IBA) has also been added to CLEM workflow leading to volume information at a micrometer resolution or topology, density, and chemical composition at a submicrometer resolution (Handschuh, Baeumler, Schwaha, & Ruthensteiner, 2013; Le Trequesser et al., 2014). Two decades ago, TEM and AFM have been combined to observe freeze-fracture replicas of tissue, bringing the 3-D surface information missing in the TEM images (Kordylewski, Saner, & Lal, 1994). Recently, AFM has been associated with CLEM to study osteolysis (Shemesh, Addadi, Milstein, Geiger, & Addadi, 2016). In this study, bones were observed before and after culture of fluorescent osteoclasts on its surface by light microscope and air SEM or environmental SEM. Then, AFM was used after removal of cells to observe the changes of topology on the bone surface due to osteolysis. AFM has also been combined with superresolution microscopy for the observation of in vitro actin filaments or tagged proteins inside live cells (Odermatt et al., 2015).

In the present article, we describe the correlative light atomic force electron microscopy technique [CLAFEM (Lafont, 2014)], in which we associated three different microscopy techniques to observe a single cell: superresolution light microscopy, AFM, and TEM or SEM (Fig. 1). We applied this approach to fixed cells infected by *Yersinia pseudotuberculosis* that we have previously demonstrated capable of hijacking the autophagy pathway to replicate inside host cells (Ligeon et al., 2014; Moreau et al., 2010). We also investigated actin distribution in living cells. We show that all these techniques bring complementary information on the sample. Light microscopy allows the identification of multiple cellular objects (molecules and organelles) even at the nanoscale with superresolution techniques on fixed and living cells, AFM provides data such as topography and elasticity also on fixed and living cells, and electron microscopy brings details at the ultrastructure level, however, only on fixed cells.

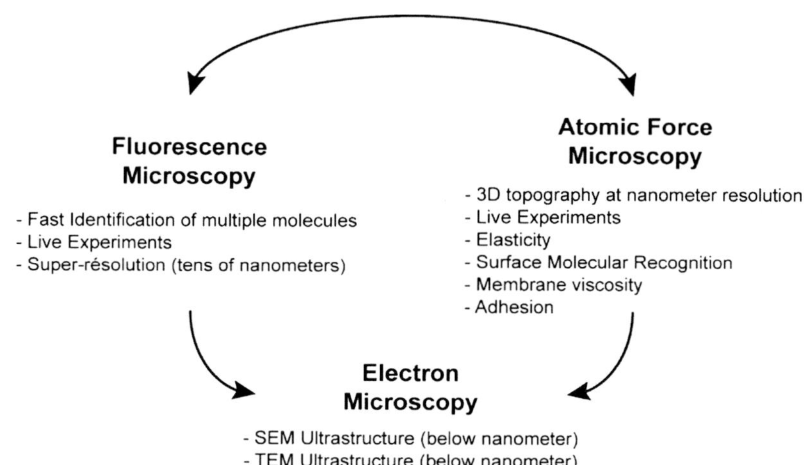

FIGURE 1

Principle of CLAFEM. Fluorescence and atomic force microscopy (AFM) are acquired with an integrated AFM-light microscope, for fixed and live samples. After identification and localization of molecules of interest by (superresolution) fluorescence microscopy, the area is scanned by AFM for biomechanical and topographical information. Then, the sample is prepared for TEM or SEM to obtain ultrastructural details.

1. MATERIALS AND METHODS

1.1 CELL CULTURE

Ptk2 cells (ATCC CCL-56) were cultured on photo-etched 25-mm-diameter coverslips (Bellco Biotechnology, 1916-91025, USA) in minimum essential medium (Gibco-Life technologies, 21090-022, France) supplemented with 10% heat-inactivated foetal bovine serum (Gibco-Life technologies, 10270, France), sodium pyruvate (Gibco-Life technologies, 11360-039, France), and L-glutamate (Gibco-Life technologies, 25000-024, France). Cells were incubated at 37°C in a humidified incubator with 5% CO_2 for 24–48 h.

1.2 SAMPLE PREPARATION—LIVE-IMAGING OF ACTIN CYTOSKELETON

Ptk2 cells were labeled for actin with 50 nM SiR actin (Spirochrome SC001, Switzerland) in medium at 37°C in a humidified incubator with 5% CO_2 for 2 h before imaging with fluorescence and AFM.

1.3 SAMPLE PREPARATION—LC-3 POSITIVE YERSINIA VACUOLES AND MICROTUBULES

Ptk2 cells were infected with the *Yersinia pseudotuberculosis* IP2777 strain with an MOI of 10 at 37°C in a humidified incubator with 5% CO_2. After 30 min of infection, 10 μg/mL of gentamycin was added to the medium to kill extracellular bacteria. After 24 h of infection, cells were labeled with 50 nM SiR tubulin (Spirochrome SC002, Switzerland) in medium for 2 h at 37°C in the incubator. Cells were then fixed with 4% paraformaldehyde and 0.1% glutaraldehyde in PBS 1X for 15 min. For LC3 immunolabeling, cells were first permeabilized with 0.2% triton X-100 in PBS 1X for 5 min, then labeled with a primary rabbit antibody against human LC3 (MBL, PM036, US-MA), at a 1/500 dilution and at 4°C overnight. This was followed by incubation with a secondary antirabbit antibody coupled with STAR 488 (Abberior, 2-0012-006-5, Germany), at a 1/200 dilution, at room temperature for 1 h. Finally, bacteria and cell nuclei were labeled with DAPI at 10 μg/mL for 5 min.

1.4 CORRELATION TECHNIQUE

The correlation between the three microscopies is made by two means. The first one, between STED and AFM, is handled by the AFM software. It consists in precisely calibrating the STED image proportions and the AFM tip position given the very accurate movements of the AFM x-y piezo scanners. This is performed semiautomatically by moving the AFM tip on a 3 × 3 (or 9 × 9), 30 μm^2 area matrix. The AFM controller sequentially sends a transistor–transistor logic (TTL) signal to the STED controller to trigger these acquisitions of the AFM tip. The acquisition is performed in the reflection mode using laser excitation (640 nm, 0.1 μW,

0.6 mW/cm^2) and the photomultiplier tube (PMT) detector. The size and resolution of calibration images must match the following experiment images. These images are saved on the AFM PC where the AFM software automatically detects them. The user then positions the tip location on one image and an algorithm automatically detects them on the following ones. The precision can be theoretically as low as the size of one pixel if the stage motors do not move. As a result, the AFM DirectOverlay software has the ability to import and transform every STED image to be in the right orientation and correct size for the AFM scanning.

The second mean of the correlation between the AFM/STED and the EM is performed by the gridded etched coverslip. As soon as the AFM/STED imaging is finished, two brightfield images of the area are acquired: one with the cell visible (100× objective), another one where the square number is visible (20× objective). The AFM head is then removed, and the sample is processed for EM. To compensate for EM preparation stretching/compressing, EM images are deformed linearly in x-y directions or not linearly using multiple notable features of the sample thanks to dedicated software (ec-CLEM, Paul-Gilloteaux et al., 2017, see also chapter: eC-CLEM: A Multidimension, Multimodel Software to Correlate Intermodal Images With a Focus on Light and Electron Microscopy by Heiligenstein, Paul-Gilloteaux, Raposo, & Salamero, 2017).

Another way for correlating these techniques is the use of fiducial markers (e.g., TetraSpeck fluorescent beads), but it has several issues that will be discussed below.

1.5 ATOMIC FORCE MICROSCOPY

Experiments were performed with a JPK NanoWizard III Ultra AFM and the Abberior Instruments STED optics mounted on an IX83 Olympus microscope. AFM calibration was performed before STED acquisition (see above for details) followed by AFM acquisition. The gridded glass coverslips were assembled on the JPK BioCell and kept in 500 μL imaging buffer, at 37°C when performing live experiment or at room temperature on fixed samples. We used Olympus BioLever mini cantilevers (BL-AC40TS-C2) that have a low spring constant (0.1 N/m), high resonant frequency (25 kHz in water), tall and slightly sharp indenter (r = 10 nm). The choice of the cantilever is critical and depends on the indentation behavior of the sample and the object to be observed. The spring constant was calibrated prior to every experiment by using the Sader method implemented in the JPK software version 6.0 (Sader, Chon, & Mulvaney, 1999). Acquisitions were performed in force mapping (QI mode), where the AFM tip raster scans while indenting into the sample at each pixel, hence providing mechanical information of the sample in 2-D or 3-D. In this mode force curves are performed at high speed (with constant speed during indentation) across the scanning area, making it better suited for either live processes or increase in resolution. We used the following parameters: scan size 30 μm^2, 512 × 512 pixels, [1–3] μm ramp size, 300 μm/s cantilever speed, [2–10] nN trigger force (higher force is needed when indenting fixed samples).

Elasticity analysis was performed either on the JPK Data Analysis 6.0 or in-house software for 3-D elasticity. "Piezo height" is the height of the AFM z-piezo at the end of the indentation, hence inside the cell. "Topography" is a zero force image corresponding to the point of contact in the force curve; it is therefore the position of the cell membrane. "Elasticity" is the Young's modulus calculated by fitting the indentation curve with the paraboloid indenter formula at each pixel. "Elasticity tomogram" is a 2-D slice of elasticity in the y-z plane obtained by calculating Young's modulus on 40 nm-cut segments of the indentation curve (Roduit et al., 2009).

1.6 STIMULATED EMISSION DEPLETION MICROSCOPY

STED acquisitions were driven by the ImSpector software (Abberior). The system is equipped with a Patented Abberior Instruments QUAD scanner (with four galvo mirrors), and illumination was done through an Olympus 100× (NA 1.4, oil) lens. SiR-actin and SiR-tubulin were excited by a 640 nm pulsed laser (PicoQuant, 45 VA max, 50/60 Hz, 0.14 μW, 7.2 mW/cm^2), and depleted by a 775 nm pulsed laser (2 W, imaging power, 32.6 mW, 1.66 kW/cm^2). Star 488 was excited with a 485 pulsed laser (PicoQuant, 45 VA max, 50/60 Hz, 9.40 μW, 485 mW/cm^2) and depleted with a 592 nm CW laser (2 W, imaging power 111 μW, 5.70 W/cm^2). To improve the quality of records, we used a gating of 2 ns for the CW 592 nm depletion and 780 ps for pulsed 775 nm depletion. 30 μm × 30 μm images were recorded with a pixel size of 40 nm, a 10 μs pixel dwell time and a line accumulation of 2. Fluorescence signals were collected, through a 25 μm pinhole, on two APDs (avalanche photodiode) before which we placed either a GFP band filter (525/50 nm) or a Cy5 band-pass filter (685/70 nm). Finally, to do the correlation, an image was taken in brightfield mode on a CCD camera (the Imaging Source DMK 41AU02, 1280 × 960 pixels) with a 20× objective.

1.7 TRANSMISSION ELECTRON MICROSCOPY

After cell imaging of microtubules and the LC3-positive bacteria vacuole by STED microscopy and AFM, cells were imaged in optical brightfield mode at 100× and 20× magnification for localization purposes. Cells were fixed anew with 1% glutaraldehyde in 0.1 mM sodium cacodylate buffer for minimum 30 min. After washing with water, cells were sequentially stained with 1% osmium tetroxide reduced with 1.5% potassium hexacyanoferrate(III) for 1 h, 1% thiocarbohydrazide for 30 min, 1% osmium tetroxide for 1 h, and finally 1% uranyl acetate for 1 h. All stains were made in water, at room temperature in the dark and were also washed with water. After staining, cells were dehydrated in graded ethanol solutions, infiltrated with epoxy resin and cured, for flat embedding on the coverslip, at 60°C for 48 h (Nguyen et al., 2011). After separation of the resin from the glass, the cell of interest was relocated with the imprinted-alphanumerical grid at the surface of the resin (Hodgson, Nam, Mantell, Achim, & Verkade, 2014). A small block of resin containing the cell of interest was prepared for sectioning parallel to the resin surface. Serial sections of 80 nm thickness were set down on carbon/formvar-coated slot grids.

Sections were observed with a Hitachi H7500 TEM (Elexience, France), and images were acquired with a digital camera from AMT (Elexience, France).

1.8 SCANNING ELECTRON MICROSCOPY

After cell live-imaging of the actin cytoskeleton by STED microscopy and AFM, cells were imaged in optical brightfield mode at 100× and 20× magnification for localization purposes. Cells were immediately fixed and permeabilized with 1% triton X-100 and 0.25% glutaraldehyde in 100 mM PIPES pH 6.9, 1 mM EGTA and 1 mM $MgCl_2$ buffer for 20 min. After washing, cells were fixed with 2% glutaraldehyde in 0.1 mM sodium cacodylate buffer for 1 h. Without washing, cells were incubated with 0.1% tannic acid in water for 30 min. After washing with water, cells were incubated with 0.2% uranyl acetate in water for 30 min. Cells were dehydrated with increasing ethanol concentration baths. After two pure ethanol baths, cells were air-dried with HMDS. Finally, dry coverslips were mounted on stubs and coated with 5 nm platinum (Quorum Technologies Q150T, Elexience, France). Cells were imaged with a Zeiss Merlin Compact VP FE-SEM (Zeiss, France) at 2 kV by an in-chamber secondary electron detector and at 10 kV by an in-lens duo detector in the secondary electron detection mode.

2. RESULTS AND DISCUSSION

For the CLAFEM experiments, PtK2 cells were cultured on free 25 mm diameter photo-etched coverslips, which fit perfectly into the JPK BioCell mounted on the STED microscope. After light and AFM acquisitions, fixed or living cells are processed for either TEM or SEM (Fig. 2). Correlation between STED and AFM was done through the calibration of the tip-cantilever position in relation to the fluorescent image. Further correlation with EM was done through the photo-etched coverslip. For TEM, the inverted imprinting of the coordinate system at the surface of the resin allows to relocate and isolate the cell of interest in a small block of resin for ultrathin sectioning. For SEM, the photo-etched coverslip is directly observed to relocate the cell of interest.

In the first CLAFEM experiment, the microtubule cytoskeleton has been labeled in live cells with the membrane permeable fluorescent probe SiR coupled to the drug docetaxel, which targets tubulin proteins. Cells were then permeabilized to indirectly immuno-label LC3 with the STAR 488 fluorophore. LC3 is a hallmark of autophagy. It is a cytosolic protein that on autophagy activation is processed to be conjugated to phosphatidylethanolamine for integration into the isolation membrane during the autophagosome elongation (Mizushima, Yoshimori, & Ohsumi, 2011). Finally, cell and bacteria DNA were labeled with DAPI before being mounted on the AFM/STED. We observed the microtubule cytoskeleton and the LC3-positive *Yersinia*-containing vacuoles by STED, AFM, and TEM in fixed cells (Fig. 3). First, triple fluorescence labeling confirms the presence of DAPI-stained bacteria in a

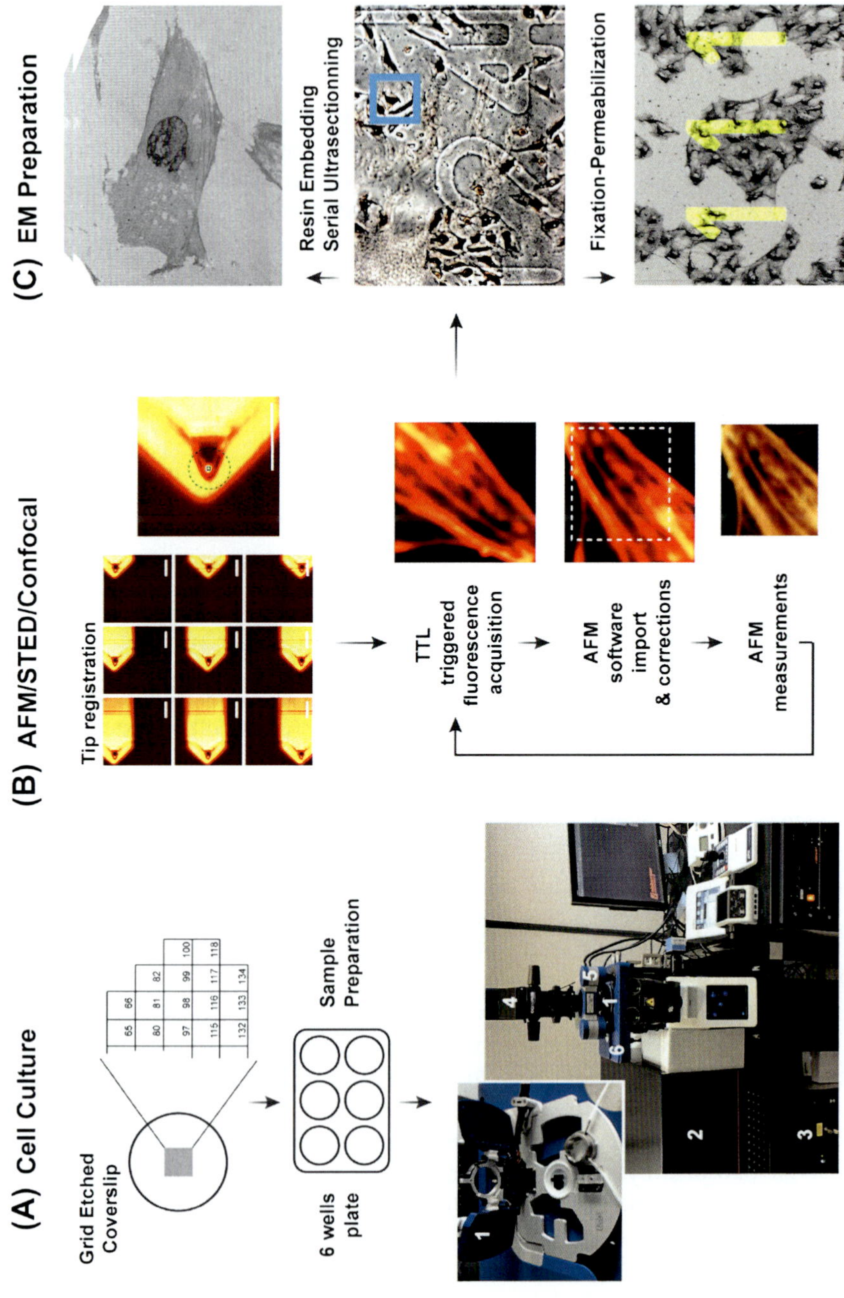
(A) Cell Culture
Grid Etched Coverslip
6 wells plate
Sample Preparation
(B) AFM/STED/Confocal
Tip registration
TTL triggered fluorescence acquisition
AFM software import & corrections
AFM measurements
(C) EM Preparation
Resin Embedding Serial Ultrasectionning
Fixation-Permeabilization

FIGURE 2

Sample preparation and correlation of techniques. (A) Cells are grown on photo-etched 25 mm diameter glass coverslips in 6-wells plate. Cells are stained and one coverslip is then transferred to the atomic force microscopy (AFM) stage for stability and temperature control during experiments. The AFM/stimulated emission depletion (STED) microscope and the BioCell are shown: (1) JPK BioCell, (2) STED optics, (3) STED lasers, (4) Inverted microscope, (5) AFM head, and (6) AFM stage. (B) The correlation between AFM and STED is performed by the AFM software: it performs a precise piezo-defined tip movement (3×3 or 9×9 positions); imports the corresponding confocal images of the AFM tip (scale bar 7.5 μm), and calculate a transformation matrix for the following acquisitions of fluorescence images. The user then has the ability to scan by AFM the area selected by fluorescence imaging and to overlay the two images. (C) The correlation between AFM/STED and electron microscopy is performed by high and low magnification brightfield imaging of the area and identification of the area number. For transmission electron microscopy observation, the cells are embedded in resin on the coverslip. After coverslip removal, the cell of interest is localized with the imprinted area number. Only a small resin block containing the cell of interest is serially sectioned (*blue box*). For scanning electron microscopy (SEM) observation, the coverslip is dried and directly introduced into the SEM chamber after metallization. The area number is visible with the SEM.

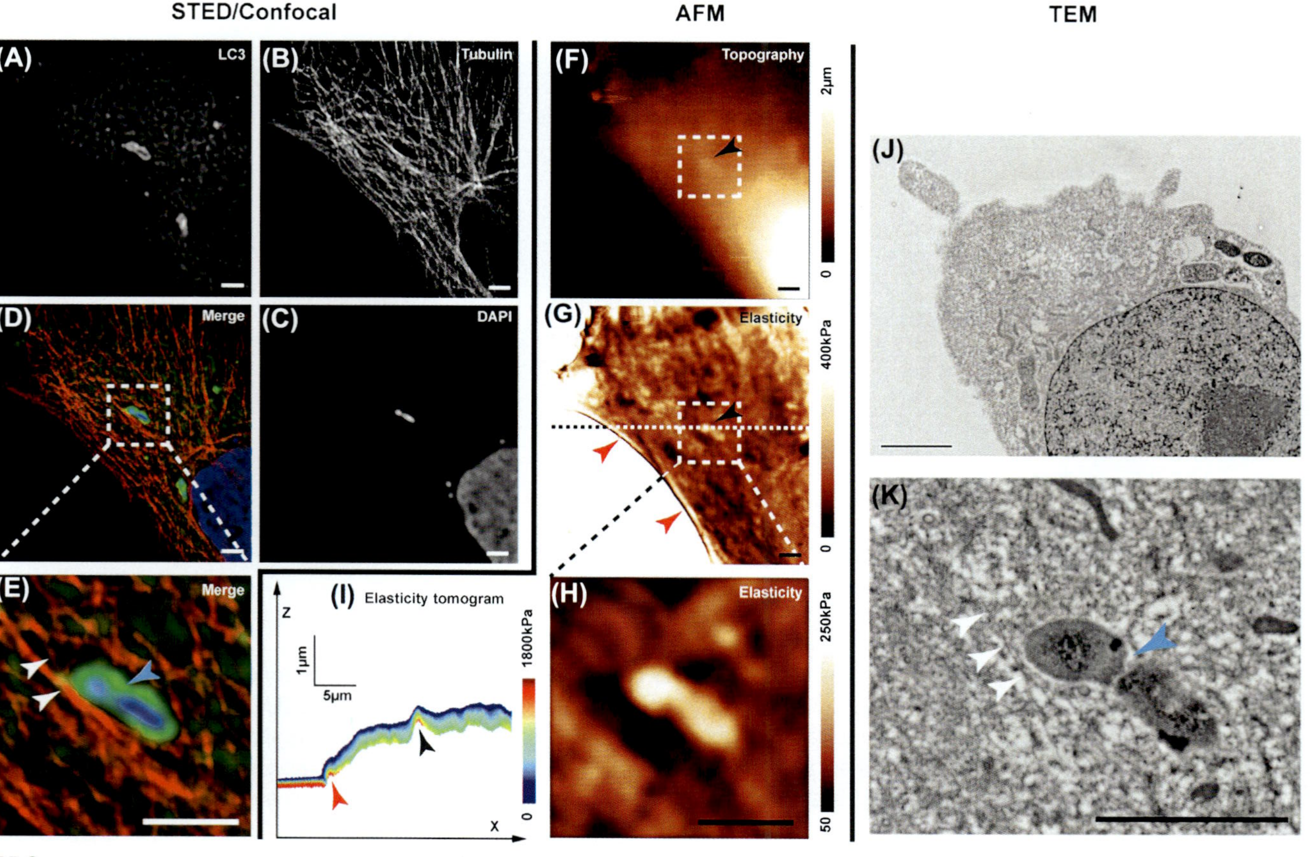

FIGURE 3

CLAFEM on fixed sample in the context of infection, PtK2 cells infected by *Yersinia pseudotuberculosis*. (A–E) Fluorescence images (STED/Confocal): (A) LC3 immunostaining, (B) tubulin labeling, (C) DAPI staining, (D): merged image, and (E) zoom of the *dashed box* in (D). (F–I) Atomic force microscopy (AFM): (F) topography, (G) elasticity map, (H) elasticity map corresponding to the *dashed box* in (G), and (I) elasticity tomogram profile corresponding to the *dotted line* in (G). (J and K) transmission electron microscopy (TEM): (J) TEM section situated at the top of the cell. (K) TEM section through the bacteria of interest. *Black arrowhead*: bacterium, *white arrowheads*: tubulin fiber, *blue arrowhead*: septum, *red arrowhead*: actin stress fiber. Scale bars 2.5 μm.

vacuole positive for LC3 (Ligeon et al., 2015, 2014), also called light chain three of microtubule-associated protein 1, surrounded by microtubules (Fig. 3A–E). DAPI and LC3-immunostained images were acquired in the confocal mode. The image of the tubulin cytoskeleton was acquired in the STED mode and shows the microtubules distributed all around the vacuole. Second, the topography image of the cell, given by the AFM, shows the height of the cell in a color-coded manner (Fig. 3F). A slight relief at the surface of the cell may indicate the bacteria-containing vacuole, but the sole topography image is not sufficient to convincingly demonstrate the presence of this vacuole. However, the elasticity image of the cell indicates stiff material corresponding to the bacteria-containing vacuole where the relief is localized (Fig. 3G and H). A slice into the elasticity tomogram passing through the vacuole shows that the bacteria-containing vacuole is harder than the surrounding area of the cell and is bumping at the cell surface (Fig. 3I). The AFM shows also two other very hard objects; the glass coverslip and the actin cytoskeleton at the edge of the cell (red arrowheads), but the tubulin cytoskeleton could not be detected in this experimental acquisition setup probably due to the localization deeper into the cell below the indentation limit. In addition, AFM does not allow detecting easily bacteria in vacuoles when close to the nucleus probably because of proximity with stiff elements. Third, a TEM image of a section at the top of the cell demonstrates that the relief is due to the bacteria-containing vacuole (Fig. 3J). TEM image of a section in the middle of the cell shows the bacteria but not the expected limiting membrane (Fig. 3K). The tubulin cytoskeleton is also visible although partially because TEM sections are thinner than optical sections. Therefore, to see the totality of the cytoskeleton network in EM, CLAFEM could be associated to 3-D electron microscopy such as focused ion beam or serial block face with a SEM (Armer et al., 2009; Beckwith et al., 2015; Felts et al., 2010; Shami et al., 2014). Because cells have been permeabilized for the immune-labeling of LC3, the TEM showed the loss of all cellular membranes during the sample preparation. Despite the presence of 0.1% glutaraldehyde in the first fixation, it was not sufficient to maintain the membranes. While we tried a higher percentage of glutaraldehyde (0.5%), we unfortunately obtained high autofluorescence background during STED acquisition (data not shown). To avoid permeabilization, we could have used cells expressing GFP-tagged LC3. GFP, widely used with classical microscopy, has been also observed with STED (Neupane et al., 2015; Rankin et al., 2011; Willig et al., 2006). However, green fluorescent proteins such as GFP and STAR 488 are less reliable probes to use with STED than with confocal microscopy. The higher power and the depletion laser used for STED acquisition, compared with confocal acquisition, lead to a fast bleaching of the green probes. It is difficult to make several STED images of these green probes, unlike red probes. Fast bleaching and loss of spatial resolution can be resolved with protected STED, in which the use of several off–state transitions of photoswitchable fluorophores allows bleaching protection and contrast enhancement (Danzl et al., 2016).

In the second CLAFEM experiment, the actin cytoskeleton has been labeled with the membrane permeable fluorescent probe SiR coupled to jasplakinolide, which targets F-actin proteins. Then, the actin cytoskeleton was observed by STED and AFM in live cells, before being further processed after fixation and permeabilization for SEM (Fig. 4). Two sequences of STED and AFM acquisitions have been done. The time of 5 min between the two STED acquisitions is the time corresponding to the AFM acquisition (5 min, 150 × 150 pixels, 14 × 14 μm). During this time, due to internal dynamics, the actin cytoskeleton slightly reorganized as shown by STED and AFM images (t = 0 vs. t = 5 min, Fig. 4). Because of its stiffness, the actin cytoskeleton is well detected in the elasticity image (Fig. 4C and F) as well as in the piezo height image (Fig. 4B and E) when compared with the STED image (Fig. 4A and D). After the second AFM acquisition, cells were chemically fixed again and also permeabilized at the same time to further observe the actin cytoskeleton by SEM. SEM images show actin stress fibers that did not reorganize (white/black arrows in Fig. 4D–H) or reorganized (blue arrowheads in Fig. 4D–F and I) during live observation. Some fibers are clearly visible in the STED images, started to fade during AFM acquisition and are not visible anymore in SEM images.

The preparation method of the sample for SEM is based on the one used for the observation of the cytoskeleton platinum replica with TEM (Svitkina, 2007). Instead of being platinum- and carbon-coated then transferred from glass coverslip to EM grid, the cytoskeleton was simply coated with platinum on the gridded coverslip and then directly observed with an FE-SEM. Thus, manipulation of the fragile replica such as separation from the glass coverslip by flotation, washing, and setting down on fragile support-film of EM grids, is avoided here. In the FE-SEM, high-resolution imaging of the cytoskeleton, capable of discerning clathrin coats such as in TEM, is obtained with an in-lens secondary electron detector (Fig. 4J).

AFM is an interesting tool for cell biology as it offers both topography and biomechanical quantitative data (adhesion, elasticity, and viscosity). However, if used alone to analyze cells, it may provide not sufficient information as the specificity of the identification of organelle and proteins requires complementary approaches. For this reason, one should prefer to associate AFM with other microscopy techniques to unambiguously identify intracellular objects that are under investigation. In the present article, we associated AFM with CLEM, light microscopy (confocal and superresolution), and EM (TEM and SEM). AFM can be used on intact, fixed, or living cells. Especially, one can thus obtain quantitative information on topography and elasticity in living cells. Intracellular compartments are also accessible to elasticity analysis using a "stiffness tomography" approach (Roduit et al., 2008) although with limitations in indentation depth. AFM thus easily detects stiff intracellular elements such as for instance internalized bacteria and actin stress fibers but failed to detect very soft objects. In addition, AFM could not detect small or deep objects in the cell because of the presence of the plasma membrane, cytoskeleton, and the cytoplasm above them. To analyze these objects by AFM, one possibility is to expose them by removing the plasma membrane and the cytoplasm (Sato, Asakawa, Fukuma, & Terada, 2016; Usukura, Narita, Yagi, Ito, & Usukura, 2016). Thus, after unroofing and fixation, AFM may allow directly and clearly analyze the actin and tubulin cytoskeletons as well as the clathrin coat.

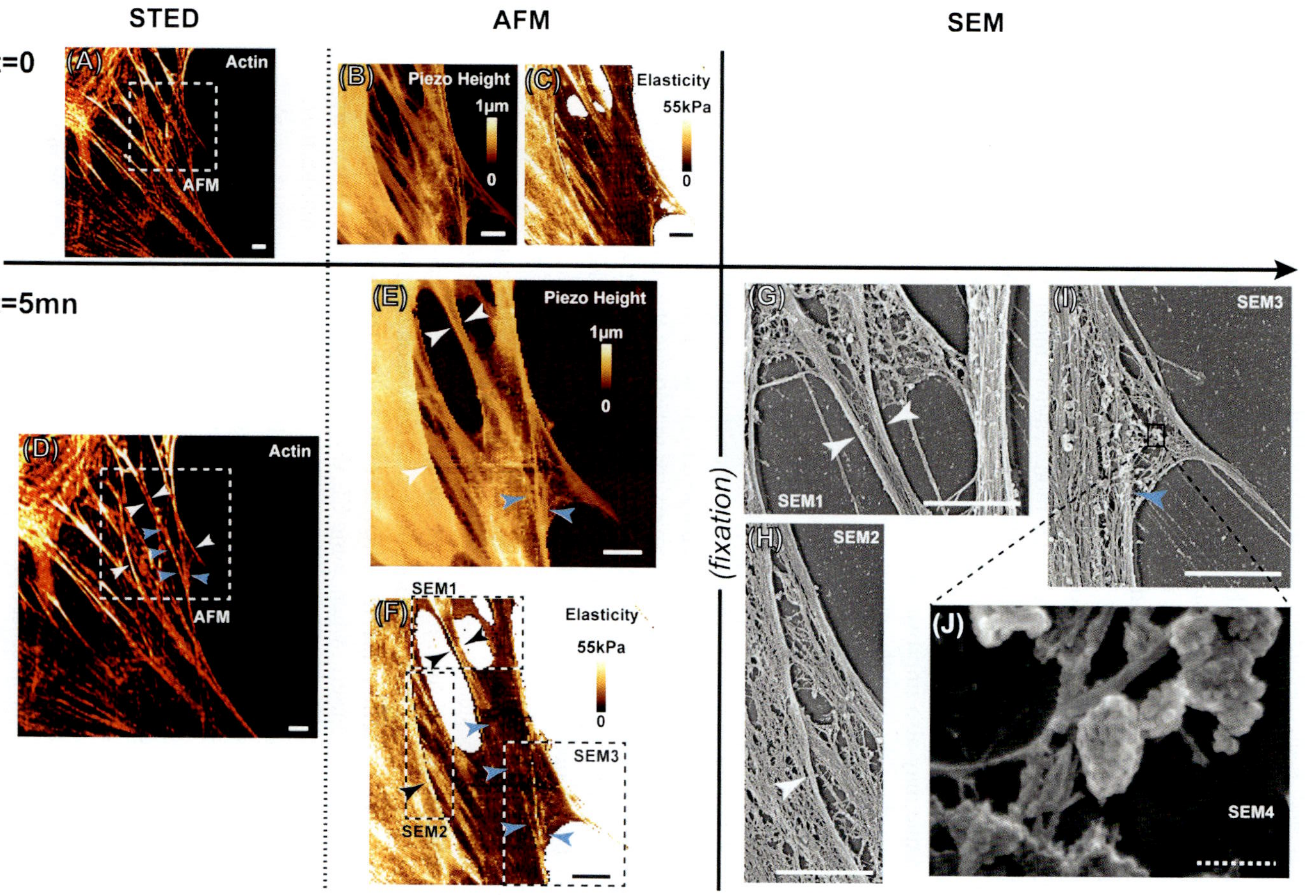

FIGURE 4

Live CLAFEM. (A) stimulated emission depletion (STED) actin imaging, (B) atomic force microscopy (AFM) piezo height and (C) elasticity at t = 0. (D) STED actin imaging, (E) AFM piezo height, (F) elasticity, and (G–J) scanning electron microscopy (SEM) images at t = 5 min. The *dashed boxes* in (A) and (D) correspond to the AFM scanning area. The *dashed boxes* in (F) correspond to the SEM images. (G and H) *Arrowheads* indicate actin fibers visible in STED, AFM, and SEM. (I) *Blue arrowheads* indicate actin fibers visible in STED, disappearing in AFM and not visible in SEM. (J) Clathrin coat visible by SEM. Scale bars: black/white 2 μm, *dashed white*: 100 nm.

Unfortunately, this technique may not be ideally suited for any cellular objects such as organelles and bacteria-containing vacuoles that could be damaged by the procedure.

We showed herein that AFM can be used on living cells. With an acquisition time of several minutes, it is not actually suitable to observe fast intracellular phenomena. However, since several years, speed improvements have been made through cantilever and system developments and the coming years should allow integration with superresolution optics. While one STED image and one AFM image are obtained sequentially, it would be also possible to scan lines after lines sequentially with STED and AFM. In this case, cellular modifications might be closely related between the two images. One important improvement for CLEM has been the development of the rapid cryofixation technique, which permits the fixation of the sample few seconds after its observation with a light microscope (Verkade, 2008). The use of this technique may also improve the CLAFEM when passing to EM. The small coverslip supporting the sample used for cryofixation (few millimeters diameter) could be adapted on the microscope but keeping in mind that space above has to be largely open for the accessibility of the cantilever of the AFM.

Correlation between STED and AFM is done through the calibration of the tip-cantilever position in relation to the fluorescent image in the integrated AFM/STED microscope. We do not need a coordinate system unlike the correlation with EM images as TEM and SEM microscopes are separated from the AFM/STED microscope. In the CLAFEM, we used photo-etched coverslips to relocate the cell of interest but this method may not be sufficient for locating the object of interest with high precision. Such high precision correlation has been proposed based on the use of fiducial markers, such as fluorescent beads, added to the sample and visible in fluorescence microscopy and EM (Kukulski et al., 2011; Schellenberger et al., 2014; Schorb & Briggs, 2014). However, the use of such beads is limited in AFM because they are not visible when located at the surface of cells. They are visible when located on the glass but it requires high-resolution scanning, thus slow scanning, which is not suitable for live imaging. In addition, their localization on the glass is not useful when looking at TEM sections localized higher in the cell, unless doing 3-D EM. On the contrary, they may be useful when doing CLAFEM with SEM. Fiducial markers, as well as cellular markers present in the three images, may serve to realign these images with high precision using registration software (Paul-Gilloteaux et al., 2017). Such software can also take into account deformations during TEM process such as the shrinkage of the sample during ethanol dehydration and the slight compression of sections during cutting.

We have shown herein that AFM can be combined with CLEM to observe either fixed or live cells, using an integrated AFM/STED microscope followed by SEM or TEM. The method provided functional dynamic of fluorescent markers involved in autophagy hijacking during infection (i.e., LC3) while cell elasticity can be monitored, and finally, electron microscopy adding the ultrastructural data level. This method can also be applied to cell-to-cell interaction in which quantitative force interaction is examined. Depending on the dynamics of the system analyzed, other modes of superresolution optic imaging can be used such as PALM/STORM. Finally, the versatility of the application makes this correlative approach applicable to many physiopathological cellular events in which functional cell dynamics and cell biomechanics at the highest resolution to date can be monitored in time lapse. Processing further the sample for ultrastructural acquisition allows to intimately analyze the cellular components involved with an unprecedented global overview of the phenomenon. Although challenging technically and requiring several know-how and important investment, the development of CLAFEM within technological facilities will be beneficial to many different fields in the future at different scales ranging from molecule to tissue investigation.

ACKNOWLEDGMENTS

We thank the BioImaging Center Lille for access to equipment. We acknowledge the financial support of the ANR (10-EQPX-04-01) and the EU-FEDER (12,001,407) to FL.

REFERENCES

Abbe, E. (1873). Beiträge zur Theorie des Mikroskops und der mikroskopischen Wahrnehmung. *Archiv für mikroskopische Anatomie, 9*(1), 413–418.

Agronskaia, A. V., Valentijn, J. A., van Driel, L. F., Schneijdenberg, C. T., Humbel, B. M., van Bergen en Henegouwen, P. M., ... Gerritsen, H. C. (2008). Integrated fluorescence and transmission electron microscopy. *Journal of Structural Biology, 164*(2), 183–189.

Ando, T., Kodera, N., Takai, E., Maruyama, D., Saito, K., & Toda, A. (2001). A high-speed atomic force microscope for studying biological macromolecules. *Proceedings of the National Academy of Sciences of the United States of America, 98*(22), 12468–12472.

Armer, H. E. J., Mariggi, G., Png, K. M., Genoud, C., Monteith, A. G., Bushby, A. J., ... Collinson, L. M. (2009). Imaging transient blood vessel fusion events in zebrafish by correlative volume electron microscopy. *PLoS One, 4*(11).

Beckwith, M. S., Beckwith, K. S., Sikorski, P., Skogaker, N. T., Flo, T. H., & Halaas, Ø. (2015). Seeing a mycobacterium-infected cell in nanoscale 3D: Correlative imaging by light microscopy and FIB/SEM tomography. *PLoS One, 10*(9), 1–19.

Binnig, G., & Quate, C. F. (1986). Atomic force microscope. *Physical Review Letters, 56*(9), 930–933.

Binning, G., Rohrer, H., Gerber, C., & Weibel, E. (1982). Surface studies by scanning tunneling microscopy. *Physical Review Letters, 49*(1), 57–61.

Betzig, E., & Trautman, J. K. (1992). Near-field optics: Microscopy, spectroscopy, and surface modification beyond the diffraction limit. *Science (New York, NY), 257*(5067), 189–195.

Bloch, D. P., Morgan, C., Godman, G. C., Howe, C., & Rose, H. M. (1957). A correlated histochemical and electron microscopic study of the intranuclear crystalline aggregates of adenovirus (ri-apc virus) in hela cells. *Journal of Biophysical and Biochemical Cytology, 3*(1), 1–8.

de Boer, P., Hoogenboom, J. P., & Giepmans, B. N. G. (2015). Correlated light and electron microscopy: Ultrastructure lights up! *Nature Methods, 12*(6), 503–513.

Bos, E., Hussaarts, L., van Weering, J. R., Ellisman, M. H., de Wit, H., & Koster, A. J. (2014). Vitrification of Tokuyasu-style immuno-labelled sections for correlative cryo light microscopy and cryo electron tomography. *Journal of Structural Biology, 186*(2), 273–282.

Burlaud-Gaillard, J., Sellin, C., Georgeault, S., Uzbekov, R., Lebos, C., Guillaume, J. M., & Roingeard, P. (2014). Correlative scanning-transmission electron microscopy reveals that a chimeric flavivirus is released as individual particles in secretory vesicles. *PLoS One, 9*(3), 1–6.

Danzl, J. G., Sidenstein, S. C., Gregor, C., Urban, N. T., Ilgen, P., Jakobs, S., & Hell, S. W. (2016). Coordinate-targeted fluorescence nanoscopy with multiple off states. *Nature Photonics, 10*(2), 122–128.

Deerinck, T. J., Martone, M. E., Lev-Ram, V., Green, D. P., Tsien, R. Y., Spector, D. L., … Ellisman, M. H. (1994). Fluorescence photooxidation with eosin – A method for high-resolution immunolocalization and in-situ hybridization detection for light and electron-microscopy. *Journal of Cell Biology, 126*(4), 901–910.

van Engelenburg, S. B., Shtengel, G., Sengupta, P., Waki, K., Jarnik, M., Ablan, S. D., … Lippincott-Schwartz, J. (February 2014). Distribution of ESCRT machinery at HIV assembly sites reveals virus scaffolding of ESCRT subunits. *Science (New York, NY), 343*, 653–656.

Felts, R. L., Narayan, K., Estes, J. D., Shi, D., Trubey, C. M., Fu, J., … Subramaniam, S. (2010). 3D visualization of HIV transfer at the virological synapse between dendritic cells and T cells. *Proceedings of the National Academy of Sciences of the United States of America, 107*(30), 13336–13341.

Gaietta, G., Deerinck, T. J., Adams, S. R., Bouwer, J., Tour, O., Laird, D. W., … Ellisman, M. H. (April 2002). Multicolor and electron microscopic imaging of connexin trafficking. *Science, 296*, 503–507.

Gustafsson, M. G. (1999). Extended resolution fluorescence microscopy. *Current Opinion in Structural Biology, 9*(5), 627–634.

Handschuh, S., Baeumler, N., Schwaha, T., & Ruthensteiner, B. (2013). A correlative approach for combining microCT, light and transmission electron microscopy in a single 3D scenario. *Frontiers in Zoology, 10*(1), 44.

Heilemann, M., Herten, D. P., Heintzmann, R., Cremer, C., Muller, C., Tinnefeld, P., … Sauer, M. (2002). High-resolution colocalization of single dye molecules by fluorescence lifetime imaging microscopy. *Analytical Chemistry, 74*(14), 3511–3517.

Heilemann, M., van de Linde, S., Schüttpelz, M., Kasper, R., Seefeldt, B., Mukherjee, A., … Sauer, M. (2008). Subdiffraction-resolution fluorescence imaging with conventional fluorescent probes. *Angewandte Chemie – International Edition, 47*(33), 6172–6176.

Heiligenstein, X., Paul-Gilloteaux, P., Raposo, G., & Salamero, J. (2017). eC-CLEM: A multidimension, multimodel software to correlate intermodal images with a focus on light and electron microscopy. In T. Mueller-Reichert, & P. Verkade (Eds.), *Correlative light and electron microscopy III* (Vol. 140, pp. 335–352).

Hell, S. W., & Wichmann, J. (1994). Breaking the diffraction resolution limit by stimulated emission: Stimulated-emission-depletion fluorescence microscopy. *Optics Letters, 19*(11), 780.

Hodgson, L., Nam, D., Mantell, J., Achim, A., & Verkade, P. (2014). *Retracing in correlative light electron microscopy: Where is my object of interest?* (1st ed.). Elsevier Inc.

Hübner, B., Cremer, T., & Neumann, J. (2013). Correlative microscopy of individual cells: Sequential application of microscopic systems with increasing resolution to study the nuclear landscape. *Methods in Molecular Biology, 1042*, 299–336.

Iijima, H., Fukuda, Y., Arai, Y., Terakawa, S., Yamamoto, N., & Nagayama, K. (2014). Hybrid fluorescence and electron cryo-microscopy for simultaneous electron and photon imaging. *Journal of Structural Biology, 185*(1), 107–115.

Jun, S., Ke, D., Debiec, K., Zhao, G., Meng, X., Ambrose, Z., … Zhang, P. (2011). Direct visualization of HIV-1 with correlative live-cell microscopy and cryo-electron tomography. *Structure, 19*(11), 1573–1581.

Kanemaru, T., Hiratab, K., Takasuc, S.-i., Isobed, S.-i., Mizukie, K., Matakaf, S., & Nakamura, K.-i. (2009). A fluorescence scanning electron microscope. *Ultramicroscopy, 109*(4), 344–349.

Kim, D., Deerinck, T. J., Sigal, Y. M., Babcock, H. P., Ellisman, M. H., & Zhuang, X. (2015). Correlative stochastic optical reconstruction microscopy and electron microscopy. *PLoS One, 10*(4), 1–20.

Klar, T. A., Jakobs, S., Dyba, M., Egner, A., & Hell, S. W. (2000). Fluorescence microscopy with diffraction resolution barrier broken by stimulated emission. *Proceedings of the National Academy of Sciences of the United States of America, 97*(15), 8206–8210.

Kordylewski, L., Saner, D., & Lal, R. (June 1994). Atomic force microscopy of freeze-fracture replicas of rat atrial tissue. *Journal of Microscopy, 173*, 173–181.

Kukulski, W., Schorb, M., Welsch, S., Picco, A., Kaksonen, M., & Briggs, J. A. (2011). Correlated fluorescence and 3D electron microscopy with high sensitivity and spatial precision. *Journal of Cell Biology, 192*(1), 111–119.

Lafont, F. (2014). Correlative light atomic force electron microscopy (CLAFEM) combining force measurements and CLEM. In *ELMI 2014* (p. 78). Oslo: Imaging the Future.

Le Trequesser, Q., Devès, G., Saez, G., Daudin, L., Barberet, P., Michelet, C., … Seznec, H. (2014). Single cell in situ detection and quantification of metal oxide nanoparticles using multimodal correlative microscopy. *Analytical Chemistry, 86*(15), 7311–7319.

Ligeon, L. A., Barois, N., Werkmeister, E., Bongiovanni, A., & Lafont, F. (2015). Structured illumination microscopy and correlative microscopy to study autophagy. *Methods, 75*, 61–68.

Ligeon, L. A., Moreau, K., Barois, N., Bongiovanni, A., Lacorre, D. A., Werkmeister, E., … Lafont, F. (2014). Role of VAMP3 and VAMP7 in the commitment of Yersinia pseudotuberculosis to LC3-associated pathways involving single- or double-membrane vacuoles. *Autophagy, 10*(9), 1588–1602.

Liv, N., Zonnevylle, A. C., Narvaez, A. C., Effting, A. P. J., Voorneveld, P. W., Lucas, M. S., … Hoogenboom, J. P. (2013). Simultaneous correlative scanning electron and high-NA fluorescence microscopy. *PLoS One, 8*(2), 1–9.

Mizushima, N., Yoshimori, T., & Ohsumi, Y. (2011). The role of Atg proteins in autophagosome formation. *Annual Review of Cell and Developmental Biology, 27*(1), 107–132.

Moreau, K., Lacas-Gervais, S., Fujita, N., Sebbane, F., Yoshimori, T., Simonet, M., & Lafont, F. (2010). Autophagosomes can support *Yersinia pseudotuberculosis* replication in macrophages. *Cellular Microbiology, 12*(8), 1108–1123.

Neupane, B., Jin, T., Mellor, L. F., Loboa, E. G., Ligler, F. S., & Wang, G. (2015). Continuous-wave stimulated emission depletion microscope for imaging actin cytoskeleton in fixed and live cells. *Sensors, 15*(9), 24178–24190 (Electronic Resource).

Nguyen, J. V., Sotoa, I., Kimb, K.-Y., Bushongb, E. A., Oglesbyc, E., Valiente-Sorianod, F. J., … Marsh-Armstrong, N. (2011). Myelination transition zone astrocytes are constitutively phagocytic and have synuclein dependent reactivity in glaucoma. *Proceedings of the National Academy of Sciences of the United States of America, 108*(3), 1176–1181.

Odermatt, P. D., Shivanandan, A., Deschout, H., Jankele, R., Nievergelt, A. P., Feletti, L., … Fantner, G. E. (2015). High-resolution correlative microscopy: Bridging the gap between single molecule localization microscopy and atomic force microscopy. *Nano Letters, 15*(8), 4896–4904.

Ohnesorge, F. M., Hörber, J. K., Häberle, W., Czerny, C. P., Smith, D. P., & Binnig, G. (October 1997). AFM review study on pox viruses and living cells. *73*, 2183–2194.

Paul-Gilloteaux, P., Heiligenstein, X., Belle, M., Domart, M.-C., Larijani, B., Collinson, L., … Salamero, J. (2017). eC-CLEM: Flexible multidimensional registration software for correlative microscopies. *Nature Methods, 14*(2), 102–103.

Polishchuk, R. S., Polishchuk, E. V., Marra, P., Alberti, S., Buccione, R., Luini, A., & Mironov, A. A. (2000). Correlative light-electron microscopy reveals the tubular-saccular ultrastructure of carriers operating between golgi apparatus and plasma membrane. *Journal of Cell Biology, 148*(1), 45–58.

Punge, A., Rizzoli, S. O., Jahn, R., Wildanger, J. D., Meyer, L., Schönle, A., … Hell, S. W. (2008). 3D reconstruction of high-resolution STED microscope images. *Microscopy Research and Technique, 71*(9), 644–650.

Rankin, B. R., Moneron, G., Wurm, C. A., Nelson, J. C., Walter, A., Schwarzer, D., … Hell, S. W. (2011). Nanoscopy in a living multicellular organism expressing GFP. *Biophysical Journal, 100*(12), 63–65.

Rayleigh. (1896). XV. On the theory of optical images, with special reference to the microscope. *Philosophical Magazine Series 5, 42*(255), 167–195.

Roduit, C., van der Goot, F. G., De Los Rios, P., Yersin, A., Steiner, P., Dietler, G., … Kasas, S. (2008). Elastic membrane heterogeneity of living cells revealed by stiff nanoscale membrane domains. *Biophysical Journal, 94*(4), 1521–1532.

Roduit, C., Sekatski, S., Dietler, G., Catsicas, S., Lafont, F., & Kasas, S. (2009). Stiffness tomography by atomic force microscopy. *Biophysical Journal, 97*(2), 674–677.

Sader, J. E., Chon, J. W. M., & Mulvaney, P. (1999). Calibration of rectangular atomic force microscope cantilevers. *Review of Scientific Instruments, 70*(10), 3967–3969.

Sato, F., Asakawa, H., Fukuma, T., & Terada, S. (2016). Semi-in situ atomic force microscopy imaging of intracellular neurofilaments under physiological conditions through the "sandwich" method. *Microscopy, 65*(4), 316–324.

Schellenberger, P., Kaufmanna, R., Sieberta, C. A., Hagena, C., Wodrichc, H., & Grünewald, K. (2014). High-precision correlative fluorescence and electron cryo microscopy using two independent alignment markers. *Ultramicroscopy, 143*, 41–51.

Schorb, M., & Briggs, J. A. G. (2014). Correlated cryo-fluorescence and cryo-electron microscopy with high spatial precision and improved sensitivity. *Ultramicroscopy, 143*, 24–32.

Shami, G., Cheng, D., Henriquez, J., & Braet, F. (2014). Assessment of different fixation protocols on the presence of membrane-bound vesicles in Caco-2 cells: A multidimensional view by means of correlative light and 3-D transmission electron microscopy. *Micron, 67*, 20–29.

Shemesh, M., Addadi, S., Milstein, Y., Geiger, B., & Addadi, L. (2016). Study of osteoclast adhesion to cortical bone surfaces: A correlative microscopy approach for concomitant imaging of cellular dynamics and surface modifications. *ACS Applied Materials & Interfaces, 8*(24), 14932–14943.

Sochacki, K. A., Shtengel, G., van Engelenburg, S. B., Hess, H. F., & Taraska, J. W. (2014). Correlative super-resolution fluorescence and metal-replica transmission electron microscopy. *Nature Methods, 11*(3), 305–308.

Svitkina, T. (2007). Electron microscopic analysis of the leading edge in migrating cells. *Methods in Cell Biology, 79*, 295–319.

Usukura, E., Narita, A., Yagi, A., Ito, S., & Usukura, J. (January 2016). An unroofing method to observe the cytoskeleton directly at molecular resolution using atomic force microscopy. *Scientific Reports, 6*, 27472.

Verkade, P. (2008). Moving EM: The rapid transfer system as a new tool for correlative light and electron microscopy and high throughput for high-pressure freezing. *Journal of Microscopy, 230*(2), 317–328.

Watanabe, S., Punge, A., Hollopeter, G., Willig, K. I., Hobson, R. J., Davis, M. W., … Jorgensen, E. M. (2011). Protein localization in electron micrographs using fluorescence nanoscopy. *Nature Methods, 8*(1), 81–85.

Willig, K. I., Kellner, R. R., Medda, R., Hein, B., Jakobs, S., & Hell, S. W. (2006). Nanoscale resolution in GFP-based microscopy. *Nature Methods, 3*(9), 721–723.

Woehl, T. J., Kashyap, S., Firlar, E., Perez-Gonzalez, T., Faivre, D., Trubitsyn, D., … Prozorov, T. (2014). Correlative electron and fluorescence microscopy of magnetotactic bacteria in liquid: Toward in vivo imaging. *Scientific Reports, 4*, 6854.

Zonnevylle, A. C., Van Tol, R. F., Liv, N., Narvaez, A. C., Effting, A. P., Kruit, P., & Hoogenboom, J. P. (2013). Integration of a high-NA light microscope in a scanning electron microscope. *Journal of Microscopy, 252*(1), 58–70.

CHAPTER

10 Correlative light–electron microscopy in liquid using an inverted SEM (ASEM)

Chikara Sato[*,§,1], **Takaaki Kinoshita**[¶], **Nassirhadjy Memtily**[*,§,||], **Mari Sato**[*], **Shoko Nishihara**[¶], **Toshiko Yamazawa**[#], **Shinya Sugimoto**[#]

[*]National Institute of Advanced Industrial Science and Technology (AIST), Tsukuba, Japan
[§]University of Tsukuba, Tsukuba, Japan
[¶]Soka University, Hachioji-shi, Japan
[||]Traditional Uyghur Medicine Institute of Xinjiang Medical University, Urumqi, China
[#]The Jikei University School of Medicine, Minato-ku, Japan
[1]Corresponding author: E-mail: ti-sato@aist.go.jp

CHAPTER OUTLINE

Methods in Cell Biology, Volume 140, ISSN 0091-679X, http://dx.doi.org/10.1016/bs.mcb.2017.03.015

Abstract

In atmospheric scanning electron microscope (ASEM), the inverted scanning electron microscope (SEM) observes the wet sample from below, while an optical microscope observes it from above simultaneously. The ASEM sample holder has a disposable dish shape with a silicon nitride film window at the bottom. It can be coated variously for the primary-culture of substrate-sensitive cells; primary cells were cultured in a few milliliters of culture medium in a stable incubator environment. For the inverted SEM observation, cells and the excised tissue blocks were aldehyde-fixed, immersed in radical scavenger solution, and observed at minimum electron dose. Neural networking, axonal segmentation, proplatelet-formation and phagocytosis, and Fas expression in embryonic stem cells were captured by optical or fluorescence microscopy, and imaged at high resolution by gold-labeled immuno-ASEM with/without metal staining. By exploiting optical microscopy, the region of interest of organ can be found from the wide area, and the cells and organelle were successfully examined at high resolution by the following scanning electron microscopy. We successfully visualized islet of Langerhans, blood microvessels, neuronal end-plate, and bacterial flora on stomach epidermal surfaces. Bacterial biofilms and the typical structural features including "leg complex" of mycoplasma were visualized by exploiting CLEM of ASEM. Based on these studies, ASEM correlative microscopy promises to allow the research of various mesoscopic-scale biological phenomena in the near future.

INTRODUCTION

The correlative light–electron microscopy (CLEM) of a sample in liquid is highly desirable not only for cell researches but for tissue studies and material science. To realize electron microscopy in solution, environmental-capsule electron microscopy

(EC-EM) has been developed (Abrams & McBrain, 1944; Daulton, Little, Lowe, & Jones-Meehan, 2001). In EC-EM, a sample in solution or gas is placed in a capsule sealed by electron-permeable thin film windows and directly imaged in situ by transmission EM (TEM) (de Jonge & Ross, 2011), scanning EM (SEM) (Thiberge et al., 2004), or scanning TEM (STEM) (de Jonge, Peckys, Kremers, & Piston, 2009). Exploiting the possibilities of EC-EM has led to important findings in various fields; ligand- and affinity-labeling studies in cell biology (de Jonge et al., 2009; Thiberge et al., 2004) and electrochemistry (Ross, 2007). However, the limited space around the sample capsule of standard TEM, precludes the simultaneous observation using optical microscopy (OM). SEM has a relatively large sample holder, but the high-resolution objective lens with large numerical aperture (NA) is not easy to be coupled. The small dimensions of the environmental capsule (capacity <20 μL) also seem to limit the culturable cell types and probably the dual-labeling for CLEM and washing efficiency achievable during the sample preparations. We have developed atmospheric SEM (ASEM) with inverted SEM to allow the use of an open sample container sandwiched by OM and SEM to overcome these limitations (Nishiyama et al., 2010). The ASEM has already been applied in bioscience and material science, and it is expected to be adapted for various research fields including clinical diagnosis.

1. INSTRUMENT DESIGN AND SAMPLE GEOMETRY OF THE ATMOSPHERIC SCANNING ELECTRON MICROSCOPY

1.1 CONFIGURATION OF THE ATMOSPHERIC SCANNING ELECTRON MICROSCOPY

The ASEM, named ClairScope (JEOL Ltd.) for commercial use, has an inverted SEM configuration to realize SEM of a sample in liquid under atmospheric pressure in a readily accessible, open container called the ASEM dish (Nishiyama et al., 2010). This SiN film-windowed dish seals the inverted SEM column from the top (Fig. 1). An OM positioned above the ASEM dish is designed for CLEM and allows observing wide areas of the sample from above, while the SEM to image specified smaller regions from below through the SiN film in the base of the dish (Fig. 1). The optical axes of both microscopes are aligned and fixed to ensure that correlative images are recorded, while the specimen stage can be moved two-dimensionally (x-y) for targeting.

The electron dose applied in the imaging, except Fig. 5I, was within the maximum dose of 47 electrons/Å^2 permitted in low-dose cryoelectron microscopy aiming at atomic-resolution single-particle reconstructions (SPRs); while Fig. 5I was imaged at the maximum magnification of 100,000×, and the electron dose applied was 150 electrons/Å^2, which is almost the maximum of total dose for tilt-series cryo-TEM tomography. The dose applied for tissues was especially small, <4 electrons/Å^2, which is less than 10% of the dose for cryo-EM SPR. For ASEM observation, an aldehyde-fixed biological sample is immersed in radical scavenger solution, that is, 10 mg/mL D-glucose or 10 mg/mL ascorbic acid

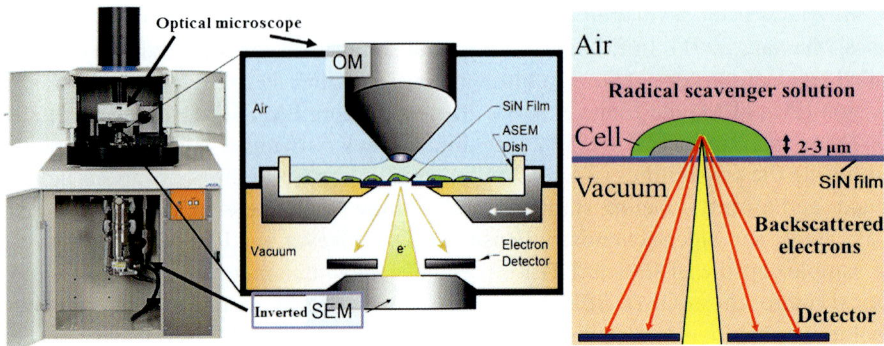

FIGURE 1

Configuration of the atmospheric scanning electron microscopy (ASEM) (Clairscope). The scanning electron microscopy (SEM) is completely inverted; the electron gun is at the bottom. At the top, a 2 mL-capacity dish with a SiN film window in its base seals the column. The body of the ASEM dish is made of plastic and ASEM dish is detachable. An optical microscope (OM) is positioned above the dish and directly opposite the inverted SEM (Nishiyama et al., 2010). A biological sample is aldehyde fixed and stained. The sample immersed in radical scavenger solution was observed by the OM and the inverted SEM. The observable specimen thickness from the SiN film was measured to be 2–3 μm at an acceleration voltage of 30 kV (Maruyama et al., 2012). The resolution measured between gold particles near the SiN film was 8 nm (Nishiyama et al., 2010).

From Maruyama, Y., Ebihara, T., Nishiyama, H., Suga, M., & Sato, C. (2012). Immuno EM-OM correlative microscopy in solution by atmospheric scanning electron microscopy (ASEM). Journal of Structural Biology, 180, *259–270 (Fig. 1a).*

solution. The resolution in aqueous solution of the inverted SEM was estimated to be 8 nm at a magnification of ×100,000, according to the measured distance between two distinguishable gold particles near SiN membrane (Nishiyama et al., 2010).

1.2 THE ATMOSPHERIC SCANNING ELECTRON MICROSCOPY DISH

The disposable ASEM dish has a thin film window in the center of a silicon chip surrounded by polystyrene body (Nishiyama et al., 2014). In the standard version, (Fig. 2A and B) there is one 0.25 × 0.25 mm, 100 nm-thick SiN film window in the center (Nishiyama et al., 2014). Starting from this, a dish with eight such windows has also been developed (Memtily et al., 2015; Nishiyama et al., 2014). The inside of the dish base can be coated with various kinds of coating reagent including poly-L-lysine or proteins (Kinoshita et al., 2014; Maruyama, Ebihara, Nishiyama, Suga, & Sato, 2012; Sato et al., 2012). The dish can hold a few milliliters of culture medium, and allows various types of cells, to be cultured in a stable environment provided by a CO_2 incubator. The culturable cell types include primary cells obtained directly from animal organs. For example, delicate neurons from mammalian brain can be cultured (Hirano et al., 2014;

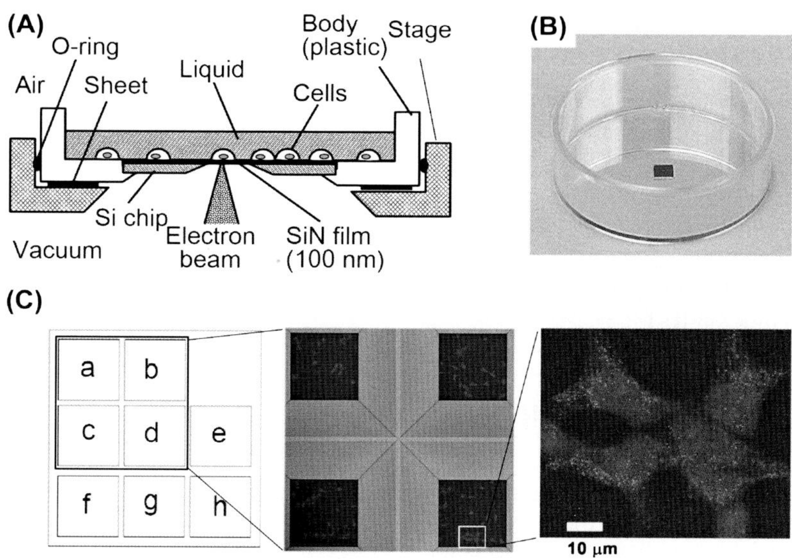

FIGURE 2

Atmospheric scanning electron microscopy (ASEM) dish. (A) Configuration diagram and (B) photograph of the standard ASEM dish. It has the shape of a Petri dish of 35 mm in diameter and is suitable for cell culture in a CO_2 incubator. The SiN film is manufactured by etching the Si side of a Si–SiN bilayered chip. To construct the ASEM dish, the chip is glued to the windowed base of the dish with the SiN side facing up so that the inner surface of the dish base is flat (Nishiyama et al., 2014). Thus, the top of the silicon chip is completely covered by SiN film. (C) Schematic diagrams of an 8-window ASEM dish. Gold-labeled cells are imaged by ASEM. All windows are 250 × 250 μm.

From Suga, M., Nishiyama, H., Konyuba, Y., Iwamatsu, S., Watanabe, Y., Yoshiura, C., ... Sato, C. (2011). The atmospheric scanning electron microscope with open sample space observes dynamic phenomena in liquid or gas. Ultramicroscopy, 111, *1650–1658 (Fig. 2a and b); Memtily, N., Okada, T., Ebihara, T., Sato, M., Kurabayashi, A., Furihata, M., ... Sato, C. (2015). Observation of tissues in open aqueous solution by atmospheric scanning electron microscopy: Applicability to intraoperative cancer diagnosis.* International Journal of Oncology, 46, *1872–1882 (Fig. 10b and c).*

Maruyama et al., 2012), which is critical for the study of neural network formations under close-to-native conditions. Embryonic stem cells have been successfully cultured, and Fas internalization into endosomal compartments in a clathrin-dependent manner in primitive endoderm cells is visualized by mathematically measuring image-blurring of the tagged gold particles using ASEM (Kinoshita et al., 2014).

The dish can be modified in various ways. It was modified to allow electrochemistry by mounting electrodes on the SiN window (Suga et al., 2011). Further, a temperature-regulated titanium ASEM dish was manufactured mainly for material science applications that require the observation of temperature-dependent phenomena, e.g., solder melting (Suga et al., 2011). This dish has a potential to be applied to the observation of thermophilic bacteria.

1.3 COLUMN PROTECTION SYSTEM IN CASE OF ACCIDENTAL SiN FILM BREAKAGE

The thin 100-nm thick SiN film window/windows in the base of the sample dish can be damaged mechanically, i.e., penetrated by a pipet tip, although the SiN film has sufficient resistance to withstand the pressure differential encountered. In such a case, a three-component protection system prevents contamination of the ASEM-column and the electron gun (Nishiyama et al., 2014). This is comprised of a specialized shutter, an inner pipe, and an air-leak valve in the chamber beneath the ASEM dish (Nishiyama et al., 2014). The ASEM system can be recovered by exchanging the BEI detector, shutter, and inner pipe (Nishiyama et al., 2014).

1.4 LABELING AND STAINING FOR CORRELATIVE LIGHT—ELECTRON MICROSCOPY

The 35 mm ASEM dish allows efficient staining and washing. We employed various traditional heavy metal EM staining protocols for ASEM observation in solution (Fig. 3A). Immunolabeling with FluoroNanogold for CLEM can be performed with standard laboratory protocols (Section 2.5). The required time and effort are similar with those for fluorescence immunolabeling (Fig. 3B).

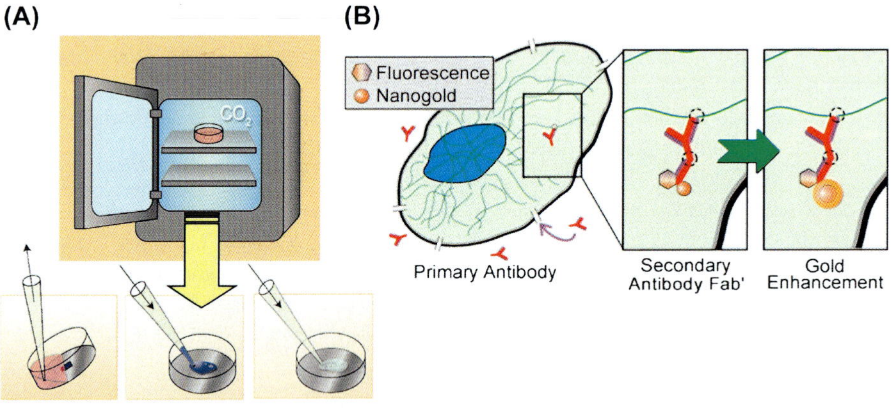

FIGURE 3

Cell culture, staining/washing, and labeling using the atmospheric scanning electron microscopy (ASEM) dish. (A) The ASEM dish, with various coatings, can be used for cell culture in a CO_2 incubator. Cell fixation, washing, labeling, and staining are easy to perform with standard laboratory equipment. (B) Primary antibody labeling is followed by secondary labeling tagged with FluoroNanogold (Nanoprobes). After fluorescence microscopy, the Nanogold is enlarged by gold enhancement.

From Maruyama, Y., Ebihara, T., Nishiyama, H., Suga, M., & Sato, C. (2012). Immuno EM-OM correlative microscopy in solution by atmospheric scanning electron microscopy (ASEM). Journal of Structural Biology, 180, *259–270 (Fig. 1b).*

1.5 IMMUNOLABELING METHOD FOR CORRELATIVE LIGHT–ELECTRON MICROSCOPY

Cells were fixed with 4% paraformaldehyde (PFA) in PBS (pH 7.4) at room temperature, for 15 min and perforated with 0.1% or 0.5% Triton X-100 in PBS at room temperature for 15 min, to label internal targets. After washing, they were blocked with 1% skimmed milk in PBS for 30 min. For primary labeling, the cells were incubated with antibodies or tagged ligands, e.g., phalloidin-biotin in the blocking solution (Maruyama et al., 2012). For examples, antibodies were mouse anti-α-tubulin antibody (Invitrogen, 4–0.5 μg/mL in the blocking solution), mouse anti-PDI antibody (Invitrogen, 1/200–400 dilution), or rabbit anti-STIM1 (C-terminal) antibody (Sigma Aldrich, 1/200 dilution). For secondary labeling against primary antibody, we used goat Fab against rabbit or mouse IgG, doubly conjugated with 1.4 nm Nanogold and fluorescent Alexa Fluor 488 or 594 dye (Nanoprobes, 1/100–1/400 dilution in the blocking solution).

For EM observation, the bound antibodies were fixed with 1% glutaraldehyde (GA) in PBS for 15 min. After washing with double distilled water (DDW), nanogold particles were enhanced by gold sedimentation using GoldEnhance EM (Nanoprobes) at room temperature for 5 min. To observe the surrounding structures of the epitopes, counter staining using metal solution can be applied. In case of phosphotungstic acid (PTA), cells were stained with 2% PTA in DDW for 15 min.

2. THE APPLICATION OF CORRELATIVE LIGHT–ELECTRON MICROSCOPY USING ATMOSPHERIC SCANNING ELECTRON MICROSCOPY

Starting with the lateral (landscape) TEM (LEM 2000; Akashi Seisakusho Ltd., Tokyo, Japan) specifically developed for CLEM in the 1980s, the move recently has been to develop CLEM instruments adopting standard EMs; e.g., Agronskaia et al. (2008) successfully developed a completely new system. Cryo-TEM together with FM has also been developed for correlative imaging (Sartori et al., 2007). Since biological samples usually need to be specifically labeled for CLEM, a range of dual-labeling methods have been developed to enable the comparison of FM mapping and higher-resolution EM images, e.g., fluorescence-Nanogold (Fluoro-Nanogold)-labeled antibodies and probes (Powell et al., 1997; Robinson & Vandre, 1997), fluorescent semiconductor particles (quantum dots) (Giepmans, Deerinck, Smarr, Jones, & Ellisman, 2005; Smith & Nie, 2009), and fluorescence conversion protocols (Gaietta et al., 2002) are now in commercial use.

In particular, ASEM to image samples in solution at atmospheric pressure and the direct link between OM and SEM (Nishiyama et al., 2010), promises to drastically increase the number of applications of CLEM.

2.1 ENDOPLASMIC RETICULUM

The endoplasmic reticulum (ER), storing Ca^{2+} ions, is an important organelle for cell excitation and neuroplasticity. CLEM was used to find a metal stain that labels the ER for subsequent ASEM imaging to precisely localize this physiologically important tubular organelle (Nishiyama et al., 2010). It was first labeled with fluorescence-tagged antibodies against the protein disulfide isomerase (PDI) present in the ER of COS7 cells cultured on an ASEM dish. After the FM, the cells were counterstained in situ using platinum blue and compared with ER-specific fluorescence-labeling pattern (Fig. 4). Platinum blue solution stains ER and also the surrounding even finer structures

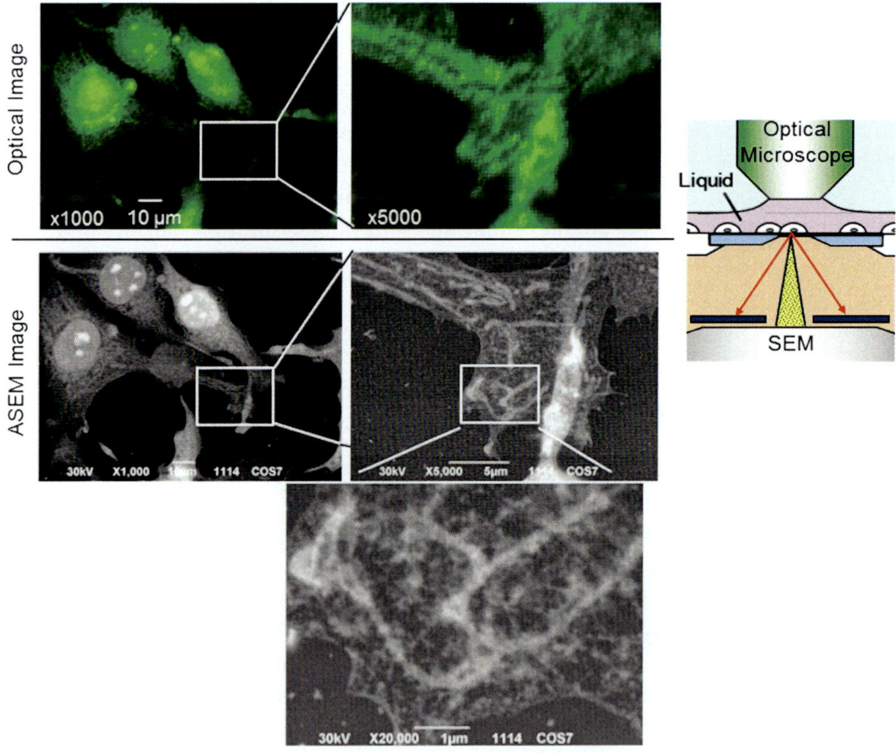

FIGURE 4

Endoplasmic reticulum (ER). The protein disulfide isomerases of ER of COS7 cells were labeled with Alexa Fluor 488 for fluorescence microscopy (FM) and then in situ with Pt blue for atmospheric scanning electron microscopy (ASEM). Only the gross structure of the ER was recognized by FM (top). The FM image was blurred at the magnification of ×5,000. ASEM of the same area revealed the presence of many fine structures (second row and bottom) (Nishiyama et al., 2010). *SEM*, scanning electron microscope.

From Nishiyama, H., Suga, M., Ogura, T., Maruyama, Y., Koizumi, M., Mio, K., ... Sato, C. (2010). Atmospheric scanning electron microscope observes cells and tissues in open medium through silicon nitride film. Journal of Structural Biology, 172, *191–202 (Fig. 7).*

(Fig. 4). Platinum blue also stains nucleic acids (Fig. 4, bottom left), and, especially in combination with 1% PFA fixation, specifically stains nuclei (Nishiyama et al., 2010).

2.2 SUPER MOLECULAR COMPLEX FORMATION OF STIM1 BY SENSING Ca^{2+}

Ca^{2+} storage in the ER plays an essential role in various physiological functions, including neuronal transmission, immune signaling, and embryogenesis. Here, we studied signaling mechanism in the immune system. STIM1 is the Ca^{2+} sensor subunit of the calcium-release–activated calcium (CRAC) channel of the ER distributed in various cells. Mutations in the CRAC system are known to cause severe immunodeficiency. When the Ca^{2+} store is depleted, STIM1 is believed to gather into puncta near the cell surface, forming higher-order complexes with Orai1 Ca^{2+} channel subunit to form functional CRAC channel in the plasma membrane mainly by OM study. However, the resolution was limited. Labeled with anti-STM1 antibody and further with FluoroNanogold, ASEM revealed that STIM1 expressed in COS7 cells (white) was usually distributed throughout the ER (Fig. 5, compare (B) and (C)). When the Ca^{2+} store is depleted by thapsigargin (Fig. 5E–I), high-resolution observation using ASEM revealed a dynamic string-like gathering of STIM1 on the ER near puncta, which was not observed using FM (Fig. 5; Maruyama et al., 2012). The result was further confirmed using Jurkat T cells (Maruyama et al., 2012).

2.3 PROPLATELET FORMATION OF MEGAKARYOCYTES

Platelets are necessary to stop bleeding and highly related to both myocardial and cerebral infarction under pathological conditions. Mature megakaryocytes (MKs) generate beaded cell projections called proplatelets and shed off platelets (Junt et al., 2007). The growth of MKs cultured in the open ASEM dish was monitored by OM and, at a critical moment, MKs were fixed, stained with heavy metal solutions in situ, and observed at high resolution using the inverted SEM. The pseudopodia extended beaded strings (Fig. 6A and B). After fixation and perforation of mature MKs, the P-selectin proteins were immunolabeled, counterstained by metal solution (Hirano et al., 2014) and observed (Fig. 6C). The results indicated that the proplatelets contained vesicles, and some of the vesicles might be α-granules, which express this adhesion protein on their surface (Fig. 6C). After monitoring independently cultured cells using phase-contrast OM (Fig. 6D), additional labeling of α-tubulin imaged at high resolution using ASEM suggests that the vesicles are transported on the microtubules of proplatelets (Fig. 6E and F) (Hirano et al., 2014).

2.4 AXONAL SEGMENTATION

CLEM of isolated *Drosophila* primary neurons grown on a poly-DL-ornithine–coated ASEM dish gave a detailed picture of BP102 localization in axial fibers

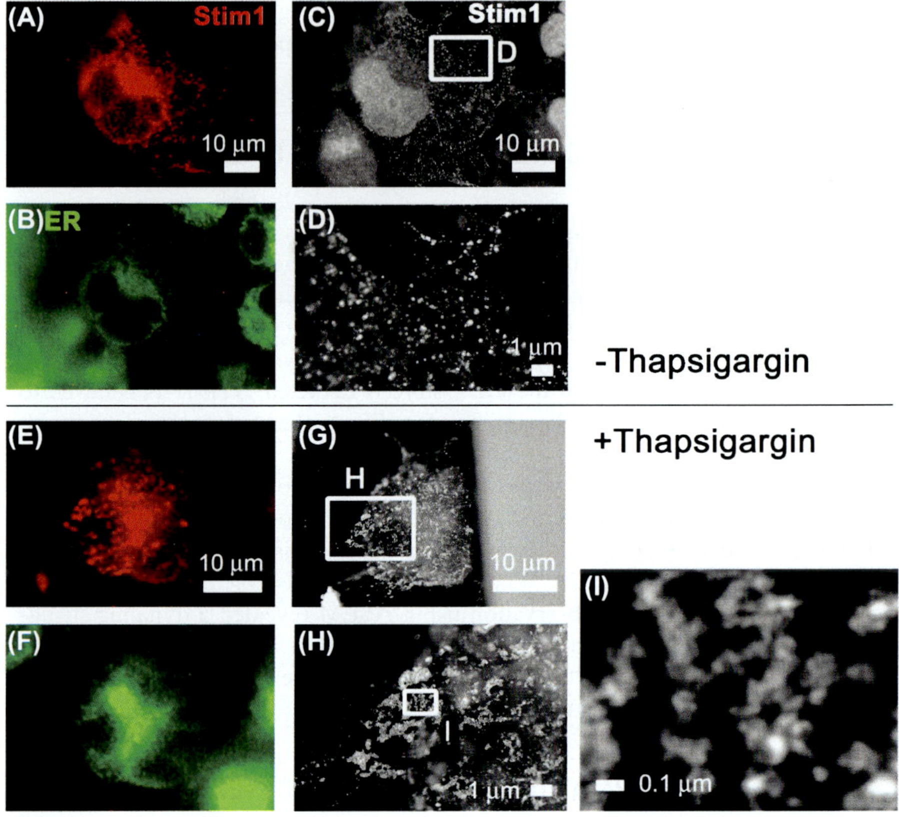

FIGURE 5

Rearrangement of STIM1 in the endoplasmic reticulum (ER) in response to Ca^{2+} store depletion. To investigate this, COS7 cells that express STIM1 were cultured on an atmospheric scanning electron microscopy (ASEM) dish and treated with thapsigargin to deplete their Ca^{2+} store, or left untreated. The cells were fixed, and perforated. STIM1 was labeled with Fluoro(Alexa594)Nanogold, gold-enhanced, and observed using ASEM. The ER was labeled with anti-PDI antibody (Alexa488; green). (A–D) Without store depletion, STIM1 was distributed on the ER. (E–I) With store depletion, STIM1 proteins gathered, forming strings on the ER (Maruyama et al., 2012).

From Maruyama, Y., Ebihara, T., Nishiyama, H., Suga, M., & Sato, C. (2012). Immuno EM-OM correlative microscopy in solution by atmospheric scanning electron microscopy (ASEM). Journal of Structural Biology, 180, *259–270 (Fig. 6b–d and h–k).*

(Kinoshita et al., 2014). FM showed that while the neuron marker HRP is present in the whole axial fiber (Fig. 7A), BP102 is specifically localized in the proximal region (Fig. 7B). Higher-resolution views of the proximal regions delivered by ASEM, revealed a polygonal frame-like structure of BP102 at the boundaries of most intraaxonal segments (Fig. 7C, arrowhead) that was not revealed by FM.

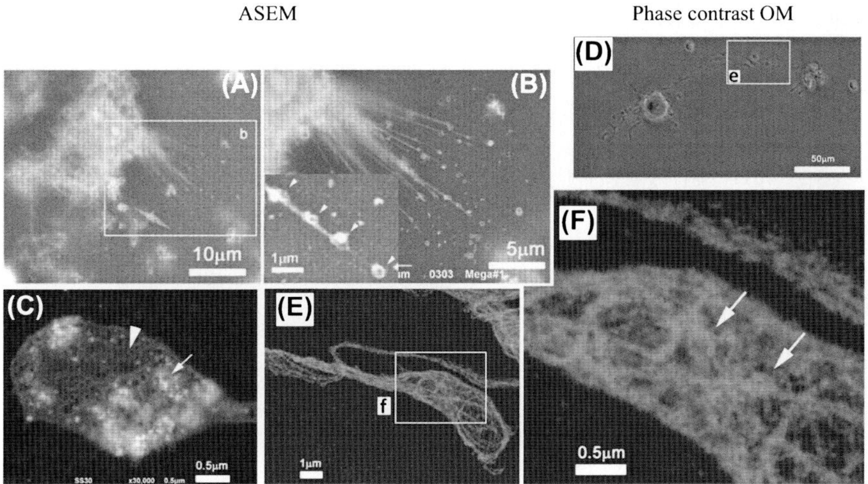

FIGURE 6

Atmospheric scanning electron microscopy (ASEM) of platelet generation by MKs. Primary MKs with proplatelet formation cultured on an ASEM dish. (A and B) Fixed and stained with Ti-blue. Beaded proplatelets extend from pseudophilia. (C) Fixed and gold tagged for P-selectin. A proplatelet bead was imaged. The *arrows* and *arrowhead* indicate putative α-granules and dense granules, respectively. (D–F) Fixed, gold tagged for P-selectin and further for α-tubulin to detect microtubules. (D) A proplatelet bead imaged using phase-contrast OM. (E and F) ASEM image at higher magnification. *Arrows* indicate putative α-granules (Hirano et al., 2014).

From Hirano, K., Kinoshita, T., Uemura, T., Motohashi, H., Watanabe, Y., Ebihara, T., ... Sato, C. (2014). Electron microscopy of primary cell cultures in solution and correlative optical microscopy using ASEM. Ultramicroscopy. 143, *52–66 (Figs. 6a,b,d, 7i, and 8a,c,d).*

Further, the axonal segmentation of neurons could be correlated to specific cytoskeletal structures by CLEM. For example, some tubulin bundles (microtubules) running alongside one another, appeared to make contact at the intraaxonal boundary and possibly elsewhere (Kinoshita et al., 2014). Microtubules in eight of the 10 axons examined made such contacts. The immunolabeling was disconnected at the intraaxonal boundaries of the other two axons, possibly because labeling was prevented by proteins bound to the α-tubulins in these regions.

2.5 SYNAPSE FORMATION INDUCED BY INDUCER-COATED FLUORESCENT MAGNETIC BEADS

Synapse formation is critical for brain development and functions. Glutamate receptor delta 2 (GluRdelta2), which is predominantly expressed in cerebellar Purkinje cells, can mediate synapse formation through binding to proteins expressed

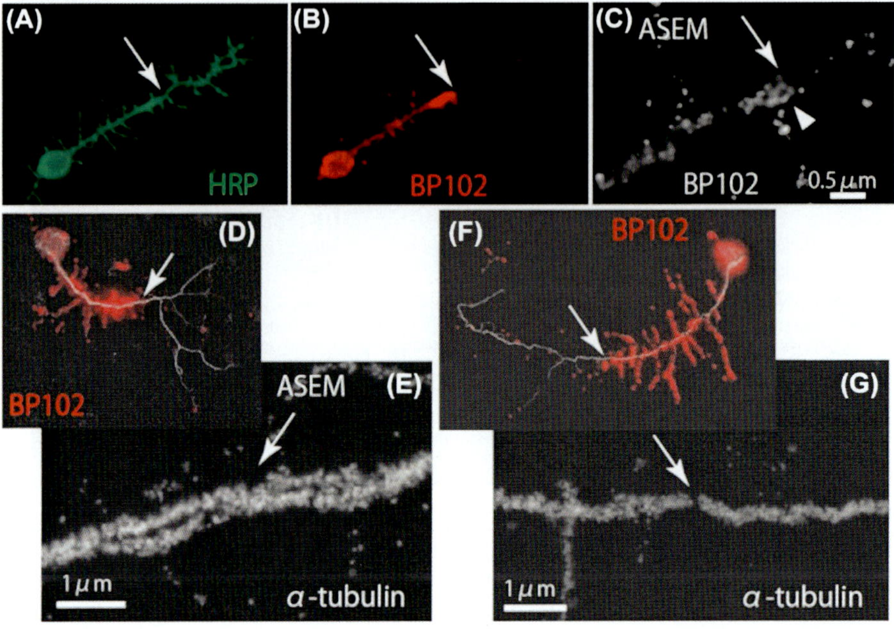

FIGURE 7

Axonal segmentation. Neurons were immunolabeled with antibodies against HRP and BP102, with Alexa488 (green) against the HRP antibody and then with Fluoro(Alexa594) Nanogold against the BP102 antibody for atmospheric scanning electron microscopy (ASEM) (A–C) or with Alexa594 (red) against BP102 and Fluoro(Alexa488)Nanogold against α-tubulin (D–G). (A) Localization of HRP epitopes (green). (B) Localization of BP102 epitopes (red). (C–G) ASEM (white). (C) ASEM of BP102. BP102 accumulates at the boundary (*arrows*) of the intraaxonal segment, forming a special structure that looks like polygonal frames (*arrowhead*). (D–G) Immunolabeled microtubule bundles. Most appeared to make contact (E) at the intraaxonal boundary (*arrows*), others seemed to be disconnected (G), possibly because immunolabeling was inhibited (Kinoshita et al., 2014).

From Kinoshita, T., Mori, Y., Hirano, K., Sugimoto, S., Okuda, K., Matsumoto, S., ... Sato, C. (2014). Immuno-electron microscopy of primary cell cultures from genetically modified animals in liquid by atmospheric scanning electron microscopy. Microscopy and Microanalysis, 20, *469–483 (Figs. 2a,b,f and 3c,e–g).*

in presynapse, thus essential for synapse formation in vivo. The extracellular N-terminal domain (NTD) of GluRdelta2 interacts with presynaptic neurexins through cerebellin-1 precursor protein. However, the precise mechanisms involved in the neuron to neuron interaction are not fully understood, since synaptic structures are too small to be observed by OM. Mouse cerebellar neurons were cultured on a poly-L-lysine- and laminin-coated ASEM dish. On DIV6, both Sindbis-EGFP virus were inoculated to visualize neurons, and fluorescent magnetic beads coated with GluRdelta2-NTD-Fc were added. After culture, the cells were fixed, immunostained against the presynaptic protein markers VGluT1 and/or Bassoon (Fig. 8). Presynaptic fibers leading to the magnetic beads and presynaptic Bassoon accumulated on the

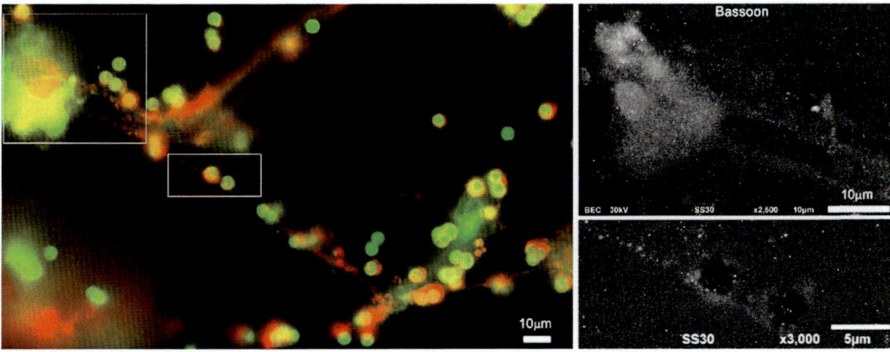

FIGURE 8

Distribution of Bassoon and VGluT1 (red) in relation to GluRdelta2-NTD-Fc-coated magnetic beads in mouse primary cerebellar cultures. The cytoplasm of EGFP-expressing neurons transfected by Sindbis-EGFP virus (green) and the magnetic beads (bright round features) are both imaged clearly. (Right) Atmospheric scanning electron microscopy of Bassoon (gray-scale image); magnetic beads appear dark, while Bassoon is observed as a gathering of *white dots* presumably where axons connect to the beads (Hirano et al.).

From Hirano, K., Kinoshita, T., Uemura, T., Motohashi, H., Watanabe, Y., Ebihara, T., ... Sato, C. (2014). Electron microscopy of primary cell cultures in solution and correlative optical microscopy using ASEM. Ultramicroscopy, 143, *52–66 (Figs. 4d,i and S2d).*

beads at the site of the contact were revealed using FM. Subsequent inverted SEM observation suggested that presynaptic proteins, presumably of an indistinguishable neurite branch, surrounded each bead when a neurite was attached to it (Hirano et al., 2014). Controlling time and space for synaptogenesis by the use of magnetic beads, is expected to allow the molecular mechanism of intercellular interaction to be analyzed in the ASEM in the near future.

2.6 shRNA (SMALL HAIRPIN RNA)-INDUCED SUPPRESSION OF SPECIFIC GENE EXPRESSION IN CULTURED CELLS

shRNA inhibition is widely applied in various researches in biology. shRNA inhibition in cells in aqueous solution was visualized at high resolution by CLEM using ASEM. The binding of Vsp9-ankyrin-repeat protein (Varp) to two GTPase Rabs, Rab32/38, is known to regulate trafficking of melanogenic enzymes in melanocytes. CLEM was employed to find out how knockdown of Varp affects Tyrp 1 trafficking in Melan-a cells (Nishiyama et al., 2014). For this, cells were transfected with both shVarp RNA and mStr plasmid. The red fluorescence of mStr allowed transfected cells to be identified (Fig. 9A, red). Immunolabeling against Tryp 1 and FM revealed the suppression of this protein (green) in their peripheral regions (Nishiyama et al., 2014). This was confirmed by correlative higher-resolution ASEM (Fig. 9B and C), which also revealed round vesicles containing Tyrp 1 distributed near the nucleus. These proteins could be untransported Tyrp 1.

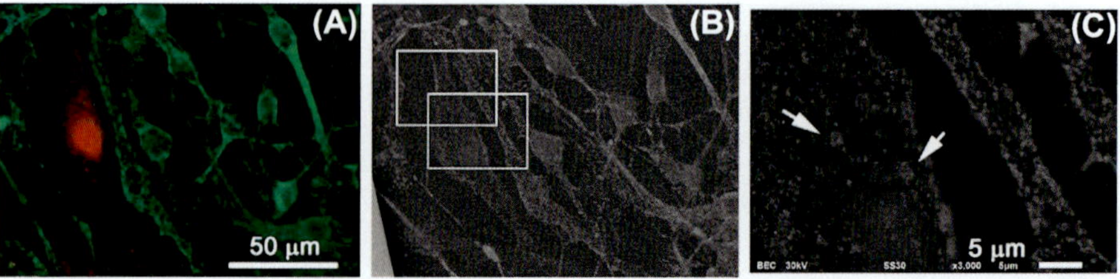

FIGURE 9

Visualization of *Varp* suppression of Tyrp 1 trafficking induced in melan-a cells by shRNA. Melan-a cells were transfected with a *Varp* shRNA expression vector. The mStr expression vector was also cotransfected as a transfection marker (red). After fixation and perforation, cells were immunostained with anti-Tyrp1 antibody and further with secondary Fab′-FluoroNanogold (green and white). (A) FM, (B and C) ASEM. siRNA-mediated knockdown of Varp dramatically reduced Tyrp1-staining (Nishiyama et al., 2014).

From Nishiyama, H., Koizumi, M., Ogawa, K., Kitamura, S., Konyuba, Y., Watanabe, Y., ... Sato, C. (2014). Atmospheric scanning electron microscope system with an open sample chamber: Configuration and applications. Ultramicroscopy, 147, *86–97 (Fig. 7c–e).*

In the future the use of OM to identify a fluorescent transfection marker and correlative ASEM imaging, promises to allow not only the knockdown effects of endogenous proteins, but also the effects of exogenous proteins genes to be studied at high resolution.

2.7 *MYCOPLASMA MOBILE* BACTERIA IN SOLUTION

Mycoplasma in solution was monitored by fluorescence, and imaged at high-resolution by inverted SEM (Fig. 10). Fixed and stained with metal soluion, characteristic easily recognizable views of *Mycoplasma mobile* were obtained by ASEM (Sato et al., 2012). Each cell was seen to have a bulb-shaped body with a protrusion (Fig. 10E). A strongly stained area in bulb can be attributed to DNA, according to accompanying FM. It is separated from a cap-like density at the other end by a variably shaped strings that might be cytoskeleton of *M. mobile* (Nakane & Miyata, 2007). This was suggested to, at least partly, colocalize with the so-called "leg" structure by immunolabeling with a monoclonal antibody tagged with FluoroNanogold, MAb7, and subsequent FM (Fig. 10A). The distribution was further imaged at high resolution using ASEM (Fig. 10B and C).

The ease with which *M. mobile* could be recognized makes ASEM a potential diagnosis tool for mycoplasma-related diseases, e.g., pneumonia and animal diseases.

2.8 BACTERIAL BIOFILM

Biofilms are highly organized microbial communities on surfaces, such as the surfaces of medical implants and host organisms, causing chronic infectious diseases. Within a biofilm, microbes are embedded in a self-produced soft

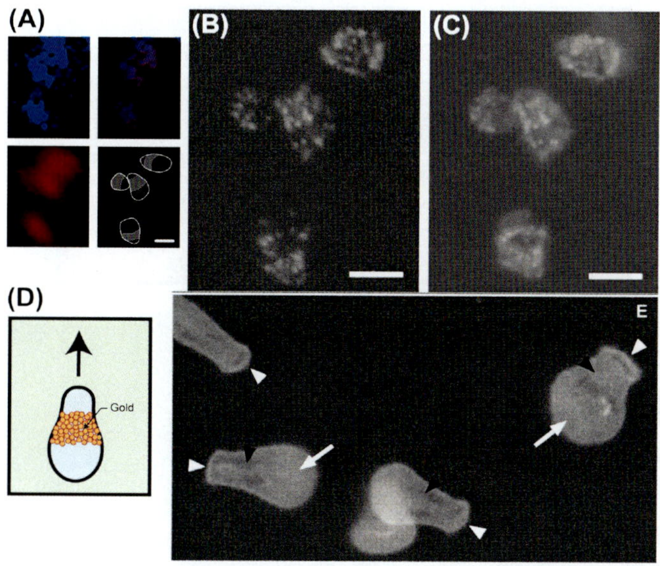

FIGURE 10

Correlative light–electron microscope observation of *Mycoplasma mobile* in solution using atmospheric scanning electron microscopy (ASEM). (A) Localization of the leg protein Gli349 labeled with dually tagged secondary antibody (red) visualized by FM; DAPI labeling of chromatin (blue). (B) Diagram indicating the location of Gli349 in the ASEM image. (C) Counterstained by heavy metal solution, the antibody is found to label the cell "neck" as schematically illustrated in (D). (E) *M. mobile* cells showed typical structure after fixation, and staining with five different heavy metals. DNA (*arrow*) and a ring-like structure (*arrowhead*) were observed at opposite ends, with another structure of variable shape (*black arrowhead*) in between. Scale bars: 0.5 μm (Sato et al., 2012).

From Sato, C., Manaka, S., Nakane, D., Nishiyama, H., Suga, M., Nishizaka, T., ... Maruyama, Y. (2012). Rapid imaging of mycoplasma in solution using atmospheric scanning electron microscopy (ASEM). Biochemical and Biophysical Research Communucations, 417, *1213–1218 (Figs. 1b and 3a,c,d).*

extracellular matrix (ECM), consisting of polysaccharides, proteins, and/or extracellular DNA (eDNA). The ECM has diverse functions to maintain the structural integrity of the biofilm and adapt to surrounding environments. For example, the ECM protects microbes from the host immune system and antibiotics. Therefore, the study of biofilm and its ECM is critical to understand the pathological mechanism of chronic diseases including pneumonia. However, the nanoscale visualization of soft, delicate biofilms in liquid is challenging. The biofilm of *Staphylococcus aureus* MR23 (MRSA23) was cultured on ASEM dish (Fig. 11, upper), fixed and stained. Imaging using inverted SEM revealed the presence of nanostructures in biofilms of MR23 (Fig. 11). These included small spheres and branched fibrils connecting bacteria. Careful observation also provided clear pictures of spheres associated with bacterial cells and seem to be snapshots of the budding of membrane vesicles as reveled (Sugimoto et al., 2016).

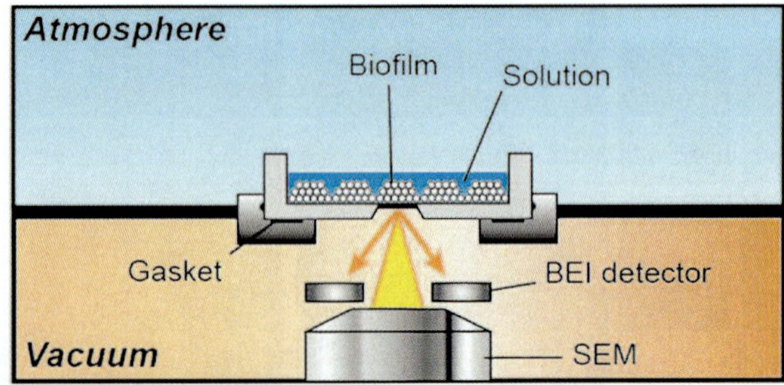

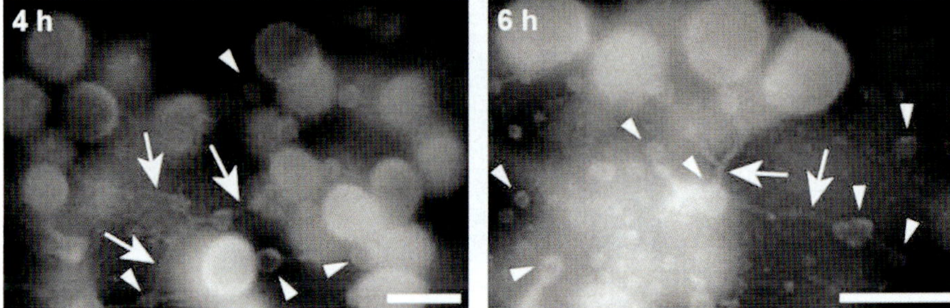

FIGURE 11 Heavy metal staining of *Staphylococcus aureus* biofilms and their monitored thin tubes between cell bodies.

(Upper) Diagram of the atmospheric scanning electron microscopy (ASEM) showing the inverted scanning electron microscopy (SEM), the detector, and the specimen dish, which separates the atmosphere (above) and the column vacuum (below). (Lower) Biofilm formation of *S. aureus* MR23 on ASEM dishes. ASEM images of an MR23 biofilm at the indicated culture times. *Arrows* and *arrowheads* indicate filamentous and spherical structures, respectively. Scale bars, 1 μm.

From Sugimoto, S., Okuda, K., Miyakawa, R., Sato, M., Arita-Morioka, K., Chiba, A., ... Sato, C. (2016). Imaging of bacterial multicellular behaviour in biofilms in liquid by atmospheric scanning electron microscopy. Scientific Reports, 6, *25889 (Figs. 1b and 2c).*

2.9 OBSERVATION OF A LARGE AREA OF TISSUE BY EXPLOITING THE OPTICAL MICROSCOPY AND OPEN CONFIGURATION OF THE ATMOSPHERIC SCANNING ELECTRON MICROSCOPY SAMPLE HOLDER

Film-sealed EC SEM has successfully observed wet tissue blocks excised from various organs, e.g., heart and kidney (Thiberge et al., 2004), kidney (Nyska et al., 2004), brain tumor (Barshack et al., 2004), avoiding dehydration artifacts. However, the samples are not readily accessible for manipulation or additional staining due to the basically closed sample holder. Using the ASEM, OM monitoring under white light also helps finding the region of interest of organ, and target the region for the following SEM. Upper OM exactly aligned to the optical axis of the inverted SEM, helps finding the region of interest of sectioned organ. Large areas of tissues can be observed by manually sliding a tissue across the ASEM dish

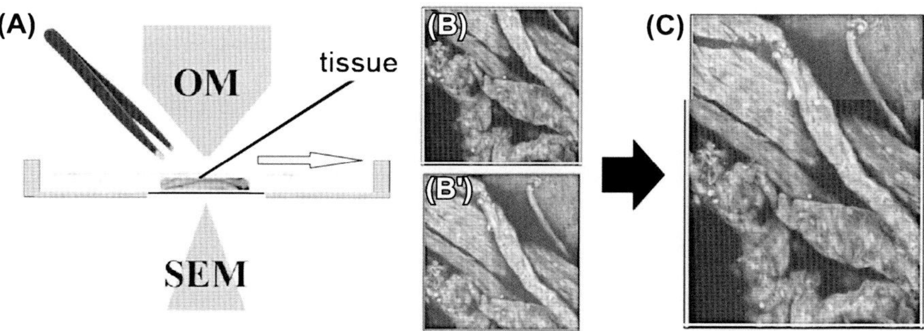

FIGURE 12

Large area observation by sliding an excised tissue slab of spinal cord across the atmospheric scanning electron microscopy (ASEM) dish. (A) Schematic representation. (B′, B) Tissue imaged before and after being moved, respectively, and (C) the combined overlapping images (Memtily et al., 2015). *OM*, optical microscopy; *SEM*, scanning electron microscope.

From Memtily, N., Okada, T., Ebihara, T., Sato, M., Kurabayashi, A., Furihata, M., ... Sato, C. (2015). Observation of tissues in open aqueous solution by atmospheric scanning electron microscopy: Applicability to intraoperative cancer diagnosis. International Journal of Oncology, 46, *1872–1882 (Fig. 10a,e, and f).*

(Fig. 12) (Memtily et al., 2015; Yamazawa, Nakamura, Sato, & Sato, 2016). The sliding also enables the imaging of larger area than the window size using the inverted SEM, by merging the images before and after the sliding (Fig. 12B, B′, and C).

2.10 OBSERVATION OF SYMBIOTIC BACTERIA ON STOMACH LUMEN MUCOSA

Symbiotic bacteria in our body play significant roles. Bacterial flora in digestive tract are known to affect our health conditions and even our mental characters. Bacteria in stomach is critically important for the carcinogenesis. Mucosa of the stomach lumen were observed using ASEM together with symbiotic bacteria in situ (Fig. 13); bacteria colonies can be seen in Fig. 13A (arrow). Staining by the modified NCMIR method instead of with PTA, delivered clear images of the symbiotic bacteria revealing their different shapes (Fig. 13B).

2.11 ISLET OF LANGERHANS

Since diabetes mellitus is clinically important, the endocrine tissues of the pancreas, islets of Langerhans, have been widely and extensively studied. Fluorescence immunolabeling of mitochondria protein (Tom20) exhibited intense signals in the islet of Langerhans, while signals were markedly fewer in exocrine cells of the pancreas (Fig. 14A), reflecting the fact that mitochondria are rich in islets of Langerhans. Because islets of Langerhans occupy only 1%–2% of the mouse pancreas by volume (Fig. 14B), they were identified by OM, and then imaged at

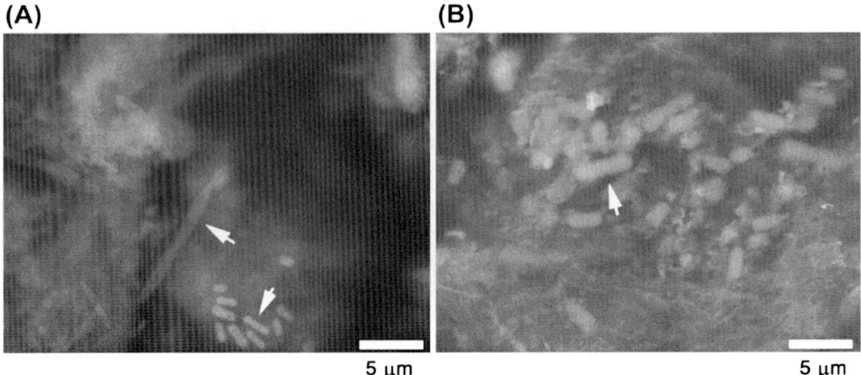

FIGURE 13

Atmospheric scanning electron microscope (ASEM) images of digestive tract. (A) Mucosal side of stomach stained with PTA. Commensalism of bacteria (*arrow*) is revealed. (B) Another area of stomach stained by the modified NCMIR method. Symbiotic bacteria of relatively large size (*arrow*) are brightly observed.

From Memtily, N., Okada, T., Ebihara, T., Sato, M., Kurabayashi, A., Furihata, M., ... Sato, C. (2015). Observation of tissues in open aqueous solution by atmospheric scanning electron microscopy: Applicability to intraoperative cancer diagnosis. International Journal of Oncology, 46, *1872–1882 (Fig. 6b and d).*

high resolution by ASEM (Fig. 14C–G; Yamazawa et al., 2016). Each cell in the imaged islet was surrounded by a white border (Fig. 14C–F). Blood capillaries contained clearly visualized erythrocyte queues, which ran between cells all over the islet (Fig. 14D–F, white arrows) working as secretion pathways for endocrine cells. At higher magnification, bright granules of various sizes were brightly visible in the cytoplasm of endocrine cells and close to the microvessels inside the islet of Langerhans cells (Fig. 14D–G). Because β-cells are found inside islets of Langerhans, the white granule-like vesicles in the cytoplasm could contain insulin.

2.12 SUBCUTANEOUS MICROVESSELS

Many diabetic patients suffer from microvascular complications. Using obesity mouse strain, the subcutaneous peripheral blood system of 10-week-old *ob/ob* mouse was observed by ASEM. Microvessels were easily detected in subcutaneous intraperitoneal adipose tissue, looking like lines of erythrocytes at low magnification (Fig. 15A, arrow) (Yamazawa et al., 2016). At higher magnification, bright capillary-like structures (Fig. 15B) including blood cells were visualized (Fig. 15B); the erythrocytes were elongated in narrower regions since the capillaries are thinner than erythrocytes. In the images, blood cells are visualized as if the microvessels are translucent tubes that are somewhat comparable to images obtained by OM, although the magnification can be far higher. Nuclei, presumably of endothelial cells, were also visualized in the walls of capillaries (Fig. 15A, star) and the

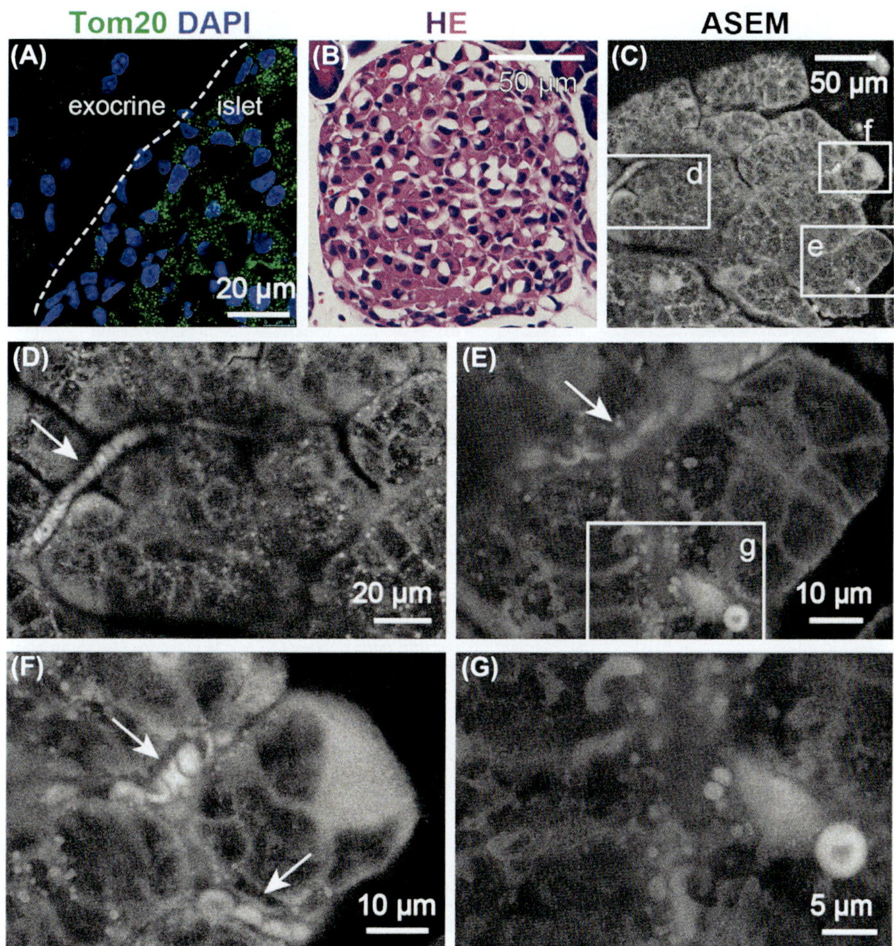

FIGURE 14 Pancreatic islets of Langerhans stained with PTA.

(A) Fluorescence microscopy of pancreas immunolabeled by antimitochondria antibody (Tom20). Many mitochondria were identified in the endocrine cells of islets of Langerhans. In contrast, fewer mitochondria were found in exocrine acinar cells. (B) Optical microscopy of hematoxylin and eosin (HE) stained islet of Langerhans cells. (C–G) Atmospheric scanning electron microscope (ASEM) observation of islet of Langerhans cells. (C) Low magnification image. (D–G) Higher magnification of the correspondingly annotated squares in preceding panels. *Arrows* in (D) to (F) indicate blood capillaries surrounding erythrocyte queues (Yamazawa et al., 2016).

From Yamazawa, T., Nakamura, N., Sato, M., & Sato, C. (2016). Secretory glands and microvascular systems imaged in aqueous solution by atmospheric scanning electron microscopy (ASEM). Microscopy Research and Technique, 79, *1179–1187 (Fig. 6a–g).*

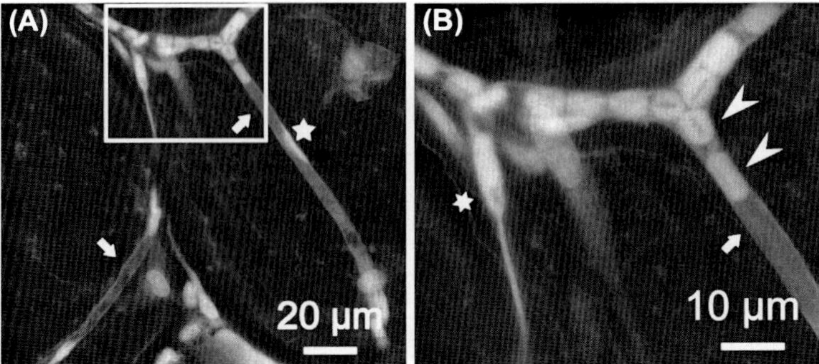

FIGURE 15

Atmospheric scanning electron microscopy (ASEM) images of capillaries of intraperitoneal tissue. Intraperitoneal adipose tissue of 10-week-old female *ob/ob* mice was fixed with GA, stained with PTA, and imaged in aqueous solution by ASEM. (A and B) Capillaries (*arrows*) in intraperitoneal adipose tissue. (A) Branching is evident; the nucleus, presumably of an endothelial cell, is indicated by a *star*. Erythrocytes (*arrowheads*) can be seen in the capillaries. (B) Higher magnification image of the rectangle in (A); a pericyte is indicated by *star* (Yamazawa et al., 2016).

From Yamazawa, T., Nakamura, N., Sato, M., & Sato, C. (2016). Secretory glands and microvascular systems imaged in aqueous solution by atmospheric scanning electron microscopy (ASEM). Microscopy Research and Technique, 79, *1179–1187 (Fig. 7b–d).*

capillaries branched in places. The cells attached to them might be pericytes (Fig. 15B, star).

2.13 OBSERVATION OF LUNG TISSUE METASTASIZED BY BREAST CANCER, AIMING AT INTRAOPERATIVE CANCER DIAGNOSIS

Intraoperative cancer diagnosis requires both accuracy and speed and significant application of microscopy. The lungs are known to be organs easily metastasized by cancer. Indeed, according to the World Health Organization statistics, lung cancer leads to the most deaths from cancer in 2012 (http://www.who.int/mediacentre/factsheets/fs297/en/). At present, OM using cryo-thin-sectioning is usually applied to make the diagnosis. Now, comparison of lung tissue excised from normal mice and from mice with tumors by ASEM indicates that this instrument could become an important less time-consuming diagnostic tool. In agreement with OM of lung, thin sections from other mice (Fig. 16A and D), ASEM of normal lung tissue revealed typical thin-wall structures with alveoli, alveolar ducts, a vein system, and trachea (Fig. 14B and C). In contrast, only faint traces of the regular alveoli and alveolar ducts were discernable for metastasized lung (Fig. 16E and F). Nuclei close to the surface of the tissue slabs appeared as bright cores and were larger in the breast cancer cells, which also have a different shape.

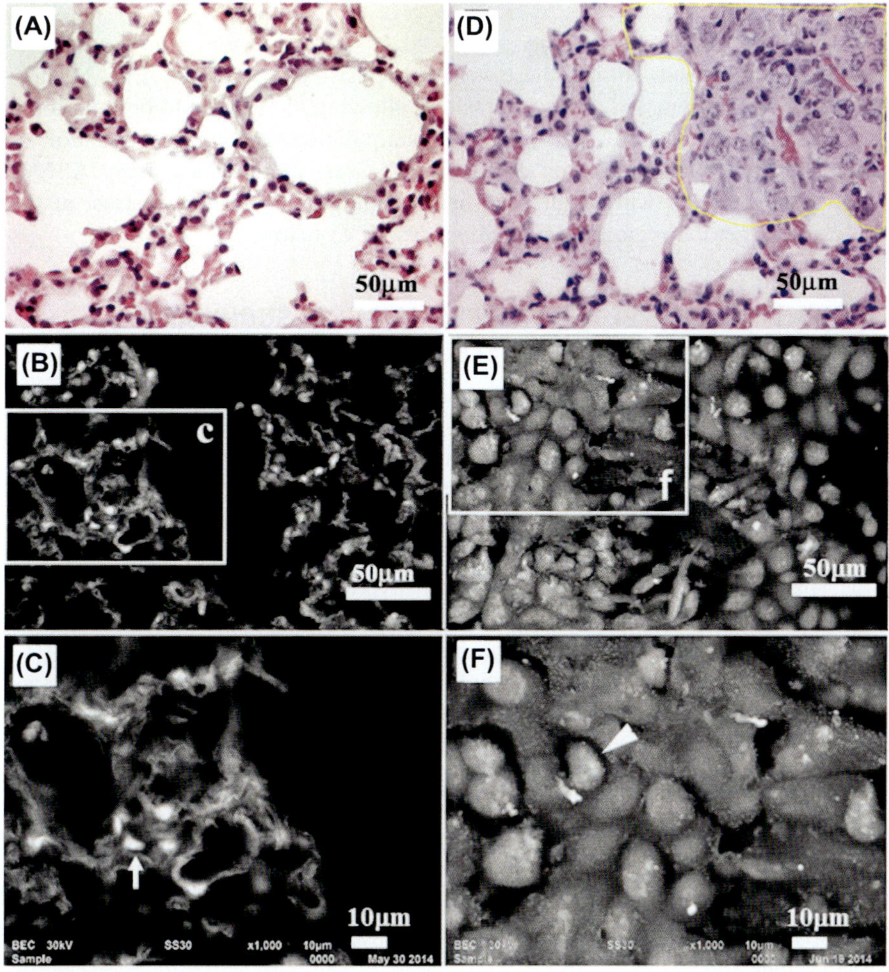

FIGURE 16

Comparison of normal lung and lung metastasized by breast cancer cells. (A) Optical microscopy of a thin-section of normal lung: hematoxylin-eosin (HE) stained; nuclei are blue, cytoplasm is red. (B) Low magnification atmospheric scanning electron microscopy (ASEM) image of an independently prepared slab of normal lung tissue stained with platinum-blue (Pt-blue) and phosphotungstic acid. Alveoli with alveolar ducts, a vein system, and trachea can be discerned. (C) Higher magnification images of the indicated fields. *Arrow*: normal sized nuclei. (D–F) Comparative observation of tissue excised from a lung metastasized with breast cancer cells. (D) OM. (E–F). ASEM. The cells are a different shape and their nuclei are larger (*arrowhead*), i.e., they are cancer cells.

From Memtily, N., Okada, T., Ebihara, T., Sato, M., Kurabayashi, A., Furihata, M.,...Sato, C. (2015). Observation of tissues in open aqueous solution by atmospheric scanning electron microscopy: Applicability to intraoperative cancer diagnosis. International Journal of Oncology, 46, *1872–1882 (Fig. 8a–f).*

3. DISCUSSION

Most of the applications outlined here were only possible because the ASEM dish is open to the atmosphere and easily accessible, allowing the sample to be monitored by OM. The addition of reagents, sample manipulation, and the investigation of phenomena that involves volume change can be easily performed using ASEM. OM monitoring under white light also helps finding the region of interest of the sample and target the region for CLEM, as described below.

3.1 TIGHT LINKAGE BETWEEN OPTICAL MICROSCOPY AND ATMOSPHERIC SCANNING ELECTRON MICROSCOPY

ASEM features in situ CLEM in solution and also in air. In case of quasiconcurrent observation, the shift from the OM to the inverted SEM is very quick: OM can be immediately shifted to SEM with x-y alignment error less than 8 μm, which is advantageous for high throughput and/or high time–resolution correlation microscopy. ASEM can capture a critical moment. For example, OM observes live cells from above and, at the critical moment, the cells are fixed and stained on the specimen stage and observed by the inverted SEM of ASEM. In case of simultaneous CLEM observation using cathodoluminescence, the fluorescence emitted by the electron-beam radiation can be detected (Nishiyama et al., 2014) using the upper OM system with high NA water immersion objective lens (Nishiyama et al., 2010). However, a real-time, image-based X-Y alignment algorithm for further exact alignment is not yet available.

In case of CLEM, the gap in observable specimen thickness between OM and SEM, i.e., the imageable specimen thickness from the SiN film is 1–3 μm using ASEM at an acceleration voltage of 10–30 kV (Suga, Nishiyama, Ebihara, Ogura, & Sato, 2009), while fluorescence microscope can observe through cell thickness. This gap should be overcome by introducing confocal FM, or superresolution OMs capable of controlling observable thickness (Manley et al., 2008) to the ASEM system.

3.2 ASEM FOR THREE-DIMENSIONAL MICROSCOPY

The observable specimen depth of ASEM can be decreased with a decrease of acceleration voltage (Suga et al., 2009). It indicates the possibility of three-dimensional reconstruction by subtracting images at different voltages, though the image resolution is significantly degraded due to electron scattering in the sample. The SEM with tilted beam radiation system developed by Ushiki's group, which is applied to CLEM (Koga, Kusumi, Shodo, Dan, & Ushiki, 2015) has a potential to realize the tomography in solution using ASEM without the specimen stage tilting. Multiphoton excitation FM can observe the bottom of a thick sample such as a tissue from the top, and it indicates the possibility of correlating the bottom surface view with the ASEM. Inversely, when the structure surrounding fluorescence is unknown, ASEM in combination with heavy metal staining can effectively visualize them.

3.3 WIDE AREA OBSERVATION PERPENDICULAR TO CULTURE SUBSTRATE

ASEM is suitable to visualize thin or flat cell structures. ASEM visualized entire structure of extended filopodia, including neural axons and dendrites perpendicular to the culture substrate (Hirano et al., 2014; Kinoshita et al., 2014) and long pseudopodia of proplatelets of MK (Hirano et al., 2014), which is rather hard to actually perform using Epon-embedded thin-sectioning EM. The thin and wide observable zone abve the SiN film of ASEM is widely applicable to cell biology, for example, counting of the synaptic formations between neurons on the culture, which is usually adopted for in vitro study of neural networking and neuroplasticity using OM (Okabe, Miwa, & Okado, 2001). The limited widow size of the ASEM dish is a shortcoming of the ASEM system, limiting the observable area; an ASEM dish has eight windows of SiN film (0.25×0.25 mm) (Memtily et al., 2015). However, the limit can be exceeded by CLEM: wide area imaging by manual shifting a tissue on the ASEM dish under the precise monitoring of the shift using OM, and the SEM images can be merged (Memtily et al., 2015). This was made possible by the open ASEM dish configuration and the axis-aligned OM (Fig. 1A). In future, a scan using a SiN window of slender rectangular shape can further expand the limit of the observable area of ASEM. The resolution of ASEM is up to 8 nm resolution close to the SiN film (Nishiyama et al., 2010). Based on beam broadening in water calculated by Monte Carlo simulations, the resolution is predicted to deteriorate to almost 200 nm at a depth of 1 μm from the SiN film and further to 500 nm at a depth of 2 μm from the SiN film (Maruyama et al., 2012).

3.4 LABELS FOR IN-SOLUTION CORRELATIVE LIGHT–ELECTRON MICROSCOPY USING ATMOSPHERIC SCANNING ELECTRON MICROSCOPY

Among fluorescent labels for CLEM, quantum dots have been shown to be effective tags with a semiconductor body of relatively simple structure (Giepmans et al., 2005). Without staining, ASEM and STEM successfully observed quantum dots in solution (Dukes, Peckys, & de Jonge, 2010; Nishiyama et al., 2010). It clearly suggests that quantum dots could be useful for various kinds of labeling for EM observation in liquid. Since ASEM directly observes a wet sample in an open dish, the effort and time required for ligand- or antibody-labeling is not large, and comparable with that required for multifluorescence labeling for OM. The success rate of labeling was 100% for more than 50 different antibodies used for fluorescence immunocytochemistry, which can be attributable to the antigen protection under fully hydrophilic condition. High-throughput ASEM can be extended to the relatively difficult culture including primary culture and can thus be applied to visualize the time course of the cell process by capturing snapshots of fixed cells; the timing of which is determined by OM. High-throughput ASEM is further applicable to drug development and screening like standard OM. The magnetic beads (2–4.5 μm) with a polystyrene core coated with magnetite further with polystyrene and finally with protein-A, have green and red fluorescence and are imaged dark in aqueous solution by ASEM. Because the position of the beads can be controlled by a magnet from the bottom of the ASEM dish during cell culture, the protein labeled beads are a fascinating system for the cell to cell interaction study using CLEM.

3.5 CULTURABILITY

The coated ASEM dish facilitates culture of different types of primary cells (Hirano et al., 2014; Kinoshita et al., 2014; Maruyama et al., 2012). These coating protocols originate from glass-coating techniques, which is attributable to the glass-like character of the surface of SiN film, including Si_3O_4. Film-coating protocols should be more extensively developed to realize various kinds of primary cultures. The primary culture technique of the ASEM dish had been applied not only to wild-type animals but also to genetically modified mice (Maruyama et al., 2012) and *Drosophila* (Kinoshita et al., 2014). Recently, genetically modified animals have been extensively studied; genetic researches in mice, mostly via target mutagenesis, have revealed the functions for 7229 genes (White et al., 2013) and successfully established various kinds of disease model mice (White et al., 2013). For delicate primary cells, a few milliliters of culture medium of the ASEM dish assure to maintain stable osmotic pressure and pH in a CO_2 incubator. Furthermore, the large gas-exchange surface of the culture medium enables stable aerobic respiration of the cells in ASEM dish. The large culture bottom of the ASEM dish also enables the culture of extended cells, e.g., neurons (Hirano et al., 2014; Kinoshita et al., 2014; Kinoshita, Sato, Fuwa, & Nishihara, 2017; Maruyama et al., 2012) or cells with filamentous connections (Sugimoto et al., 2016).

CONCLUSION

The development of ASEM allows in situ CLEM of samples in solution at atmospheric pressure. By SiN film–coating techniques, primary culture from animal tissues, the critical moments of cell activities were directly observed in natural aqueous conditions. Since the sample space of the ASEM is open to the air, micromanipulation, addition of reagent, and electrical stimulation are easy with the help of monitoring using OM. Magnetic beads used here should be exploited as time- and spatial-regulating mediator for samples in solution studied by CLEM. The quasisimultaneous or simultaneous correlative observation by ASEM should shorten the lag time when high time resolution is required for the studies, including stimulus-induced cell reactions. Further, the ASEM realized high-throughput monitoring without pretreatment in nonbiofield including wet materials (Nishiyama et al., 2014; Suga et al., 2011). We conclude that the

ASEM will be a valuable tool not only in biology, medicine, and diagnosis, but also in various fields including material- and nanoscience including process control in the biological and nonbiological industries.

ACKNOWLEDGMENTS

We thank Dr. Toshihiko Ogura at the National Institute of Advanced Industrial Science and Technology (AIST) for valuable discussions in the development of the ClairScope. This work was supported by Grant-in-Aid for Scientific Research on Innovative Areas, Sparse modeling (to C.S.), by CREST (to C.S.), by a Grant-in-Aid for Scientific Research from JSPS (15K14499) (to C.S.), by grants from the Ministry of Education, Culture, Sports, Science, and Technology (MEXT) (to C.S.), by Grant-in-Aid from CANON (to C.S.) and by Grant-in-Aid from AIST (to C.S).

REFERENCES

Abrams, I. M., & McBrain, J. W. (1944). A closed cell for electron microscopy. *Journal of Applied Physics, 15*, 607–609.

Agronskaia, A. V., Valentijn, J. A., van Driel, L. F., Schneijdenberg, C. T., Humbel, B. M., van Bergen en Henegouwen, P. M., … Gerritsen, H. C. (2008). Integrated fluorescence and transmission electron microscopy. *Journal of Structural Biology, 164*, 183–189.

Barshack, I., Polak-Charcon, S., Behar, V., Vainshtein, A., Zik, O., Ofek, E., … Nass, D. (2004). Wet SEM: A novel method for rapid diagnosis of brain tumors. *Ultrastructural Pathology, 28*, 255–260.

Daulton, T. L., Little, B. J., Lowe, K., & Jones-Meehan, J. (2001). In situ environmental cell-transmission electron microscopy study of microbial reduction of chromium(VI) using electron energy loss spectroscopy. *Microscopy and Microanalysis, 7*, 470–485.

Dukes, M. J., Peckys, D. B., & de Jonge, N. (2010). Correlative fluorescence microscopy and scanning transmission electron microscopy of quantum-dot-labeled proteins in whole cells in liquid. *ACS Nano, 4*, 4110–4116.

Gaietta, G., Deerinck, T. J., Adams, S. R., Bouwer, J., Tour, O., Laird, D. W., … Ellisman, M. H. (2002). Multicolor and electron microscopic imaging of connexin trafficking. *Science, 296*, 503–507.

Giepmans, B. N., Deerinck, T. J., Smarr, B. L., Jones, Y. Z., & Ellisman, M. H. (2005). Correlated light and electron microscopic imaging of multiple endogenous proteins using quantum dots. *Nature Methods, 2*, 743–749.

Hirano, K., Kinoshita, T., Uemura, T., Motohashi, H., Watanabe, Y., Ebihara, T., … Sato, C. (2014). Electron microscopy of primary cell cultures in solution and correlative optical microscopy using ASEM. *Ultramicroscopy, 143*, 52–66. Retrieved from http://www.who.int/mediacentre/factsheets/fs297/en/.

de Jonge, N., Peckys, D. B., Kremers, G. J., & Piston, D. W. (2009). Electron microscopy of whole cells in liquid with nanometer resolution. *Proceedings of the National Academy of Sciences of the United States of America, 106*, 2159–2164.

de Jonge, N., & Ross, F. M. (2011). Electron microscopy of specimens in liquid. *Nature Nanotechnology, 6*, 695–704.

Junt, T., Schulze, H., Chen, Z., Massberg, S., Goerge, T., Krueger, A., … von Andrian, U. H. (2007). Dynamic visualization of thrombopoiesis within bone marrow. *Science, 317*, 1767–1770.

Kinoshita, T., Mori, Y., Hirano, K., Sugimoto, S., Okuda, K., Matsumoto, S., … Sato, C. (2014). Immunoelectron microscopy of primary cell cultures from genetically modified animals in liquid by atmospheric scanning electron microscopy. *Microscopy and Microanalysis, 20*, 469–483.

Kinoshita, T., Sato, C., Fuwa, T. J., & Nishihara, S. (2017). Short stop mediates axonal compartmentalization of mucin-type core 1 glycans. *Scientific Reports, 7*, 41455.

Koga, D., Kusumi, S., Shodo, R., Dan, Y., & Ushiki, T. (2015). High-resolution imaging by scanning electron microscopy of semithin sections in correlation with light microscopy. *Microscopy, 64*(6), 387–394.

Manley, S., Gillette, J. M., Patterson, G. H., Shroff, H., Hess, H. F., Betzig, E., & Lippincott-Schwartz, J. (2008). High-density mapping of single-molecule trajectories with photoactivated localization microscopy. *Nature Methods, 5*, 155–157.

Maruyama, Y., Ebihara, T., Nishiyama, H., Suga, M., & Sato, C. (2012). Immuno EM-OM correlative microscopy in solution by atmospheric scanning electron microscopy (ASEM). *Journal of Structural Biology, 180*, 259–270.

Memtily, N., Okada, T., Ebihara, T., Sato, M., Kurabayashi, A., Furihata, M., … Sato, C. (2015). Observation of tissues in open aqueous solution by atmospheric scanning electron microscopy: Applicability to intraoperative cancer diagnosis. *International Journal of Oncology, 46*, 1872–1882.

Nakane, D., & Miyata, M. (2007). Cytoskeletal "jellyfish" structure of *Mycoplasma mobile*. *Proceedings of the National Academy of Sciences of the United States of America, 104*, 19518–19523.

Nishiyama, H., Koizumi, M., Ogawa, K., Kitamura, S., Konyuba, Y., Watanabe, Y., … Sato, C. (2014). Atmospheric scanning electron microscope system with an open sample chamber: Configuration and applications. *Ultramicroscopy, 147*, 86–97.

Nishiyama, H., Suga, M., Ogura, T., Maruyama, Y., Koizumi, M., Mio, K., … Sato, C. (2010). Atmospheric scanning electron microscope observes cells and tissues in open medium through silicon nitride film. *Journal of Structural Biology, 172*, 191–202.

Nyska, A., Cummings, C. A., Vainshtein, A., Nadler, J., Ezov, N., Grunfeld, Y., … Behar, V. (2004). Electron microscopy of wet tissues: A case study in renal pathology. *Toxicologic Pathology, 32*, 357–363.

Okabe, S., Miwa, A., & Okado, H. (2001). Spine formation and correlated assembly of presynaptic and postsynaptic molecules. *The Journal of Neuroscience, 21*, 6105–6114.

Powell, R. D., Halsey, C. M., Spector, D. L., Kaurin, S. L., McCann, J., & Hainfeld, J. F. (1997). A covalent fluorescent-gold immunoprobe: Simultaneous detection of a pre-mRNA splicing factor by light and electron microscopy. *The Journal of Histochemistry and Cytochemistry, 45*, 947–956.

Robinson, J. M., & Vandre, D. D. (1997). Efficient immunocytochemical labeling of leukocyte microtubules with FluoroNanogold: An important tool for correlative microscopy. *The Journal of Histochemistry and Cytochemistry, 45*, 631–642.

Ross, F. M. (2007). *In situ transmission electron microscopy*. Springer.

Sartori, A., Gatz, R., Beck, F., Rigort, A., Baumeister, W., & Plitzko, J. M. (2007). Correlative microscopy: Bridging the gap between fluorescence light microscopy and cryo-electron tomography. *Journal of Structural Biology, 160*, 135–145.

Sato, C., Manaka, S., Nakane, D., Nishiyama, H., Suga, M., Nishizaka, T., ... Maruyama, Y. (2012). Rapid imaging of mycoplasma in solution using atmospheric scanning electron microscopy (ASEM). *Biochemical and Biophysical Research Communications, 417*, 1213–1218.

Smith, A. M., & Nie, S. (2009). Next-generation quantum dots. *Nature Biotechnology, 27*, 732–733.

Suga, M., Nishiyama, H., Ebihara, T., Ogura, T., & Sato, C. (2009). Atmospheric electron microscope: Limits of observable depth. *Microscopy and Microanalysis, 15*, 924–925.

Suga, M., Nishiyama, H., Konyuba, Y., Iwamatsu, S., Watanabe, Y., Yoshiura, C., ... Sato, C. (2011). The atmospheric scanning electron microscope with open sample space observes dynamic phenomena in liquid or gas. *Ultramicroscopy, 111*, 1650–1658.

Sugimoto, S., Okuda, K., Miyakawa, R., Sato, M., Arita-Morioka, K., Chiba, A., ... Sato, C. (2016). Imaging of bacterial multicellular behavior in biofilms in liquid by atmospheric scanning electron microscopy. *Scientific Reports, 6*, 25889.

Thiberge, S., Nechushtan, A., Sprinzak, D., Gileadi, O., Behar, V., Zik, O., ... Moses, E. (2004). Scanning electron microscopy of cells and tissues under fully hydrated conditions. *Proceedings of the National Academy of Sciences of the United States of America, 101*, 3346–3351.

White, J. K., Gerdin, A. K., Karp, N. A., Ryder, E., Buljan, M., Bussell, J. N., ... Steel, K. P. (2013). Genome-wide generation and systematic phenotyping of knockout mice reveals new roles for many genes. *Cell, 154*, 452–464.

Yamazawa, T., Nakamura, N., Sato, M., & Sato, C. (2016). Secretory glands and microvascular systems imaged in aqueous solution by atmospheric scanning electron microscopy (ASEM). *Microscopy Research and Technique, 79*, 1179–1187.

CHAPTER

11

Relocation is the key to successful correlative fluorescence and scanning electron microscopy

Delfine Cheng*,[a], Gerald Shami*,[a], Marco Morsch[§], Minh Huynh*, Patrick Trimby*, Filip Braet*,[1]

The University of Sydney, Sydney, NSW, Australia

[§]*Macquarie University, Sydney, NSW, Australia*

[1]*Corresponding author: E-mail: filip.braet@sydney.edu.au*

CHAPTER OUTLINE

[a]Both authors contributed equally.

Methods in Cell Biology, Volume 140, ISSN 0091-679X, http://dx.doi.org/10.1016/bs.mcb.2017.03.013

Abstract

In this chapter the authors report on an automated hardware and software solution enabling swift correlative sample array mapping of fluorescently stained molecules within cells and tissues across length scales. Samples are first observed utilizing wide-field optical and fluorescence microscopy, followed by scanning electron microscopy, using calibration points on a dedicated sample-relocation holder. We investigated HeLa cells in vitro, fluorescently labeled for monosialoganglioside one (GM-1), across both imaging platforms within tens of minutes of initial sample preparation. This resulted in a high-throughput and high spatially resolved correlative fluorescence and electron microscopy analysis and allowed us to collect complementary nanoscopic information on the molecular and structural composition of two differently distinct HeLa cell populations expressing different levels of GM-1. Furthermore, using the small zebrafish animal model *Danio rerio*, we showed the versatility and relocation accuracy of the sample-relocation holder to locate fluo-tagged macromolecular complexes within large volumes using long ribbons of serial tissue sections. The subsequent electron microscopy imaging of the tissue arrays of interest enabled the generation of correlated information on the fine distribution of albumin within hepatic and kidney tissue. Our approach underpins the merits that an automated sample-relocation holder solution brings in support of results-driven research, where relevant biological questions can be answered, and high-throughput data can be generated in a rigorous statistical manner.

INTRODUCTION

The desire to correlate structural information across different imaging modalities originated as early as the inception of electron microscopy (EM) in the mid-1940s, when Porter, Claude, and Fullam (1945) performed light microscopy (LM) and transmission electron microscopy (TEM) to describe and correlate images of the endoplasmic reticulum in whole cells. This landmark study represented the realization about the additional cytoarchitectural insights that could be gained by combining multiple imaging modalities. Soon thereafter, when the first commercial electron microscopes found their way in various research settings, it became readily apparent that there was a pressing need to correlate structure—function information across different imaging modalities. More especially, an ongoing debate among the cell biology community stated that the sole application of EM imaging on a given biological sample would certainly not deliver on all outstanding questions about cellular ultrastructure, making a call to carefully compare the unknown EM world with data obtained from the more familiar microscopy techniques at that time (e.g., LM) (Moberg, 2012). As a result, from around the 1950s onwards, many cell biologists devoted significant amounts of time and efforts to devise a means of cross-correlating optical and electron microscopic information that were recorded from exactly the same sample, preferably even the same area.

The first reports using the term correlative and correlated microscopy were essentially not the typical exemplars of genuine correlative microscopy (CM) approaches as we know today but rather the first innovative attempts of those days within the exciting field of CM. The earlier work mainly took the advantage of serial sectioning of biological samples in which the different adjacent sections were specifically prepared and examined for different microscopies. Some elegant examples include the correlative assessment of: (1) histologically stained pancreas tissue for LM and TEM studies (Moses, 1956); (2) silver-impregnated thyroid glands to disclose glycoprotein (Vanheyningen, 1965); and (3) optical translucent EM-embedded lung tissue exposed to ozone for consecutive LM and TEM studies (Plopper, Dungworth, & Tyler, 1973). Coinciding with the rise of fine structure cytochemical techniques for correlated LM and EM imaging purposes (Lange, 1972), others took advantage of the unique staining patterns left behind by the dyes and/or electron dense material (i.e., molecular footprints) allowing them to beam up the same region of interest (ROI) across microscopies. Examples include the cross-correlated analysis and use of: (1) Giemsa-stained human blood smears for LM and scanning electron microscopy (SEM) (McDonald & Hayes, 1969); (2) a selective fuchsin staining approach to precisely locate neurosecretory substances within the LM, SEM, and TEM (Coates & Teh, 1978); (3) fluorescence retention methods within TEM-embedded material to unveil protein distribution in virus-infected cells (Rieder & Bowser, 1985); and (4) colloidal gold to monitor glycoproteins dynamics in platelets via LM, TEM, and SEM (Albrecht, Goodman, & Simmons, 1989).

Irrespective of all those clever cross-correlative imaging approaches, sample manipulation workflows in the early days were highly tedious, due to the laborious and time-consuming task of relocating the same ROI in a single sample, between different microscopy modalities (*for a review on this matter*, see Su et al., 2010). It was not until the late 20th century that the first practical relocation techniques emerged, allowing researchers to accurately and reproducibly relocate ROIs, thereby addressing specific research questions regarding biological fine structure. Such techniques have primarily included the use of artefactual and/or fiducial markers such as needle scratches, the deposition of fluorescent latex particle patterns postlabeling, or the use of coordinate engraved consumables (Fig. 1). While such manual relocation methods have aided in improved relocation accuracy, a prevailing drawback is low sample throughput, often resulting as a consequence of technical difficulties such as changes in sample orientation (i.e., flipping and rotation) and significant differences in the morphological appearance of biological structures as well

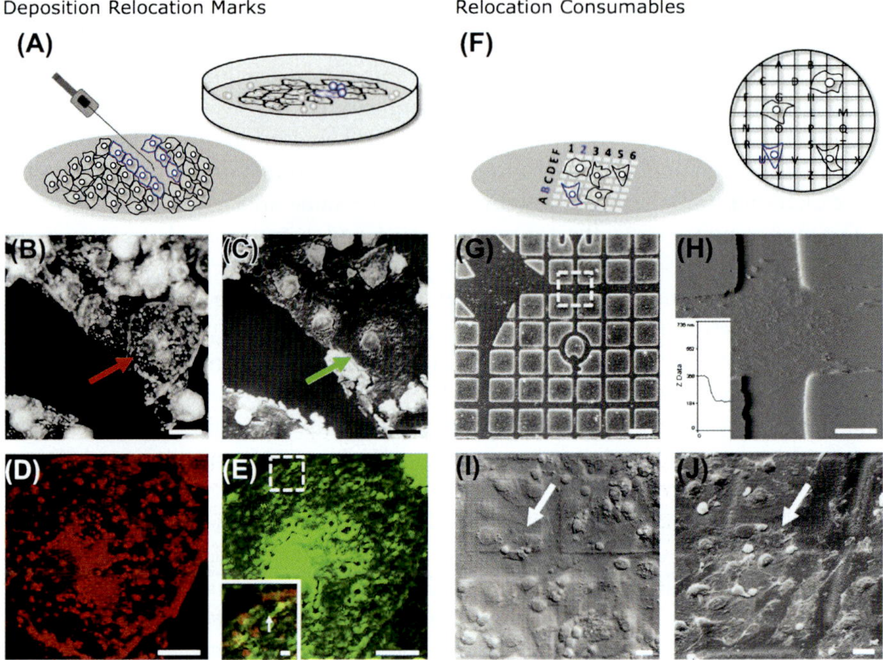

FIGURE 1

Schematic diagrams and practical examples on cultured hepatic endothelial cells outlining alternative and low-cost relocation means suitable for correlative light, fluorescence, and scanning electron microscopy purposes (Braet et al., 2002, 2007). The first group (A–E) concerns the creation of unique footprints that can be easily recognized across the imaging platforms via the addition of fluo-tagged spherical beads or via the application of artifactual marks using a fine sharp object on the surface of the sample. Outlined under (B–E) is an illustration of how cells can be easily relocated along a needle scratch (for comparative reason see *red vs. green arrows*), (B and D) represent fluorescent imaging of rhodamine-phalloidin labeled cells, and (C and E) the corresponding scanning electron microscopy information. The inset under (E) is the corresponding merged image information of a part of the porous cell surface (*white arrow*) derived from the *dotted line box*. The second group (F–J) deals with the readily available commercial relocation means such as electron microscopy finder grids and pattern-etched glass coverslips. (G and H) represent scanning electron (G) and atomic force (H) microscopy images of glass coverslips with relocation marks without the presence of sample material. Note that the alphabetical markings are easily recognizable and measurements reveal that there is ~150-nm of height difference compared with the surrounding glass cell culture surface. The *white arrows* under images (I) (differential interference contrast) and (J) (scanning electron microscopy) show the corresponding areas of interest. Scale bars: (B and C), 10 μm; (D and E), 5 μm; Inset (E), 0.5 μm; (G), 50 μm; (H), 1 μm; (I and J), 10 μm.

as the resolving power between various imaging modalities complicating data interpretation. Part of those challenges were fortunately alleviated by the rise of digital imaging means including professional image software processing packages (*for a review*, see Jahn et al., 2012).

Conversely, a historical hallmark in providing a hardware tool for genuine correlative light and electron microscopy (CLEM) studies was provided — to the best of our knowledge — by Geissinger (1974). Geissinger devised a "sample-relocation-stage" equipped with calibration spacers and micrometer screws permitting accurate relocation of the samples' ROIs between LM and SEM, and this CM tool was subsequently used by his group and collaborators in more than 20 research papers between the mid-1970s and mid-1980s (e.g., Abandowitz & Geissinger, 1975). Essentially, the sample-relocation holder utilized in this present report closely resembles Geissinger's homemade CM stage of somewhat 40 years ago. Evidently, the contemporary solution at hand is brought up-to-date with the latest advances in microelectromechanical sample stage technology, a fully controlled digital imaging environment, including a fully automated ROI relocation software interface for the different microscopy modalities (Lucas, Gunthert, Gasser, Lucas, & Wepf, 2012). Examples to-date have allowed for the swift and automated generation of correlated image arrays (Husain, Thomas, Kirschmann, Oberti, & Hahnloser, 2011; Wang et al., 2013), which most importantly fulfills a fundamental tenet of scientific research, high-throughput data generation.

In this chapter we will outline our experience utilizing an automated CLEM sample stage to rapidly and accurately explore topographical details of cultured cancer cells and to disclose the internal ultrastructure of zebrafish, all combined with relevant fluorescent biomolecular markers for the localization of specific ROIs. In so doing, we demonstrate the versatility, spatial accuracy, and universal applicability of such a hardware- and software-driven CLEM approach for the investigation of a range of biological samples.

1. RATIONALE

We will present a good-practice workflow blending the inherent strengths of classical sample preparation and labeling techniques, combined with a modern fiducial-based relocation sample stage system for the correlative investigation of both cells and tissues. More specifically, we will outline and discuss the use of a commercially available hardware and software solution that utilizes a 3-point calibration on a dedicated sample-relocation holder (i.e., Carl Zeiss Microscopy GmbH, Shuttle & Find). From this point onwards, the combined use of LM, fluorescence microscopy (FM), and SEM, inherently with different spatial- and temporal-resolution limits, and applied to the same ROI via the Shuttle & Find solution, is shortly referred to in this chapter as CLEM. The feasibility and ease of this CLEM approach will be illustrated by presenting data of HeLa human cervical carcinoma cell cultures labeled for monosialoganglioside one (GM-1) obtained via LM and FM, that were subsequently correlated with high-resolution SEM information. The benefits of employing this swift and high-throughput correlative cell imaging approach to collect novel structure–function information will be underpinned by quantitative morphometric measurements in which FM information will be directly compared with SEM. Furthermore, we will demonstrate the accuracy of the CLEM solution in which we postprocessed samples after initial CLEM investigation, followed by reinsertion of the object within the sample-relocation holder, opening up numerous future possibilities for novel and alternative CLEM approaches. In a second experimental set, we will demonstrate the benefits that the automated relocation solution brings to finding the distribution of fluorescent complexes within a large tissue sample or explore large volumes. More especially, individual tissue sections will be examined for traces of the endocytosis marker albumin conjugated to Texas Red with FM, and ROIs of within the liver and the kidney organs will be next analyzed under high-resolution SEM conditions, including the collection of volumetric information. Finally, the different approaches to analyze and model the different microscope data sets, which result in high-throughput CLEM arrays, will be addressed throughout this chapter.

2. METHODS

The overall methodological steps outlined under this section are graphically depicted in Fig. 2, which provides the sequential key steps to be undertaken in sample processing to accurately correlate LM and FM data with SEM information. Depending on the type of specimen (e.g., cells grown on coverslips or mounted tissue sections), including the type of fluorescent tag utilized, the sample preparation steps will slightly differ. Briefly, for the investigation of membrane markers on cultured cells, the processing essentially involves a rapid two-step sample preparation approach consisting of a direct FM labeling approach and subsequent sample drying for SEM purposes using hexamethyldisilazane (HMDS) (Braet, 2010; Braet, De Zanger, & Wisse, 1997). For tissue samples, the sample preparation is more time-consuming as tissue material has to be optimally prepared for fluorescence retention in resin embedding media (Moore, Cheng, Shami, & Murphy, 2016). However, from that point onwards large areas of serial sections can be successively investigated under FM for the marker of choice, and next with or without postsample preparation steps relocated and imaged under high-resolution EM imaging conditions.

As we will illustrate irrespective of the nature of the sample, the entire digital environment permits instant interpretation of the CLEM recordings and accompanying metadata. If desirable, and as illustrated herein, the data sets can be analyzed further, using the professional image analysis software ImageJ (Abramoff, Magelhaes, & Ram, 2004) and/or processed for further editing purposes using Photoshop (Sedgewick, 2008). We further outline the full methodical details in vide infra. The reader should be aware that each sample ultimately has its own specific sample manipulation needs for CLEM studies. The authors wish therefore to refer

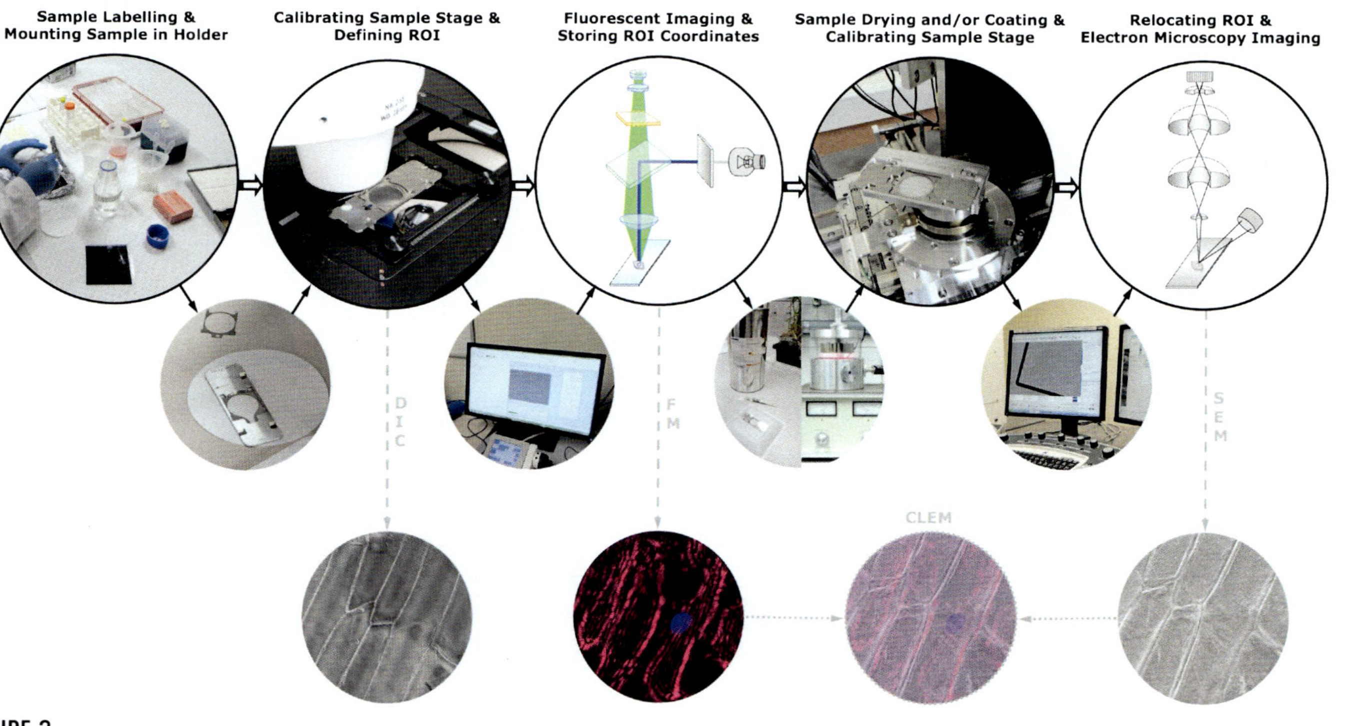

FIGURE 2

General workflow of the various experimental steps involved in correlating regions of interest (ROIs) between optical and scanning electron microscopy imaging modalities. In the top panel (*black arrows*), cultured cells or tissue sections, grown or attached to a coverslip are fluorescently labeled and mounted in the dedicated correlative light and electron microscopy (CLEM) sample relocation holder. Preceding fluorescence microscopy (FM) imaging, a 3-point calibration is performed under the light microscopy (LM) imaging mode. Next, ROIs within the FM are recorded and their coordinates embedded within the metadata of the acquired image. The sample is next prepared for scanning electron microscopy (SEM) and subsequently loaded into the SEM, after which the 3-point calibration is repeated. Note that the coverslip containing the sample can be removed if postsample processing is desired or can be held within the sample-relocation holder. In either way, accurate relocation can be achieved. Preceding the secondary calibration, the ROIs are automatically recalled, and a direct overlay of the FM and electron microscopy data can be achieved, providing the localization of areas of interest based on specific fluorescent markers, with their ultrastructural appearance. The lower panel displays an example outcome and shown herein is onion tissue labeled for actin (red color, 0.33 μM Alexa 467 labeled phalloidin solution) and DNA (blue color, 5 μg/mL DAPI). The different imaging modalities (gray arrows) employed are differential interference microscopy, FM, SEM, and the merged FM and SEM image information (i.e., CLEM).

to the continuing growing online information (Zeiss, 2013), listing numerous specific sample-relocation holder protocols covering a diverse set of biological samples and applications.

2.1 CELL CULTURE AND LABELING FOR CORRELATIVE IMAGING STUDIES

For cells, once the glass coverslips have been sterilized one can commence the routine cell culture practice that is in place within the laboratory. The following standard procedure is applicable to all cell line–based experimentation with the exception of the growth media and cell culture supplements that might vary from cell to cell line. In vide infra we will outline a characteristic experimental design that encompasses all necessary sample labeling steps that lead to the successful acquisition of FM data.

1. In this paper, we used the HeLa human cervical carcinoma cell line for acquiring cell arrays on the presence of the ganglioside GM-1. The cell line was obtained from the American Type Culture Collection (ATCC).
2. Culture cells in 75 cm^2 tissue culture flasks and maintained in complete media consisting of advanced Dulbecco's Modified Eagle Medium (DMEM), fetal bovine serum [FBS (10%)], L-glutamine (2.5 mM), and antibiotic/antimycotic solution (100 U/mL penicillin, 100 μg/mL streptomycin, 25 μg/mL amphotericin B).
3. Incubate cells at 37°C under 5% CO_2 and subculture to 70% confluency.
4. Wash cells with phosphate-buffered saline (PBS), and harvest cells by applying trypsin-EDTA for 5 min at 37°C before centrifugation at 2000 rpm for 3 min.
5. Resuspend cells in complete media and place cells into 6-well tissue culture plates containing cleaned 22 mm × 22 mm glass coverslips at confluency.
   Note: Cleaning of coverslips: Prewash square 22 mm × 22 mm glass coverslips by soaking them in 0.1 M HCl for 1 h, followed by a few washes in distilled water. The coverslips are then subsequently soaked twice in 70% ethanol and in 100% ethanol, for 1 h each. This step is essential for optimum cell adherence, including optimal imaging under LM and FM conditions, as greasy and unwanted particles are removed during this cleaning process.
6. Allow cells to attach, spread, and grow on the coverslips for at least 24–48 h at 37°C under 5% CO_2 before subsequent labeling with fluorophores (i.e., from step 7 onwards).
7. Wash the cells twice with complete media pH 7.4 at 37°C.
8. Make up a 1 μg/mL Alexa Fluor 647 cholera toxin staining solution in complete media.
9. Incubate cells for 20 min in staining solution at 37°C in a CO_2 incubator.
10. Wash cells five times with PBS at room temperature.
11. Fix cells with 4% paraformaldehyde (PFA) solution in PBS for 10 min at room temperature.

12. Wash cells five times with PBS at room temperature.
13. Mount coverslip into the sample-relocation holder (Zeiss Life Science Cover Glass 22 × 22 holder) with approximately 350 μL of PBS. Note: It is crucial that the clamping frame, which holds the coverslip has fully locked into place, if not this could result in movement of the sample during acquisition and inaccurate relocation.
14. Proceed as outlined under Section 2.3.

2.2 LABELING AND TISSUE PREPARATION FOR CORRELATIVE IMAGING STUDIES

For tissue, zebrafish larvae were used as experimental tissue model. Optically transparent larvae of 12 days postfertilization (dpf) were chosen as it was determined that at this developmental stage, the gastrointestinal system and associated digestive glands are fully functional and closely resembles the ultrastructure of human and rodent liver (Cheng, Shami, Morsch, Chung, & Braet, 2016). More specifically, *Casper* mutants (*Dario rerio*, *roy orbison nacre*), a strain of zebrafish deprived from melanocytes and iridophores, in both embryogenesis and adulthood, were used for their optically lucent skin (White et al., 2008). In this report, bovine serum albumin (BSA) conjugated with Texas Red (BSA-TxRed) was injected ortotopically into the zebrafish liver to study the uptake and distribution of the macromolecular complex within hepatic (Goresky, 1982) and kidney cells (Simpson & Shand, 1983).

1. Collect transparent *Casper* zebrafish (12 dpf) raised in a standardized clear water system (i.e., methylene blue free). Next, deprive the fish from food commencing the night before the injection. Note: Food constituents and methylene blue are well-known sources of autofluorescence (Hedrera et al., 2013; Matsui et al., 2010).
2. Anaesthetize the larvae with 0.02% tricaine until inactive and transfer subsequently into a glass bottom culture dish containing 0.8%—1.5% low-temperature melting agarose. The agarose is dissolved in egg water containing 0.02% tricaine. Via the use of a dissection microscope and a fine brush, position the fish sideways for unobstructed access to the liver with the injection needle.
3. Let the agarose set for a few minutes and then add some tricaine water to the glass bottom culture dish to keep the zebrafish hydrated and dormant.
4. Pull a glass needle and fill the needle with BSA-TxRed (5 mg/mL in PBS). Attach the glass needle to the microinjector of the micromanipulator apparatus. Calibrate the injection volume to 1.4 nL using the oil droplet technique (Rosen, Sweeney, & Mably, 2009). Note: For a volume of 1.4 nL, the injection time and pressure were adjusted to expel a drop diameter of 0.14 mm.
5. Carefully drive the tip of the needle into the liver and inject 1.4 nL of BSA-TxRed. Immediately thereafter, transfer the zebrafish larvae to the FM to assess the local distribution of BSA-TxRed using the 596/615 nm settings.

6. Euthanize the zebrafish with an overdose of tricaine at the chosen experimental time point, remove carefully the zebrafish from the surrounding agarose, and immediately transfer the zebrafish to a glass bottom culture dish that contains fixative solution. Fixative solution consists of 4% PFA and 0.1% glutaraldehyde in zebrafish buffer solution: i.e., 0.1 M sodium cacodylate buffer supplemented with 4% sucrose and 0.15 mM $CaCl_2$.
7. After 1 h initial fixation, transfer the larvae to fresh fixative solution and leave immersed overnight at 4°C, in the dark. Next, wash off excessive fixative with zebrafish buffer (4°C), and bring the sample through a series of graded ethanol solutions: 30% and 50%, 2 × 5 min each, all at 4°C.
8. Incubate the zebrafish in 2% uranyl acetate in 50% ethanol overnight, 4°C in the dark. Next, dehydrate with 70% and 80% ethanol, 3 × 5 min, all at 4°C.
9. Infiltrate and embed the sample in LR White: 25%, 50%, 75%, for 2 h each followed by 100%, 2 × 8 h, all at 4°C. Embed the zebrafish larvae in flat moulds, preferably sideways, left side down, to correspond with the side previously imaged under FM conditions. This allows later relocation for correlative purposes. Seal the moulds from air using a piece of ACLAR film and allow to polymerize at 50°C overnight. After polymerization, the resin blocks should be kept in the dark, at room temperature until sectioning. In our hands, the fluorescence signal was well retained even after 3 months following embedding. Note: Acrylate resins are renowned for their fluorescence retention capabilities following polymerization.
10. Face up the resin block to expose the sample, and trim away excess surrounding resin. Ensure the top and bottom of the block face are perfectly parallel as this is crucial to obtain straight section ribbons. For this purpose, use a trimming knife. Paint the bottom edge of the block face with a thin layer of cement glue: xylene (2:1), and allow to dry for a few minutes. In this way, the intervening layer of glue will maintain the sections together and ribbons of multiple serial sections can be produced and manipulated easily (Micheva, O'Rourke, Busse, & Smith, 2010).
11. Generate semithin sections of 500 nm with the use of an 8-mm Histo (45 degrees) jumbo diamond knife and collect on glass microscopy slides. Stain with 0.5% toluidine blue solution for orientation purposes under LM. Once the area of interest has been reached, by using the toluidine blue sections as guidance, cut ribbons of serial sections 90 nm thick (Cheng et al., 2016). The number of sections is dependent on the desired volume to be investigated and the size of the block face. If multiple ribbons are generated carefully arrange them to conserve their order.
12. Collect the ribbons of serial sections on previously cleaned coverslips (see, step 5 under Section 2.1) that are coated with poly-L-lysine (PLP) and glow charged. For this, 0.1% aqueous PLP was placed on the surface of the coverslip for 30 min and then excess PLP was removed by suction. The coverslips were air-dried for 2–3 h and next glow discharged for 5 s to render them hydrophilic. The latter step eases the collection process and prevents sections

wrinkles. The ribbons are typically retrieved by slowly dipping a coverslip into the knife trough and positioning the edges of each ribbon to the water interface. Next, slowly withdraw the coverslip from the water, blot the excess water with filter paper, and allow the sections to air-dry for ~10 min in a dark and dust-free environment at room temperature. Note: Once sections are generated they should be imaged within the FM as soon as possible. Despite the strong signal in freshly cut sections, the signal does not sustain for more than a day on the coverslip.

13. Mount coverslip into the sample-relocation holder (Zeiss Life Science cover glass 22 × 22 holder) and proceed as outlined under Section 2.3.

2.3 LOCATING ARRAYS OF INTEREST VIA LIGHT AND FLUORESCENCE MICROSCOPY

The numbered workflow under this section allows (1) the registration of the cell(s) of interest for correlative purposes using bright field imaging, (2) the instant collection of fluorescent information under low-magnification wide-field fluorescence imaging conditions as reference to assess the labeling quality, (3) followed by the acquisition of FM image data of the selected ROIs. For cells and tissue sections, the following steps were routinely executed:

1. Place the sample on the microscope stage, making sure that the sample relocation holder is held firmly between the two spring-loaded sliders.
2. Locate the "Shuttle and Find" tool within the software and select the correct correlative holder i.e., "Life Science cover glass 22 × 22".
3. The correlative sample holder has three fiducial markers (assigned numbers 1, 2, and 3). The markers consist of a large L-shape (1 mm) and a smaller L-shape (50 μm).
4. Manually move the stage to the first fiducial L-shaped marker and start the "Sample Holder Calibration Wizard." A dry objective with low magnification (5×–40×) should be used for this step to attain an overview image, easing the observation of the L-shaped markers.
5. Follow the steps as described within the calibration wizard for detection of all three fiducial markers: i.e., "3-point calibration."
6. Once the calibration has completed successfully, close the wizard.
7. Acquire LM and/or FM images. For cells, change to the 63× NA 1.3 water/glycerol immersion objective, and apply a water/glycerol mixture (1:1) between the lens and bottom surface of the coverslip. In case a wet objective is used for tissue sections, a water/glycerol mixture (1:1) is applied on the surface of the sections to match for refractive index. Note: The low viscosity of the water/glycerol mixture allows easy washing off at the end of FM imaging by gently dripping water over the coverslip. Ensure that all files are saved as ".czi" file format, which is required when relocating areas on the SEM (see Section 2.4).

2.4 RELOCATING ARRAYS OF INTEREST FOR SCANNING ELECTRON MICROSCOPY

For cells, the following steps were routinely executed:

1. Remove the sample-relocation holder from the FM stage.
2. Bring the sample-relocation holder over to a fume hood, remove excess PBS via folded filter paper, and add ~350 μL of HMDS or until the coverslip is completely covered with HMDS. Wait for 3 min.
3. Remove excess HMDS via folded filter paper at the edge of the glass coverslip, and bring next the sample-relocation holder containing the sample over to a desiccator. Let air dry for 15 min.
4. Place the sample-relocation holder into the SEM adapter and mount it on the SEM stage. Note: Ensure the arrow on the sample-relocation holder faces the arrow on the SEM adapter.
5. Initialize the SEM stage and repeat steps 4–6 as described in Section 2.3.
6. Optimize SEM imaging conditions as required within the "SmartSEM" software interface. In this case we typically used an accelerating voltage between 3 and 5 kV and utilized a conventional secondary electron detector. The working distance varied between 2.8 and 4.1 mm.
7. Load LM and FM images (i.e., ".czi file") into the "ZEN 2 SEM" software.
8. Activate "Live" mode, go to "S&F" tab and drag an LM (or FM) image into the right image container.
9. Double click on the LM (or FM) image, the SEM stage moves to exactly the same XY-position. Note: One can also define ROI by drawing rectangles directly on the FM image under the "ZEN 2 SEM" software. The SEM will then navigate to that area and automatically match magnifications between the imaging modalities.
10. To acquire images, click "Snap". Examine the samples at higher magnification or other imaging acquisition settings as required.

For tissue sections, we essentially follow the same steps as described above with the exception of some additional sample preparation steps and different SEM image acquisition settings:

1. Remove the sample-relocation holder from the FM stage, and recover the glass coverslip containing the sections. Add an orientation mark at the edge of the coverslip with the use of a sharp or solvent-proof marker as an aid to remount the coverslip correctly in the holder.
2. Poststain by covering the sections with 2% aqueous uranyl acetate solution for 10 min. Next, gently rinse the coverslip by dripping distilled water. Proceed with care to prevent sections from coming off the coverslip.
3. Let the coverslip completely dry in a dust-free environment.
4. Evaporate a thin layer of carbon (~15 nm) on the coverslip (1 s—color thickness guidance "Orange").

5. Remount and image the sample as described above under steps 4–10 (i.e., same procedure as for cells). Note: Optimize SEM imaging conditions as required within the "SmartSEM" software interface. For sections, an accelerating voltage of 12 kV, a working distance of 7 mm and a backscattered electron detector were used. Importantly, images are displayed with inverted contrast so that they resemble TEM images and ease interpretation.

2.5 POSTSAMPLE PREPARATION APPROACHES FOR CORRELATIVE IMAGING STUDIES

When desired, the sample can be retrieved from the sample-relocation holder and subject to further manipulation. Postsample preparation steps can include—but is not limited to—sputter-coating (e.g., cell cultures), or additional carbon coating, and/or even further poststaining with EM contrasting solutions (e.g., tissue sections). Those are examples that benefit image contrast and final resolution. On the other hand, even after optimizing the SEM image acquisition settings, an additional thin layer of carbon can tackle charging issues when examining tissue sections. For cells, we routinely took the advantage of investigating samples under uncoated versus coated conditions. Noteworthy, although trivial, before removing the glass coverslip for postsample processing it is essential to add an orientation marking at the edge of the coverslip (Fig. 4). This allows correct remounting of the coverslip in the right orientation on the sample-relocation holder using the holder's engraved markings, guaranteeing subsequent LM and/or SEM relocation. The following sequence of events can be considered when additional postsample preparation steps are desirable:

1. Remove the sample-relocation holder from the SEM stage and recover the coverslip.
2. Add an orientation marking at the edge of the coverslip with the use of a sharp.
3. Sputter coat the sample with 15 nm gold (2 min at 25 mA) for cells or with an additional layer of 5 nm carbon for tissue sections (<0.3 s—color thickness guidance "Indigo") or any other postsample preparation steps.
4. Remount the coverslip in the sample-relocation holder in the same orientation by using the orientation mark.
5. Mount the holder on the SEM stage and acquire SEM images as outlined in the previous section.

2.6 CORRELATIVE DATA ANALYSIS AND MORPHOMETRY

Correlative data arrays of ROIs were generated with the aid of the "ZEN 2" software (Blue Edition) equipped with the "Shuttle and Find" module (Carl Zeiss Microscopy GmbH) allowing for the automatic overlay of the LM, FM, and SEM images at matching magnification. In addition, acquiring a mosaic of the ROI was also very helpful as a larger area can be viewed and compared, and was particularly useful in more complex samples such as tissues.

For correlative image analysis, all end magnifications recorded within the individual metafiles across the different imaging platforms were transferred into ImageJ (Abramoff et al., 2004) and/or Adobe Photoshop CS6 (Sedgewick, 2008) to allow the calculation of the number of pixels per length. In doing so this permits accurate analysis of the LM, FM, and SEM data within a given CLEM experiment. Of special note, during any subsequent cross-correlation image analysis steps it is crucial to ensure that the pixel sizes are retained throughout all the subsequent image manipulation processes. The following sequence of events was performed on 8-bit images under ImageJ to extract morphometric data on cell cultures:

1. Open original image data under ImageJ, *File* → *Open*.
2. Go to, *Process* → *Find Edges*.
3. Next, *Process* → *Binary* → *Make Binary*.
4. Subsequently, *Image* → *Adjust* → *Threshold*. Note: Specific image analysis settings under steps 2 and 4 are required, and vary for each new experimental design. The primary factors that influence outcomes are the type of cell model and probes used, including the image acquisition conditions. Therefore, it is difficult to recommend universal numerical settings for each of the above image analysis steps. In general, they should be defined empirically and applied to all images within the same experiment. Using the standard default settings (or decide automatically) is a good start, and individual parameters can be adjusted to suit the desired outcome. For the interested reader we recommend the work by Russ and Neal (2015).
5. In the final step, *Analyse* → *Analyse Particles*. Similar to the remark above, settings such as size and circularity, are empirically to be defined.
6. Export measurements to an electronic spreadsheet of your choice and make calculations accordingly.

3. INSTRUMENTATION AND MATERIALS

3.1 CELL CULTURE AND LABELING FOR CORRELATIVE IMAGING STUDIES

Instrumentation: Forma direct heat CO_2 incubator (Thermo Scientific); Heraeus Labofuge 200 (Thermo Scientific); Neubauer chamber (ProSciTech).

Materials: Glass coverslips (High precision 1.5H, 22 mm × 22 mm, ZEISS); HeLa human cervical carcinoma cell line (American Type Culture Collection, Item No. PTA-5659); 75 cm^2 tissue culture flasks (In Vitro Technologies, COR430641); 6-well tissue culture plates (In Vitro Technologies, COR3516); tissue culture consumables (Corning Life Sciences); transfer pipettes.

Reagents: Advanced Dulbecco's Modified Eagle Medium (DMEM) media (Life Technologies, 12491023); Alexa-Fluor 647 cholera toxin (Subunit B, Life Technologies, C34778); antibiotic/antimycotic (Life Technologies, 15240104); distilled water; EM grade 16% paraformaldehyde (ProSciTech, C007); ethanol

(Merck, Cat No. 4.10230, CAS # 64-17-5); heat-inactivated fetal bovine serum (FBS; Life Technologies, 10100-147); hydrochloric acid (Analytical grade, Sigma Chemicals); L-glutamine (Sigma-Aldrich, G7513); paraformaldehyde (ProSciTech, C007); phosphate buffered saline (PBS), (In Vitro Technologies, IVT3001302); trypsin-EDTA (Sigma-Aldrich, 59430C).

3.2 LABELING AND TISSUE PREPARATION FOR CORRELATIVE IMAGING STUDIES

Instrumentation: Bench-top micro-manipulator (MM33, Maerzhaeuser); heat block (SBD110, Select BioProducts); heating oven (Binder WTC, Tuebingen); incubator (TEI-43G, Thermoline Scientific); microinjector (Picospritzer III, Parker Hannifin); stereoscope (SMZ 745, Nikon); ultramicrotome (Ultracut 7, Leica); vertical pipette puller (700D, KOPF).

Materials: ACLAR film (ProSciTech, GL105); capillary glass (Borosilicate with filament, GC100F-15, World Precision Instruments, Inc); *Casper* zebrafish larvae (*Dario rerio*; *roy orbison nacre*); contact cement glue (DAP Weldwood); diamond trimming knife (Diatome); filter paper (No.1 Whatman); glass bottom culture dishes (35 mm, P35G-0-20-C, MatTek); glass coverslips (High precision 1.5H, 22 mm × 22 mm, ZEISS); glass microscopy slides; histo jumbo diamond knife (8 mm, 45 degrees, Diatome); injector razor blades (carbon steel, ProSciTech, L057C); micrometer; PTFE flat embedding moulds (ProSciTech, RL090); standard paint brush (long liner, size 1); transfer pipettes.

Reagents: Bovine serum albumin conjugated with Texas Red (Molecular Probes, A23017); calcium chloride (Riedel-De Haen AG Seelze-Hannover, 820102); distilled water; egg water used for raising fish (0.6 g instant ocean sea salt in 10 L deionized water); EM grade 16% paraformaldehyde (ProSciTech, C004); EM grade 25% glutaraldehyde (ProSciTech, C001); ethanol; London Resin (LR) white catalyzed with benzoyl peroxide (medium grade, ProSciTech, C025); low-temperature melting agarose (Fisher Scientific); phosphate buffered saline; poly-L-lysine solution 0.1% (w/v) (Sigma, P8920); sodium cacodylate tri-hydrate (ProSciTech, C020); sucrose (Sigma-Aldrich, S9378); uranyl acetate (ProSciTech, C079-F); toluidine blue O (ProSciTech, C078); tricaine (0.02%, MS-222, Argent Labs); xylene.

3.3 LOCATING ARRAYS OF INTEREST VIA LIGHT AND FLUORESCENCE MICROSCOPY

Instrumentation: Filter cubes DsRed (BP 538-562, BS 570, BP 570-640), and Cy 5 (BP 625-644, BS 660, BP 665-715); Plan-Apochromat 40×/0.95 (Korr M27 420660-9970-000) & LCI Plan-NeoFluar 63×/1.30 (Imm Korr DIC M27 420882-9970-000 water/glycerol) objectives; Vision IsoStation vibration isolation table (Newport); Zeiss Axiocam 506 mono camera; Zeiss AxioObserver Z1 inverted fluorescence microscope.

Materials: ZEN 2 software (Blue Edition) equipped with the Shuttle and Find module (Carl Zeiss Microscopy GmbH).

Reagents: Distilled water; glycerol (99% Grade, Sigma-Aldrich, G9012).

3.4 RELOCATING ARRAYS OF INTEREST FOR SCANNING ELECTRON MICROSCOPY

Instrumentation: For cells, ZEISS Ultra Plus SEM (Carl Zeiss Microscopy GmbH); for tissues, carbon coater (Emitech K950C) and ZEISS Sigma HD VP SEM (Carl Zeiss Microscopy GmbH).

Materials: Graphite rods (ProSciTech, V001G); sharp; solvent-proof extra-fine marker, SmartSEM version 5.07 (Carl Zeiss Microscopy GmbH); ZEN 2 software (Blue Edition) equipped with the Shuttle and Find module (Carl Zeiss Microscopy GmbH).

Reagents: Hexamethyldisilazane (HMDS, Sigma-Aldrich, 440191); uranyl acetate (ProSciTech, C079-F).

3.5 POSTSAMPLE PREPARATION APPROACHES FOR CORRELATIVE IMAGING STUDIES

Instrumentation: Carbon coater (Emitech K950C); sputter coater (Emitech K550X).

Materials: Gold target (ProSciTech, VS57X03-AU-X); graphite rods (ProSciTech, V001G); sharp; solvent-proof extra-fine marker.

3.6 CORRELATIVE DATA ANALYSIS AND MORPHOMETRY

Instrumentation: Home-built 64-bit computer (16 Gigabytes RAM 1600 MHz, Intel i7-2700K Quad Core 3.5 GHz CPU) running Windows 7 Professional software package.

Materials: Adobe Photoshop CS6 (Sedgewick, 2008); ImageJ (v1.47, NIH) (Abramoff et al., 2004); Microsoft Excel 2011 professional package; ZEN 2 software (Blue Edition) equipped with the Shuttle and Find module (Carl Zeiss Microscopy GmbH).

4. RESULTS

4.1 CORRELATIVE ANALYSIS ON HeLa CELLS

Cell cultures were typically investigated immediately after fluorescent labeling under wet-imaging conditions in the absence of a microscope coverslip using widefield FM. Besides data capturing, coordinates were recorded via the software interface allowing the subsequent automated relocation at the SEM level (Fig. 3). The entire process of CLEM imaging, in which four randomly chosen areas were analyzed, took typically less than 45 min, including sample transfer and correlative SEM examination. In this example, HeLa human cervical carcinoma cells were labeled for GM-1 (Jahn & Braet, 2008), resulting in a heterogeneous labeling pattern of cells from which two distinct cell populations could be discerned under FM. This is indeed in line with our previous studies, in which it was demonstrated that cell lines depending on their cell culture conditions generate different GM-1 expressing cell populations (Jahn, Biazik, & Braet, 2011). As such, CLEM was first used to locate GM-1 positive cells via FM at low magnification (Fig. 3A–C), followed by subsequent high-resolution topology investigation of the GM-1 negative versus GM-1 positive cells (Fig. 3D–E). This approach did allow—besides the detection of subpopulations of GM-1 positive cells—to examine next the surface topology of those cells in more detail. It became apparent that GM-1 positive cells bearing more microvilli per square micrometer compared with the cells lacking any GM-1. These small fingerlike projections have a typical width in the order of 100 up to 130 nm which is clearly under the resolving power of widefield microscopy but can be easily picked up with SEM. Taking advantage of being able to locate and relocate numerous randomly chosen areas (i.e., at least four) that contained a high number of cells (i.e., ~400 cells/area), per experimental condition, allowed us to perform analysis on a statistically relevant cell population (Fig. 3F).

The images and their embedded metadata make them highly suitable for successive digital data analysis (Rohde, 2013; Stuurman & Swedlow, 2012). In our hands, making the images binary followed by specific thresholding settings was sufficient to extract morphometric numerical data. Often, and as mentioned by many CLEM researchers before, the arrival of CM automation tools running under a full digital environment finally permits to extend CLEM images into relevant numbers. Lastly, to test the system on its accuracy, and explore the possibility for further postsample preparation steps, including investigation at higher accelerating voltages, the coverslip with cells was removed from the holder after initial CLEM imaging (Fig. 4A) and next coated with a thin layer of gold. Subsequent transfer of the coated sample to the SEM allowed the automated relocation of the previously imaged cells with a relocation-precision in the order of about one cell (i.e., ~20 μm) (Fig. 4B). Besides the ability to readily compare uncoated versus coated imaging conditions on the same cells, after removing and reinserting the sample in the holder, the CLEM solution on offer permits further cross-correlative analysis after postsample processing at a high relocation accuracy.

4.2 CORRELATIVE ANALYSIS ON ZEBRAFISH TISSUE

The second part of the experimental data underpins the benefits of the use of an automated sample-relocation holder under large array screening conditions for fluorescent signals across multiple serial tissue sections and the successive image acquisition of selected ROI within the EM at high-resolving power (Fig. 5). Indeed, targeted ultramicroscopy in combination with precise and rapid locating of ROIs

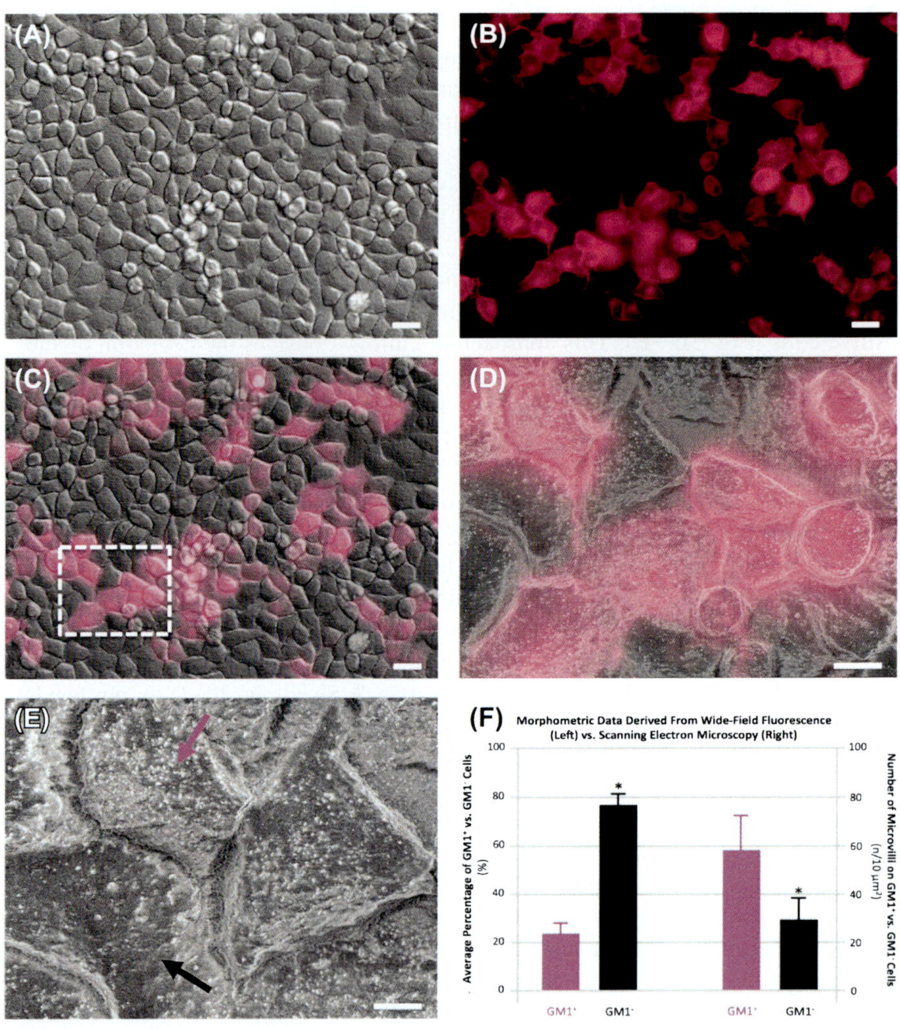

FIGURE 3

Combining differential interference contrast, wide-field fluorescence and scanning electron microscopy information on cultured HeLa human cervical carcinoma cells, followed by cross-correlative data analysis. In this CLEM example, the automated location-relocation solution allowed the swift examination of cells labeled for the cell membrane-associated ganglioside GM-1 (Jahn & Braet, 2008). Screening of large areas of cells at multiple points, and examination at relative low magnification using combined differential interference contrast and fluorescent imaging, permitted the disclosure of GM-1 positive (salmon red color in images B and C) versus GM-1 negative cells (i.e., compare images A vs. C). Subsequent sample transfer to the scanning electron microscope allowed the investigation of the topology of GM-1 positive and GM-1 negative cells at high-magnifications (D). Detailed examination disclosed that the two cell populations have a distinct different membrane surface topology has been proven advantageous for assessing fluo-tagged markers within small model organisms using correlative FM and TEM techniques (Kolotuev, Bumbarger, Labouesse, & Schwab, 2012; Kolotuev, Schwab, & Labouesse, 2009). Additionally, we also took advantage of fluorescence retention in resin embedded tissue material (Peddie et al., 2014; Van Driel, Knoops, Koster, & Valentijn, 2008), enabling us to instantly discern the fine distribution of albumin in X, Y, and Z dimensions in zebrafish larvae, from the organ up to the subcellular level. The ability to interrogate large-field volumetric CLEM data on consecutive ultrathin sections at 90 nm depth intervals also delivered a marked improvement in depth-independent fluorescent labeling because of the use of the water-miscible LR White acrylic resin. Indeed other comparable CLEM approaches to attain X, Y, and Z correlative arrays, such as in situ serial block-face SEM (Hughes, Hawes, Monteith, & Vaughan, 2014), are limited to the use of epoxy resins (e.g., Epon or Spurr) (Shami et al., 2016). Despite the fact that epoxy resins have less tendency to crumble from the knife edge and to fall back onto the block-face they do not excel in overall fluorescence retention for prolonged times, thereby impeding the collection of reliable multidimensional CLEM information. Finally, besides the stable presence of the fluorescent conjugate, autofluorescence potentially generated by the use of glutaraldehyde was absent (Fig. 5). In our hands, the combined use of glutaraldehyde (0.1%) and uranyl acetate (2%) in the primary sample preparation steps resulted in well-preserved fine ultrastructure with adequate EM contrast, including the preservation of the bright fluorescent emission properties of the dye Texas Red (Fig. 6). This is in accordance with previous studies in which various members of the fluorescent proteins have been successfully investigated in aldehyde-fixed and LR White resin–embedded tissue material (Bell, Mitchell, Paultre, Posch, & Oparka, 2013; Kolotuev et al., 2012).

Fluorescent-conjugated albumin is a widely accepted macromolecular marker (MW ~66,000 Daltons) to study vascular-related events under live small animal imaging conditions (National Center for Biotechnology Information, 2004–2013). Prolonged circulation of the conjugate adequately distributes throughout the organism allowing studying its target destinations (i.e., microvasculature) including

(E). GM-1 negative cells have a relatively smooth cell membrane surface (*black arrow*) whereas GM-1 positive cells possess numerous microvilli. Since multiple areas were screened and relocated across the different imaging platforms, a substantial amount of cross-correlative image data could be collected permitting data analysis (F). Subsequent morphometric analysis of the optical and fluorescent data revealed that about ~20% of the confluent HeLa cell cultures were positive for GM-1. Whereas SEM data examination shows that GM-1 positive cells bearing a higher average number of microvilli on their surface compared with the GM-1 negative cells. The latter is a structural detail that cannot be disclosed using optical microscopy techniques because of their limiting resolving power. Scale bars: (A–C), 25 μm; (D), 10 μm; (E), 5 μm.

(D) Reprinted from Braet, F., & Trimby, P. (2016). Micron, 89. *Front Cover, with permission from Elsevier.*

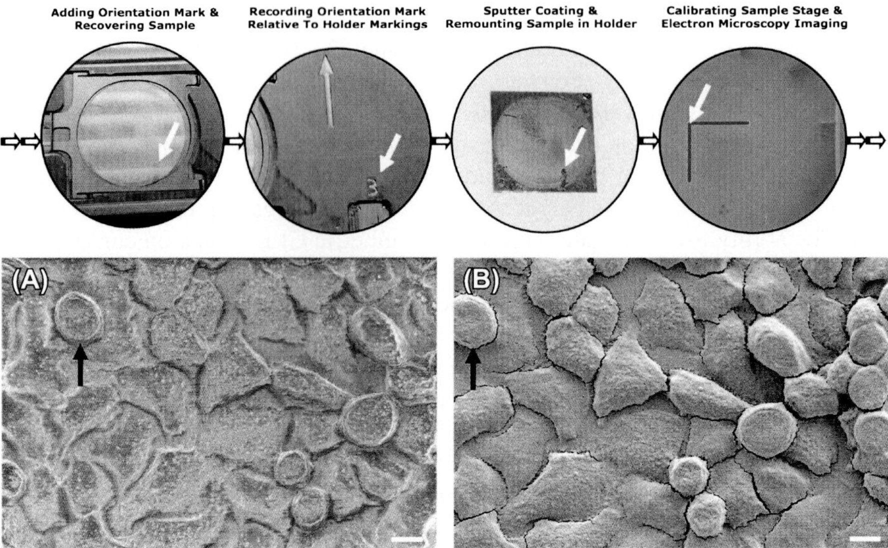

FIGURE 4

Illustration of the relocating precision of the sample-relocation holder after an initial CLEM experimentation was completed (see, Fig. 3), that was followed by sample retrieval, next sputter-coated with a thin layer of gold (i.e., 15 nm), and subsequently reinserted in the holder for scanning electron microscopy purposes. In the schematic workflow (top) the *white arrows* points to: (1) the orientation mark artificially added to the edge of the coverslip; and (2) the orientation mark engraved on the sample-relocation holder (Fig. 4). Although trivial, those markers allow reinserting the coverslip in the correct orientation. Image A depicts the observation of HeLa cell under uncoated conditions and image (B) after sputter coating. Note the remarkable difference in image information between uncoated versus the more familiar looking coated imaging conditions. Compare the location of the *black arrows* under (A and B), illustrating that the relocation precision in this experiment is in the order of ~1 cell. Scale bars: (A and B), 10 μm.

clearance pathways (e.g., receptor-mediated endocytosis). In this study, BSA-TxRed was injected intrahepatically and allowed to circulate for half an hour before CLEM sample preparation commenced. The zebrafish was sectioned transversely behind the gills—i.e., at the start of the digestive tract—and across the liver and kidney (Fig. 5A). Subsequent FM observation disclosed the target destinations of BSA-TxRed, appearing bright red under FM (Fig. 5B). When combining the LM and FM images, those destinations could be easily identified as the liver and kidney based on their unique microanatomical appearance (Fig. 5C) (Cheng et al., 2016). However, it was only following EM correlation that we could determine that BSA-TxRed accumulated within the endothelial cells of the liver capillaries (Fig. 5D and E) and the microvascular endothelium of the kidney (Fig. 5D and F). Investigation at high magnification displayed furthermore that BSA-TxRed was trapped within the space of Disse and also incorporated within the hepatocytes (Fig. 5E). Those observations are in accordance with the "liver sieving" and "endothelial massage" concepts on how macromolecular complexes get delivered to the liver parenchyma (Wisse, De Zanger, Charels, Van Der Smissen, & McCuskey, 1985). Previous biochemistry assessments combined with FM indeed confirm that BSA ends up in the reticuloendothelial system of the liver (Goresky, 1982; Smedsrod, 2004). Intriguingly, as soon as 30 min after injection, the majority of the conjugate was located in the vicinity and within the kidney (Fig. 5F). This might indicate that the excess of BSA-TxRed is rapidly cleared from the organism. Indeed, it has been reported that albumin-overload studies in rats and rabbits result in the deposition of large patches of albumin within the kidney microvasculature (Andres, Seegal, Hsu, Rothenberg, & Chapeau, 1963; Simpson & Shand, 1983). When imaging multiple consecutive sections, we could follow the course of the BSA-TxRed cell inclusions in height (Fig. 6). These round-shaped vesicles clearly resemble endosome–lysosome complexes. While the total projection of the combined CLEM information allowed us to localize "patches" of albumin (Fig. 5F), it was only after studying the individual serial sections that we could individually define the shape and course of those inclusions (Fig. 6), underpinning the strengths of correlative array tomography. Hence, we put forward that examination of LR White-embedded serial sections on glass microscope slides (Micheva & Smith, 2007) combined with the automated sample-relocation hardware solution outlined herein (Loussert & Humbel, 2015) is an attractive low-cost and time-saving CLEM alternative for similar commercial solutions allowing fully automated imaging of correlative array tomography sections (Zeiss, 2015). The approach outlined in this chapter allowed us to screen the entire organism for the presence of fluorescent-labeled complexes and the subsequent investigation of different target organs (i.e., liver and kidney) in multiple spatial dimensions at high resolution.

5. DISCUSSION

5.1 MAKING THE MOST OF THIS AUTOMATED APPROACH

Contemporarily, the difficulties associated in the past decades with specimen preparation and relocation for traditional CLEM (Fig. 1), have been largely circumvented via the recent development of a variety of commercial hardware and software systems (de Boer, Hoogenboom, & Giepmans, 2015) (Fig. 2). Those solutions have allowed for the swift and automated generation of correlated information, which most importantly fulfills the need for high-throughput CLEM data generation. One such system that is made available uses an automated sample-relocation holder, which drives the generation of specific fluorescent information on fine structures of interest utilizing FM, combined with high-resolution cytoarchitectural information, generated using SEM. Locating and relocating of ROIs at high accuracy is achieved via the use of a 3-point calibration engraved on a dedicated sample-relocation holder

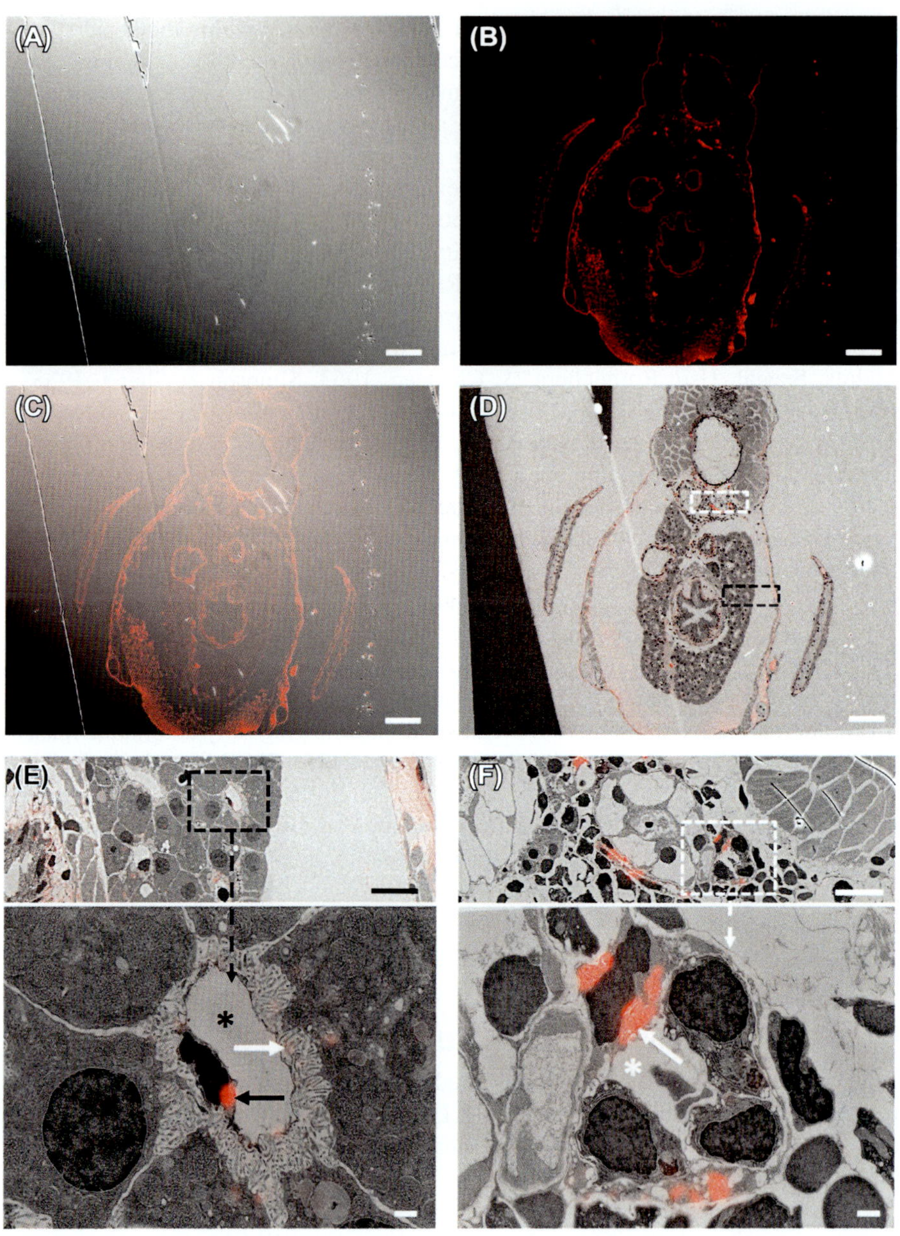

FIGURE 5

Combining differential interference contrast, wide-field fluorescence and scanning electron microscopy information of a transversely cut section (90 nm) from a zebrafish larva that was injected intrahepatically with BSA-TxRed. BSA-TxRed was allowed to circulate for 30 min before CLEM preparation commenced. Note the remarkable retention of fluorescence within

(Lucas et al., 2012), which deviates from traditional CLEM relocation, where relocation fiducials are often etched to the sample substrate or deposited on the sample, thereby potentially interfering with the ROIs. In this study, multiple ROIs could be swiftly defined and captured in a single sample, which is subsequently embedded in the metadata of each image. Additional image acquisition parameters such as magnification are also transferred, thereby allowing a direct overlay of the corresponding ROIs between the different microscope modalities at matching pixel dimensions. For completeness, it has to be noted that other and similar locate and relocate stage solutions exist such as Mi*X*croscopy (JEOL-Nikon, 2014; Loussert & Humbel, 2015).

There is no doubt that the combined use of a set of optical imaging approaches and electron microscopy techniques has been increasingly important in furthering our understanding of the morphology and functional insights of biological organisms and structures, ranging from the whole organism level down to the supramolecular level (de Boer et al., 2015). Indeed, to genuinely understand how complex biological structures function, we must correlate and integrate knowledge of their dynamic behavior and of their molecular machinery (Smith, 2012). Despite the long history of CM we are finally at the cross-roads where the researcher is fully equipped with the necessary relocation tools running under a digital environment, including the availability of fluorescent labeling strategies specifically designed for CM, to explore the cell's landscape for its constituents and structural machinery on a much higher scale than ever before (Loussert & Humbel, 2015). Automated CLEM approaches confer a plethora of advantages for the generation of complementary structure–function nanoscopic information (Lucas et al., 2012; Schorb et al., 2017). These include, the compatibility of such systems with a diverse range of biological samples, the generation of either topographic or internal ultrastructural information, high-accuracy relocation fidelity, and the ability to investigate large volumes of samples in three-dimensions.

the LR White embedded samples. At low magnification, correlating differential interference contrast (A) and wide-field fluorescence (B) information allowed the swift location of fluorescently labeled albumin (i.e., ROI) at the organismal level (C). (D) Merging the fluorescent information together with the corresponding SEM image disclosed that the liver (*black dotted line box*) and kidney (*white dotted line box*) are the primary target destinations of BSA-TxRed. (E) At higher magnification, albumin was accumulated within the endothelial lining of the liver (*black arrow*), trapped within the space of Disse (*white arrow*) and some discrete patches of albumin can be seen at the apical side of the surrounding hepatocytes. Note the microvillous surface of the hepatocytes. (F) High power magnification examination of the kidney area disclosed the presence of concentrated amounts of albumin (*white arrow*) within the endothelium of the kidney microvasculature. *Asterisks* denote the lumen of blood vessels. Scale bars: (A–D), 50 μm; (E and F), 10 μm (top) and 1 μm (bottom).

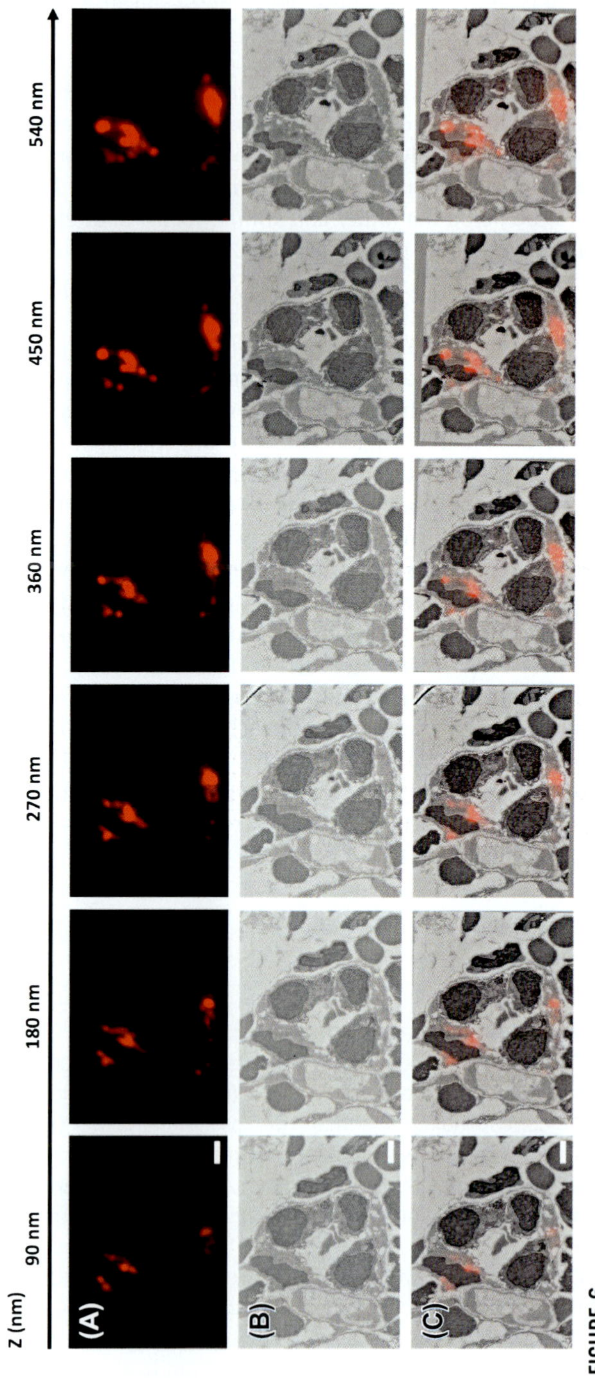

FIGURE 6

Fluorescence microscopy (A), scanning electron microscopy (B) and corresponding merged images (C) of a kidney blood vessel (for full comparison, see Fig. 5F). In this figure set, six consecutive sections with a thickness of 90 nm were examined. Note the course of the round-shaped vesicles changing in shape and size along the volume imaged (i.e., 540 nm). The vesicles typically resemble the structure and distribution of endosome–lysosome complexes. Scale bars: 2 μm.

5.2 OUTLOOK

Presently, advances in CM imaging technologies and methodologies are occurring at a prolific rate. The most exciting developments concern the correlative approaches under full cryogenic conditions in which samples are physically fixed resulting in fine structural preservation of biological specimens close to their native state. The recent concurrent advances made within the disciplines of cryo-EM and CLEM ultimately permitted the full consolidation of the best sample preparation, labeling and imaging approaches in one new world of cryo-CLEM; by which vitrified specimens light up under fluorescence excitation, and fine ultrastructure subsequently appears under the microscope's electron beam. Unquestionably, rapid cryofixation in a time window of less than a second results in the best preservation of supramolecular complexes, and delivers the best structural preservation for subsequent tomography modeling within nanometers precision. Recent Cryo-CLEM studies elegantly illustrate that this approach will be the standard for structural biologists within the foreseeable future (Bos et al., 2014; Schorb et al., 2017). Of note, rendering three-dimensional EM information using focused ion beam (FIB) cryo-SEM approaches is a space to watch out for (Schertel et al., 2013; Vidavsky et al., 2016), and the first cryo-CLEM FIB-SEM studies are a direct attest to this statement (Arnold et al., 2016; Fukuda et al., 2014). For a recent review and future prospects on this matter see, Wolff et al. (Wolff, Hagen, Grünewald, & Kaufmann, 2016).

Finally, with the advent of the proof of principle of integrated light and electron microscopy (Agronskaia et al., 2008), including cryo-ILEM (Faas et al., 2013), many started to develop fully integrated microscopy approaches by which different imaging modalities, mostly FM combined with SEM, are built into one single instrument (Timmermans & Otto, 2015). Commercial solutions are even on offer such as the integrated FM-SEM platform supplied by Delmic (i.e., SECOM). Building on the previous paragraph, one can clearly foresee what a genuine "integrated FM and cryo-FIB SEM platform" together with improved electron detection can deliver to the future generation of molecular- and structural biologists. It is the Holy Grail sought by microscopists for many decades.

5.3 CONCLUSION

This contribution outlines a CLEM approach that allows the fast transfer and automated relocation of samples between the FM and the SEM, enabling swift correlative mapping of fluorescently stained molecular complexes—via a software-driven 3-point calibration on a dedicated sample-relocation holder. Ultimately, the significance of the automated CLEM approach outlined herein is derived from its role as an adjunct to results-driven science generating complementary information, where relevant biological questions can be answered, from the micron down to the nanometer scale. High-throughput data can be swiftly generated over multiple points within the sample, allowing the extraction of relevant CLEM information in a rigorous statistical manner. We foresee that this automated FM and SEM approach—which is in part similar to the light and fluorescent histology slide scanners—can be readily applied to clinically relevant samples to better understand disease mechanism or aid diagnosis where high-resolution imaging is required such as in a variety of liver and kidney diseases.

ACKNOWLEDGMENTS

The authors acknowledge the facilities of the Australian Microscopy & Microanalysis Research Facility at the Australian Centre for Microscopy & Microanalysis of the University of Sydney and the Zebrafish Facility at Macquarie University. The authors are indebted to Mrs Susan Hart and Mr. Simon Kinder from Carl Zeiss Australia for their continuing support in this work and to Mr. Jeffrey Henriquez for his assistance in some of the sample preparation reported herein. The authors also acknowledge the use of Motifolio SmartArt objects (License 275086) that were modified and reedited to suit content in support of Figs. 1 and 2. The authors are indebted to the Australian Research Council for granting support through the "Linkage Infrastructure, Equipment and Facilities" funding scheme (LE100100030 and LE110100203). We also wish to thank the University of Sydney (Research & Innovation) and the Charles Perkins Centre (Cellular Imaging Facility) for funding some of the infrastructure reported herein.

REFERENCES

Abandowitz, H. M., & Geissinger, H. D. (1975). Preparation of cells from suspensions for correlative scanning electron and interference microscopy. *Histochemistry, 45*, 89–94.

Abramoff, M. D., Magelhaes, P. J., & Ram, S. J. (2004). Image processing with ImageJ. *Biophotonics International, 11*, 36–42.

Agronskaia, A. V., Valentijn, J. A., van Driel, L. F., Schneijdenberg, C. T., Humbel, B. M., van Bergen En Henegouwen, P. M., … Gerritsen, H. C. (2008). Integrated fluorescence and transmission electron microscopy. *Journal of Structural Biology, 164*, 183–189.

Albrecht, R. M., Goodman, S. L., & Simmons, S. R. (1989). Distribution and movement of membrane-associated platelet glycoproteins: Use of colloidal gold with correlative video-enhanced light microscopy, low-voltage high-resolution scanning electron microscopy, and high-voltage transmission electron microscopy. *American Journal of Anatomy, 185*, 149–164.

Andres, G. A., Seegal, B. C., Hsu, K. C., Rothenberg, M. S., & Chapeau, M. L. (1963). Electron microscopic studies of experimental nephritis with ferritin-conjugated antibody. Localization of antigen-antibody complexes in rabbit glomeruli following repeated injections of bovine serum albumin. *The Journal of Experimental Medicine, 117*, 691–704.

Arnold, J., Mahamid, J., Lucic, V., de Marco, A., Fernandez, J. J., Laugks, T., … Plitzko, J. M. (2016). Site-specific cryo-focused ion beam sample preparation guided by 3D correlative microscopy. *Biophysical Journal, 110*, 860–869.

Bell, K., Mitchell, S., Paultre, D., Posch, M., & Oparka, K. (2013). Correlative imaging of fluorescent proteins in resin-embedded plant material. *Plant Physiology, 161*, 1595–1603.

de Boer, P., Hoogenboom, J. P., & Giepmans, B. N. (2015). Correlated light and electron microscopy: Ultrastructure lights up! *Nature Methods, 12*, 503–513.

Bos, E., Hussaarts, L., van Weering, J. R., Ellisman, M. H., de Wit, H., & Koster, A. J. (2014). Vitrification of Tokuyasu-style immuno-labelled sections for correlative cryo light microscopy and cryo electron tomography. *Journal of Structural Biology, 186*, 273–282.

Braet, F. (2010). Evaluation of the microanatomy of the liver via a rapid sample preparation protocol and a table-top scanning electron microscope. *The Open Anatomy Journal, 2*, 98–101.

Braet, F., De Zanger, R., & Wisse, E. (1997). Drying cells for SEM, AFM and TEM by hexamethyldisilazane: A study on hepatic endothelial cells. *Journal of Microscopy, 186*, 84–87.

Braet, F., Spector, I., Shochet, N. R., Crews, P., Higa, T., Menu, E., ... Wisse, E. (2002). The new anti-actin agent dihydrohalichondramide reveals fenestrae-forming centers in hepatic endothelial cells. *BMC Cell Biology, 3*, 7.

Braet, F., Wisse, E., Bomans, P., Frederik, P., Geerts, W., Koster, A., ... Ringer, S. (2007). Contribution of high-resolution correlative imaging techniques in the study of the liver sieve in three-dimensions. *Microscopy Research Technique, 70*, 230–242.

Cheng, D., Shami, G. J., Morsch, M., Chung, R. S., & Braet, F. (2016). Ultrastructural mapping of the zebrafish gastrointestinal system as a basis for experimental drug studies. *Biomedical Research International, 2016*, 8758460.

Coates, P. W., & Teh, E. C. (1978). Demonstration of neurosecretory substance in previously scanned specimens: Adaptation of the Gomori aldehyde-fuchsin method for correlative SEM/TEM/LM histochemistry. *American Journal of Anatomy, 153*, 469–475.

Faas, F. G., Bárcena, M., Agronskaia, A. V., Gerritsen, H. C., Moscicka, K. B., Diebolder, C. A., ... Koster, A. J. (2013). Localization of fluorescently labeled structures in frozen-hydrated samples using integrated light electron microscopy. *Journal of Structural Biology, 181*, 283–290.

Fukuda, Y., Schrod, N., Schaffer, M., Feng, L. R., Baumeister, W., & Lucic, V. (2014). Coordinate transformation based cryo-correlative methods for electron tomography and focused ion beam milling. *Ultramicroscopy, 143*, 15–23.

Geissinger, H. D. (1974). A precise stage arrangement for correlative microscopy for specimens mounted on glass slides, stubs or EM grids. *Journal of Microscopy, 100*, 113–117.

Goresky, C. A. (1982). The processes of cellular uptake and exchange in the liver. *Federation Proceedings, 41*, 3033–3039.

Hedrera, M. I., Galdames, J. A., Jimenez-Reyes, M. F., Reyes, A. E., Avendaño-Herrera, R., Romero, J., & Feijóo, C. G. (2013). Soybean meal induces intestinal inflammation in zebrafish larvae. *PLoS One, 8*, e69983.

Hughes, L., Hawes, C., Monteith, S., & Vaughan, S. (2014). Serial block face scanning electron microscopy—The future of cell ultrastructure imaging. *Protoplasma, 251*, 395–401.

Husain, M., Thomas, C., Kirschmann, M., Oberti, D., & Hahnloser, R. (2011). Functional and structural investigation of songbird brain projection neurons with shuttle and find. *Microscopy and Microanalysis, 17*, 242–243.

Jahn, K. A., Barton, D. A., Kobayashi, K., Ratinac, K. R., Overall, R. L., & Braet, F. (2012). Correlative microscopy: Providing new understanding in the biomedical and plant sciences. *Micron, 43*, 565–582.

Jahn, K. A., Biazik, J. M., & Braet, F. (2011). GM1 expression in Caco-2 cells: Characterisation of a fundamental, passage-dependent transformation. *Journal of Pharmaceutical Science, 100*, 3751–3762.

Jahn, K. A., & Braet, F. (2008). Monitoring membrane rafts in colorectal cancer cells by means of correlative fluorescence electron microscopy (CFEM). *Micron, 39*, 1393–1397.

JEOL-Nikon. (2014). New JEOL-Nikon miXcroscopy correlative imaging solution. *Microscopy Today, 22*.

Kolotuev, I., Bumbarger, D. J., Labouesse, M., & Schwab, Y. (2012). Targeted ultramicrotomy: A valuable tool for correlated light and electron microscopy of small model organisms. *Methods in Cell Biology, 111*, 203–222.

Kolotuev, I., Schwab, Y., & Labouesse, M. (2009). A precise and rapid mapping protocol for correlative light and electron microscopy of small invertebrate organisms. *Biology of the Cell, 102*, 121–132.

Lange, R. H. (1972). Correlated light and electron microscopy with special reference to histochemistry. *Mikroskopie, 27*, 193–199.

Loussert, F. C., & Humbel, B. M. (2015). Correlative microscopy. *Archives of Biochemistry and Biophysics, 581*, 98–110.

Lucas, M. S., Gunthert, M., Gasser, P., Lucas, F., & Wepf, R. (2012). Bridging microscopes: 3D correlative light and scanning electron microscopy of complex biological structures. *Methods in Cell Biology, 111*, 325–356.

Matsui, A., Tanaka, E., Choi, H. S., Kianzad, V., Gioux, S., Lomnes, S. J., & Frangioni, J. V. (2010). Real-time, near-infrared, fluorescence-guided identification of the ureters using methylene blue. *Surgery, 148*, 78–86.

McDonald, L. W., & Hayes, T. L. (1969). Correlation of scanning electron microscope and light microscope images of individual cells in human blood and blood clots. *Experimental and Molecular Pathology, 10*, 186–198.

Micheva, K. D., O'Rourke, N., Busse, B., & Smith, S. J. (2010). Array tomography: Production of arrays. *Cold Spring Harbor Protocols, 2010*. pdb.prot5524.

Micheva, K. D., & Smith, S. J. (2007). Array tomography. *Neuron, 55*, 25–36.

Moberg, C. L. (2012). Entering the realm of the cell with an electron microscope. In C. L. Moberg (Ed.), *Entering an unseen world* (pp. 41–71). New York: The Rockefeller University Press (Chapter 3).

Moore, C. L., Cheng, D., Shami, G. J., & Murphy, C. R. (2016). Correlated light and electron microscopy observations of the uterine epithelial cell actin cytoskeleton using fluorescently labeled resin-embedded sections. *Micron, 84*, 61–66.

Moses, M. J. (1956). Studies on nuclei using correlated cytochemical, light, and electron microscopy techniques. *The Journal of Biophysical and Biochemical Cytology, 2*, 397–408.

National Center for Biotechnology Information, US. (2004–2013). *Molecular imaging and contrast agent database*. Retrieved from https://www.ncbi.nlm.nih.gov/books/NBK5330/.

Peddie, C. J., Blight, K., Wilson, E., Melia, C., Marrison, J., Carzaniga, R., … Collinson, L. M. (2014). Correlative and integrated light and electron microscopy of inresin GFP fluorescence, used to localise diacylglycerol in mammalian cells. *Ultramicroscopy, 143*, 3–14.

Plopper, C. G., Dungworth, D. L., & Tyler, W. S. (1973). Pulmonary lesions in rats exposed to ozone. A correlated light and electron microscopic study. *The American Journal of Pathology, 71*, 375–394.

Porter, K. R., Claude, A., & Fullam, E. F. (1945). A study of tissue culture cells by electron microscopy – Methods and preliminary observations. *The Journal of Experimental Medicine, 81*, 233–246.

Rieder, C. L., & Bowser, S. S. (1985). Correlative immunofluorescence and electron-microscopy on the same section of epon-embedded material. *The Journal of Histochemistry and Cytochemistry, 33*, 165–171.

Rohde, G. K. (2013). New methods for quantifying and visualizing information from images of cells: An overview. *Conference Proceedings of the IEEE Engineering in Medicine and Biology Society, 2013*, 121−124.

Rosen, J. N., Sweeney, M. F., & Mably, J. D. (2009). Microinjection of zebrafish embryos to analyze gene function. *Journal of Visualized Experiments*. http://dx.doi.org/10.3791/1115.

Russ, J. C., & Neal, F. B. (2015). *The image processing Handbook* (7th ed.). New York: CRC Press.

Schertel, A., Snaidero, N., Han, H. M., Ruhwedel, T., Laue, M., Grabenbauer, M., & Mobius, W. (2013). Cryo FIB-SEM: Volume imaging of cellular ultrastructure in native frozen specimens. *Journal of Structural Biology, 184*, 355−360.

Schorb, M., Gaechter, L., Avinoam, O., Sieckmann, F., Clarke, M., Bebeacua, C., … Briggs, J. A. (February 2017). New hardware and workflows for semi-automated correlative cryo-fluorescence and cryo-electron microscopy/tomography. *Journal of Structural Biology, 197*(2), 83−93.

Sedgewick, J. (2008). *Scientific imaging with photoshop: Methods, measurement, and output* (1st ed.). Berkeley, USA: New Riders Press.

Shami, G., Cheng, D., Huynh, M., Vreuls, C., Wisse, E., & Braet, F. (2016). 3-D EM exploration of the hepatic microarchitecture − Lessons learned from large-volume in situ serial sectioning. *Scientific Reports, 6*, 36744.

Simpson, L. O., & Shand, B. I. (1983). Morphological changes in the kidneys of mice with proteinuria induced by albumin-overload. *British Journal of Experimental Pathology, 64*, 396−402.

Smedsrod, B. (2004). Clearance function of scavenger endothelial cells. *Comparative Hepatology, 3*, S22.

Smith, C. (2012). Microscopy: Two microscopes are better than one. *Nature, 492*, 293−297.

Stuurman, N., & Swedlow, J. R. (2012). Software tools, data structures, and interfaces for microscope imaging. *Cold Spring Harbor Protocols, 2012*, 50−61.

Su, Y., Nykanen, M., Jahn, K. A., Whan, R., Cantrill, L., Soon, L. L., … Braet, F. (2010). Multi-dimensional correlative imaging of subcellular events: Combining the strengths of light and electron microscopy. *Biophysical Reviews, 2*, 121−135.

Timmermans, F. J., & Otto, C. (2015). Contributed review: Review of integrated correlative light and electron microscopy. *The Review of Scientific Instruments, 86*, 011501.

Van Driel, L. F., Knoops, K., Koster, A. J., & Valentijn, J. A. (2008). Fluorescent labeling of resin embedded sections for correlative electron microscopy using tomography-based contrast enhancement. *Journal of Structural Biology, 161*, 372−383.

Vanheyningen, H. E. (1965). Correlated light and electron microscope observations on glycoprotein-containing globules in the follicular cells of the thyroid gland of the rat. *The Journal of Histochemistry and Cytochemistry, 13*, 286−295.

Vidavsky, N., Akiva, A., Kaplan-Ashiri, I., Rechav, K., Addadi, L., Weiner, S., & Schertel, A. (December 2016). Cryo-FIB-SEM serial milling and block face imaging: Large volume structural analysis of biological tissues preserved close to their native state. *Journal of Structural Biology, 196*(3), 487−495.

Wang, W. J., Tay, H. G., Soni, R., Perumal, G. S., Goll, M. G., Macaluso, F. B., … Tsou, M. F. B. (2013). CEP162 is an axoneme-recognition protein promoting ciliary transition zone assembly at the cilia base. *Nature Cell Biology, 15*, 591−601.

White, R. M., Sessa, A., Burke, C., Bowman, T., LeBlanc, J., Ceol, C., … Zon, L. I. (2008). Transparent adult zebrafish as a tool for in vivo transplantation analysis. *Cell Stem Cell, 2*, 183−189.

Wisse, E., De Zanger, R. B., Charels, K., Van Der Smissen, P., & McCuskey, R. S. (1985). The liver sieve: Considerations concerning the structure and function of endothelial fenestrae, the sinusoidal wall and the space of Disse. *Hepatology, 5*, 683–692.

Wolff, G., Hagen, C., Grünewald, K., & Kaufmann, R. (2016). Towards correlative super-resolution fluorescence and electron cryo-microscopy. *Biology of the Cell, 108*, 245–258.

Zeiss. (2013). *Correlative microscopy protocols – A reference guide to correlative sample preparation*. Retrieved from http://applications.zeiss.com/.

Zeiss. (2015). *Atlas 5 array tomography*. Retrieved from http://www.zeiss.com/microscopy/us/products/scanning-electron-microscopes/upgrades/atlas-5-array-tomography.html.

CHAPTER 12

Correlative two-photon and serial block face scanning electron microscopy in neuronal tissue using 3D near-infrared branding maps

Robert M. Lees*, Christopher J. Peddie[§], Lucy M. Collinson[§], Michael C. Ashby*, Paul Verkade*[,1]

**University of Bristol, Bristol, United Kingdom*

[§]The Francis Crick Institute, London, United Kingdom

[1]Corresponding author: E-mail: p.verkade@bristol.ac.uk

CHAPTER OUTLINE

Methods in Cell Biology, Volume 140, ISSN 0091-679X, http://dx.doi.org/10.1016/bs.mcb.2017.03.007

Abstract

Linking cellular structure and function has always been a key goal of microscopy, but obtaining high resolution spatial and temporal information from the same specimen is a fundamental challenge. Two-photon (2P) microscopy allows imaging deep inside intact tissue, bringing great insight into the structural and functional dynamics of cells in their physiological environment. At the nanoscale, the complex ultrastructure of a cell's environment in tissue can be reconstructed in three dimensions (3D) using serial block face scanning electron microscopy (SBF-SEM). This provides a snapshot of high resolution structural information pertaining to the shape, organization, and localization of multiple subcellular structures at the same time. The pairing of these two imaging modalities in the same specimen provides key information to relate cellular dynamics to the ultrastructural environment. Until recently, approaches to relocate a region of interest (ROI) in tissue from 2P microscopy for SBF-SEM have been inefficient or unreliable. However, near-infrared branding (NIRB) overcomes this by using the laser from a multiphoton microscope to create fiducial markers for accurate correlation of 2P and electron microscopy (EM) imaging volumes. The process is quick and can be user defined for each sample. Here, to increase the efficiency of ROI relocation, multiple NIRB marks are used in 3D to target ultramicrotomy. A workflow is described and discussed to obtain a data set for 3D correlated light and electron microscopy, using three different preparations of brain tissue as examples.

INTRODUCTION

Light microscopy (LM) is invaluable to cell biologists as a tool for obtaining dynamic information about cellular function and structure, especially with the aid of fluorescent reporters or dyes. Multiphoton (MP) fluorescence microscopy allows the visualization of these cellular dynamics inside living tissue at greater depth than is possible with conventional fluorescence microscopy (e.g., confocal microscopy) (Zipfel, Williams, & Webb, 2003). This is achieved by employing lower energy, longer wavelength light in the near-infrared part of the spectrum (~700–1400 nm). To excite a fluorophore, near simultaneous absorption of multiple, lower energy photons are required instead of a single, higher-energy photon in single photon fluorescence microscopy. The probability of MP excitation is extremely low in comparison with single photon excitation. To overcome this, the laser is repeatedly pulsed in ultra-fast (100 fs) bursts to increase the photon density. Because photon density falls away with distance from the focal plane, the chance of obtaining MP excitation is effectively zero outside that focal plane, conferring inherently high three-dimensional (3D) resolution (Zipfel et al., 2003). MP microscopy is advantageous to a cell biologist because the near-infrared light is refracted less as it passes through the tissue. With a sample that refracts very little light, it is possible to routinely achieve imaging depths of around 300 μm, with a strongly emitted signal that raises in excess of 500 μm. Although there are other examples of MP excitation, this chapter is focused on two-photon (2P) excitation, the absorption of two photons, using an MP microscope.

Neuroscience research can utilize 2P microscopy to aid in studies of the mammalian brain that were previously limited to smaller, semitransparent model organisms (Svoboda & Yasuda, 2006). It is possible to image the cortical neurons of mice longitudinally in vivo through cranial windows or thinned skull with either genetically encoded or virally expressed fluorescent reporters (Holtmaat et al., 2009; Yang, Pan, Parkhurst, Grutzendler, & Gan, 2010). Electrophysiological studies can also be aided in the case of genetically encoded calcium indicators (GECIs) (Akerboom et al., 2012). GECIs are used to look at intracellular Ca^{2+} transients in neurons that are a proxy measure of membrane depolarization and hence neuronal cell activity. Another use of 2P microscopy in neuroscience is for the uncaging of neurotransmitter (e.g., glutamate; Shoham, O'Connor, Sarkisov, & Wang, 2005). 2P excitation is used to uncage the neurotransmitter with high spatial resolution, and the response from the cell is used to map receptors on the neuron or to induce activity in specific target cells (Ashby & Isaac, 2011). These techniques have contributed to the understanding of both structural and functional neuronal dynamics within the spatial resolution limits of 2P microscopy.

There are technical challenges to labeling and imaging multiple different fluorophores with 2P in tissue. This includes the lack of fluorophores with sufficiently separate 2P excitation spectra and the difficulty of obtaining a good signal-to-noise ratio deep in tissue. Therefore, it has become routine to fix samples and probe for extra information through immunochemistry using other forms of LM (often with resectioning of the sample). This is especially useful for identifying cell types and protein localization post hoc. However, LM becomes limited by spectral and spatial resolution when identifying, quantifying, and measuring multiple different subcellular structures that are abundant in the tissue volume (such as synapses and mitochondria). Problems also occur in penetration of the tissue with antibodies and therefore obtaining good signal from higher depths. There are advances in this area, including the reduction of light scattering in whole tissue samples by clearing (Renier et al., 2014) and serial sectioning of tissue for easier probing, reprobing and imaging by array tomography (Micheva, O'Rourke, Busse, & Smith, 2010).

Electron microscopy (EM) can be used to overcome the resolution limits of LM to be able to distinguish subcellular structures that are a few nanometers apart. Structures can be identified and have their function inferred by determining protein composition using immuno-EM. As a tool for neuroscience, EM has been used to characterize the ultrastructure of neurons for decades, resulting in a comprehensive understanding of structural identity but with no dynamic information. Pairing neuronal dynamics from 2P microscopy with the ultrastructural environment provides unprecedented insights into structure—function relationships. However, correlating 3D light and electron microscopy in the same tissue sample is difficult.

Until recently, the only option for reconstructing tissue volumes at ultrastructural resolution was through serial section transmission EM (ssTEM) methods. This involves a highly trained individual cutting and staining perfect serial sections, acquiring images, and subsequently aligning them in a laborious workflow. However, by automating the sectioning, imaging, and alignment simultaneously, time and human error can be mitigated. Both serial block face scanning electron microscopy (SBF-SEM) and focused ion beam scanning electron microscopy (FIB-SEM) achieve this goal. They work by destructively sectioning or milling off the top layer of the tissue and imaging each serial block face (Peddie & Collinson, 2014). It is conceivable to go from having a fixed brain tissue specimen to a registered 3D ultrastructural volume within a week. Because FIB-SEM and SBF-SEM are destructive techniques, they are at a disadvantage in comparison with ssTEM methods (which have also become more automated in recent years; Kasthuri et al., 2015). The choice of technique depends on the volume to be analyzed and the resolution required; at present, SBF-SEM routinely produces a much larger volume than FIB-SEM, albeit with lower Z-resolution. The obtainable voxel size for SBF-SEM is currently <20 nm in the Z-axis and <5 nm in X/Y-axis, whereas FIB-SEM can achieve an isotropic voxel size of 5 nm. In comparison, ssTEM reaches a pixel size of <0.5 nm and a reliable slice thickness of only 50 nm. The resolution limits of these systems are sufficient for identification of synaptic structures, making them attractive for use in neuroscience research, especially connectomics (Helmstaedter, 2013).

One of the biggest technical challenges with 3D correlated light and electron microscopy (CLEM) is finding a reliable and efficient method for relocating the region of interest (ROI) between imaging modalities. This problem is not novel and has been approached from many different angles. CLEM in tissue cannot benefit from finder grid coordinate systems as in cell culture. However, other potential solutions include, immunogold labeling the structures of interest, diaminobenzidine (DAB) precipitation in the fluorescently labeled cell, and the use of nearby vasculature as a fiducial marker. These techniques have their uses for particular samples but present problems for relocation of an ROI in thick tissue. Membrane structure can become compromised in immunogold labeling due to steps of permeabilization that requires detergents. Additionally, antibodies can take a long time to penetrate thicker tissue specimens; as paraformaldehyde fixation is reversible over time, there will be further loss of structural integrity during long antibody incubations. Immunogold labeling also only improves identification and not relocation of the ROI. DAB precipitation can be useful in particularly thin samples where it is possible to see the precipitate with a brightfield microscope after resin embedding. The ROI can then be targeted quickly during ultramicrotomy prior to EM (Knott, Holtmaat, Trachtenberg, Svoboda, & Welker, 2009). However,

DAB can reduce the contrast intracellularly at the ROI, which is especially bad for SBF-SEM/FIB-SEM imaging where high contrast is required. Finally, the use of vasculature as a fiducial marker to relocate an ROI can be extremely useful when coupled with X-ray imaging using a micro-CT (Karreman et al., 2016). The X-ray images show contrast where blood vessels are; this can be used to target ultramicrotomy with great success. However, vasculature is subject to physiological variability and so cannot be user defined to target the ROI. There are also efforts to produce new genetically encoded tags for CLEM that can be resolved by both LM and EM, which are covered in a previous volume of this book (Hodgson, Nam, Mantell, Achim, & Verkade, 2014). These may soon be reliable enough for LM imaging deep in tissue with 2P microscopy.

To circumvent the problems of existing approaches, a technique was developed that uses the near-infrared laser of a 2P microscope to make fiducial marks in the tissue, termed near-infrared branding (NIRB; Bishop et al., 2011; Bishop, Nikić, Kerschensteiner, & Misgeld, 2014). These branding marks can be used to target relocation by placing them near to the ROI. NIRB marks are formed by increasing laser power to the point where damage of the tissue occurs. Cumulative buildup of light exposure results in microbubbles pushing the tissue apart to form a hole over the area where the laser is scanned. The damage caused is restricted to within a few micrometers of the focal plane, for this reason branding appears to be an MP process. The edges of the branding marks are fluorescent in nature, with an emission spectrum similar to that of the tissue autofluorescence (Bishop et al., 2011). This is presumably due to the creation and accretion of autofluorescent molecules formed during the branding process.

Branding has repeatedly been shown to work in fixed neuronal tissue, as well as kidney and lymphoid tissue (Bishop et al., 2011; Grillo et al., 2013; Maco, Holtmaat, Jorstad, Fua, & Knott, 2014; Mostany et al., 2013). NIRB marks can be customized to fit any ROI through the arbitrary scan settings of most microscopy software. These marks can then be relocated in the embedded tissue during ultramicrotomy, where they appear unstained in the semithin sections in contrast to the surrounding stained tissue. This allows efficient and accurate relocation of the ROI on the block face for 3D EM. Subsequent imaging and reconstruction of the ROI has been demonstrated using both FIB-SEM and ssTEM techniques (Bishop et al., 2014; Grillo et al., 2013; Maco et al., 2014). NIRB is therefore of remarkable value when correlating volumes of light and electron images to efficiently target relocation of an ROI and to provide a fiducial marker between modalities. This chapter will discuss considerations that should be made during the workflow and describe in detail a protocol for achieving efficient relocation of an ROI using 3D NIRB maps (Fig. 1).

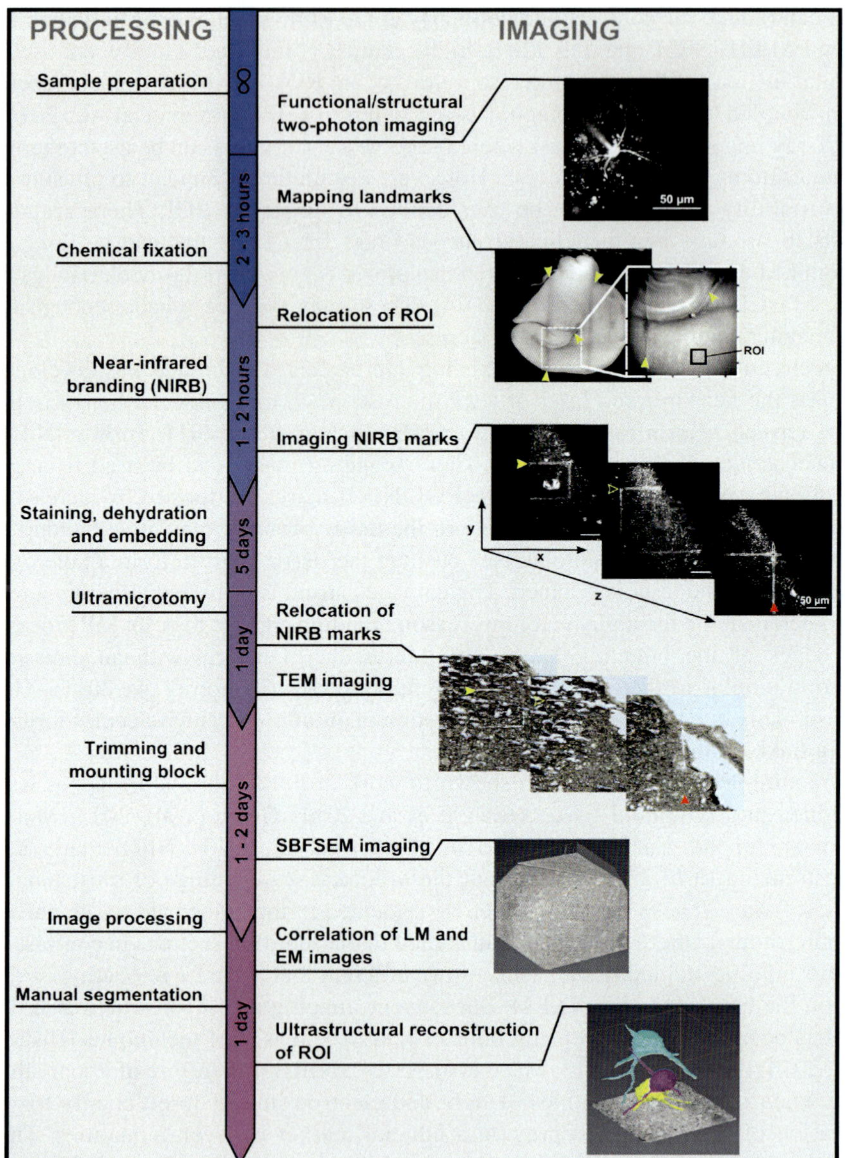

FIGURE 1

Workflow diagram outlining steps required for correlative two-photon and serial block face scanning electron microscopy using near-infrared branding. Steps to the left of the timeline are involved in processing, and to the right of the timeline are imaging-related steps. The total duration of the workflow is ~10 days. *EM*, electron microscopy; *LM*, light microscopy; *NIRB*, near-infrared branding; *ROI*, region of interest; *SBF-SEM*, serial block face scanning electron microscopy; *TEM*, transmission electron microscopy.

1. RATIONALE

2P microscopy allows cell biologists to use light microscopy as a tool to obtain structural and functional dynamics of cells inside their physiological environment. 3D EM techniques can obtain ultrastructural information from the same ROI. However, the relocation of the exact ROI for 3D EM imaging is an unattractive and tedious process. Efficient and accurate relocation can be accomplished by using 3D NIRB maps, which also act as fiducial markers for correlation of the two imaging modalities.

NIRB was first demonstrated by Bishop et al. (2011) and also documented in the previous CLEM edition of this book for use with ssTEM and FIB-SEM in brain tissue (Bishop et al., 2014; Maco et al., 2014). For relocation of a single synapse in a 1×10^9 μm^3 piece of brain tissue, NIRB marks can reliably reduce the search area to 1×10^3 μm^3, a millionfold decrease. The technique does not compromise the preservation or contrast of ultrastructure at the ROI, which is a problem with other approaches. Previously, the use of a single set of NIRB marks in one plane was used to relocate the ROI. Here, the efficiency and accuracy of relocation are increased by creating a 3D NIRB map. The map is formed of multiple fiducial NIRB marks that aid the relocation of an ROI throughout ultramicrotomy. It is therefore possible to target ultramicrotomy efficiently and is especially useful where relocation can take a long time, e.g., larger block faces and ROIs deep in the tissue.

This chapter shows that correlation of 2P microscopy with SBF-SEM imaging can be achieved with relative ease, in comparison with previous techniques, using common equipment present in most EM labs. The reduction in expert skill and time afforded by SBF-SEM should increase the appeal of this CLEM workflow to cell biologists.

2. METHODS

2.1 MULTIPHOTON MICROSCOPE SETUP

The same microscope is used for imaging and subsequent branding. The considerations that need to be made when choosing components for the setup are outlined here:

- The laser must be mode-locked (repetitive, ultrafast pulsing; required for 2P excitation), have a wavelength suitable for sample excitation, and produce enough power to create branding marks at the sample (see *Near-infrared branding*). A single, tunable laser is used here, set at 910 nm for imaging green fluorescent protein (GFP) in the sample and 800 nm for branding. The use of two laser beams, one dedicated to excitation of the sample and one for branding (set at around 800–900 nm), would make the branding process quicker by not having to tune a single laser multiple times.

- The choice of objective lens is important for high-resolution imaging and to achieve well-defined branding marks. An objective with a higher numerical aperture (N.A.), and hence smaller point spread function, allows for tighter confinement of NIRB marks. This is particularly useful for outlining smaller structures (e.g., a single dendritic spine). Additionally, the objective must have a long working distance to image and brand within the tissue at the depth of the ROI (working distance will depend on the depth of the ROI). Water-immersion objectives with an N.A. of 1.0 or more are used here, this gives enough control to make sure the marks can be clearly defined by the user. A lower-magnification objective (e.g., 4×, 0.1 N.A. air) is also required for brightfield imaging to identify landmarks on the tissue.
- Finally, the most important aspect of the imaging setup is the ability to control both the X- and Y-axis movement of the laser to achieve the correct size/shape of NIRB mark. It is preferred to have servomotor-controlled galvanometers (servo mirrors) rather than resonant. Resonant scanning can be used, but segmented linescans are not possible, as one of the mirrors cannot have its position user defined. Rotation of the field of view (FOV) can be used to create shapes with resonant mirrors. This becomes a highly laborious process in comparison with servo mirrors, which can have a user-defined scanning path. The preferred software functionality on top of normal image acquisition is an arbitrary linescan function. However, with some software this may not be possible. Therefore, as with resonant scanning, the rotation of the FOV can be used to create shapes with NIRB marks. The branding process is discussed in more detail later on (see *Near-infrared branding*).

2.2 INITIAL FUNCTIONAL AND/OR STRUCTURAL TWO-PHOTON IMAGING

2.2.1 Principle

Below, three examples of sample preparations from different experiments are used to illustrate the scope of applications for CLEM in brain tissue. The first preparation is for an electrophysiological recording (Fig. 2A). Here, an acute 400 μm thalamocortical brain slice is taken from a neonatal mouse (postnatal day 6/7) and whole-cell patch clamp electrophysiology is used to record from a single neuron while filling it with fluorescent dye from the patch pipette. In this scenario, it is necessary to correlate the ultrastructure of both the cell and the surrounding cells (particularly their synaptic inputs) to the response from the recorded cell. The second preparation is a fixed tissue slice (Fig. 2B), where a 400 μm thalamocortical slice is taken from an adult transgenic mouse expressing tdTomato in layer 4 of the barrel cortex. CLEM is required here to identify cell type—specific effects on synaptic vesicle pool size, synapse size, and number etc. The last sample preparation is for intravital imaging (Fig. 2C), in which an adult mouse is fitted with a cranial window and has fluorescent proteins expressed by viral transduction in the somatosensory cortex. This is the most common

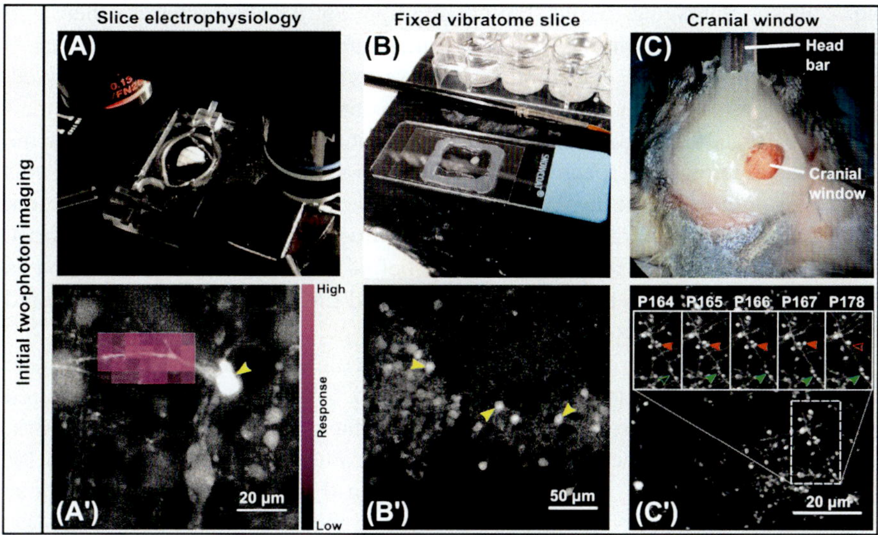

FIGURE 2

Initial functional and/or structural two-photon (2P) imaging. Representative images of sample preparation for 2P imaging of (A) slice electrophysiology, (B) fixed section from a whole brain, and (C) intravital imaging through a cranial window. (A′) Representative 2P image of patched cell (*yellow arrowhead* = soma) filled with Alexa Fluor 594 hydrazide from a glutamate uncaging experiment. Responses from the cell due to discrete glutamate uncaging along the dendrite are shown (*magenta boxes*). (B′) Representative 2P image of transgenic neurons expressing tdTomato (*yellow arrowheads*) in layer IV of the mouse barrel cortex. (C′) Representative 2P image showing axonal structural imaging of cytosolic green fluorescent protein in layer I of the mouse somatosensory cortex. Addition (*green arrowheads*) and loss (*red arrowheads*) of axonal boutons is indicated in repeated longitudinal imaging (inset).

scenario in which NIRB is used for CLEM. Typically, dendritic spines or axonal boutons that have their structure imaged longitudinally (over time) are relocated to assess changes in size and corresponding ultrastructure of both the labeled and surrounding cells. These examples are not limiting, but show a variety of cases in which NIRB can be utilized.

To identify the ROI during subsequent 2P imaging, high-magnification structural images are acquired after functional imaging and directly before fixation, if not already collected. These are also used to correlate to the final SBF-SEM data set. To ensure that the structure of interest can be picked out from the neuropil in SBF-SEM images, it is essential to optimize the sample preparation. The fluorescence labeling should not be homogenous or dense and the structure should be clearly identified from morphological features (e.g., crossing processes, branches, vicinity to other cell bodies or the vasculature). This is especially relevant for ubiquitous structures, e.g., axons, but not so important for cell bodies, as there will not be many present within the SBF-SEM imaging volume.

High-magnification 2P structural images are acquired before the tissue is fixed, as the fluorescence signal will be maximal. This is because fixation with paraformaldehyde and glutaraldehyde will introduce autofluorescence and quenching of the endogenous fluorescence, decreasing signal to noise. It is key to consider this when choosing fluorescent tags/dyes as they will be affected differently by fixation. GFP, tdTomato, and Alexa Fluor 594 maintain fluorescence well in this protocol. Where it is not possible to acquire images before fixation, or there is a need to have high temporal correlation between the LM and EM, images should be acquired as soon after fixation as possible.

Low-magnification brightfield images are collected to record the tissue topography. If the tissue is larger than the FOV, a set of images can be acquired and stitched together to create a mosaic image of the tissue. An obvious landmark, such as a ruffle/tear in the tissue or vasculature pattern, is used as the origin for coordinating the position of the ROI for relocation (Fig. 4). If there is no obvious landmark, one can be introduced through cutting of the tissue on the corner or edge.

To prevent further structural changes to the tissue, it is fixed as soon as possible after the final LM images are acquired. Fixation of whole tissue is a relatively slow process, therefore it is important to consider carefully the osmolarity and pH of any solutions being perfused to avoid compromising the preservation of ultrastructure. To avoid any potential problems, fresh fixative with the correct pH is made before use. It should be noted that $CaCl_2$ is added in this protocol to help preserve lipid membranes.

For intravital imaging, vasculature of the brain surface is used as a fiducial marker when relocating ROIs. However, cardiac perfusion removes the contrast created by red blood cells and therefore a blood vessel marker must be used. Here, DiI (D-282, Invitrogen), a lipophilic fluorescent dye that is also visible using brightfield illumination, is used during perfusion to bind to the endothelial cell membrane of the vasculature (Fig. 4E).

Regardless of fixation technique, the tissue is eventually sectioned to <400 μm thick and ideally as thin as 100 μm, as long as this still encompasses the ROI. This is because thicker tissue sections are more likely to ruffle during staining and embedding steps, decreasing the chance of successful flat embedding.

2.2.2 Materials

2.2.2.1 Equipment

1. MP microscope (e.g., Bruker Ultima Intravital, Leica TCS SP8 MP, Scientifica VivoScope etc.)
   a. Ti:sapphire laser (e.g., Spectra-Physics Mai Tai DeepSee; Newport Spectra-Physics, UK)
   b. Low-magnification objective (e.g., 4×, 0.1 N.A.)
   c. High-magnification objective (1.0 + N.A.)

2. Fume hood (for fixation)
3. Vibrating tissue slicer (e.g., Leica VT 1200; Leica Microsystems)
4. Double-edged razor blades (for vibrating tissue slicer)
5. Fine paintbrushes
6. pH meter (to pH perfusates)
7. Glass microscope slides
8. Silicone grease
9. Plastic syringe (10 mL) w/snipped 200 μL pipette tip attached (for silicone grease)
10. Coverslips (#1.0, rectangular, 22 × 40 mm)
11. Plastic Pasteur pipettes
12. Perfusion/dissection apparatus (e.g., Fine Science Tools)
    a. Small hemostats
    b. Large blunt scissors
    c. Medium sharp scissors
    d. Blunt forceps
    e. Small scoop
    f. Butterfly cannula w/tubing
    g. 10 mL syringes (for PBS and OPTIONAL DiI perfusate)
    h. 50 mL syringe (for fixative)
    i. 2/3-way tap
13. Small, precision pen-grip drill w/drill bit (e.g., dental drill; OPTIONAL—for drilling implant from cranial window imaging)

2.2.2.2 Reagents

1. PBS (0.01 M; 137 mM sodium chloride, 2.7 mM potassium chloride, and 10 mM phosphate buffer, pH 7.4)
2. Glutaraldehyde (25% EM-grade)
3. Paraformaldehyde (16% EM-grade)
4. Sodium cacodylate
5. Calcium chloride (C3306, Sigma–Aldrich)
6. ddH_2O
7. DiI (D-282, Invitrogen; OPTIONAL—for labeling vasculature)
    a. D-Glucose (OPTIONAL—for DiI diluent)
    b. Ethanol (OPTIONAL—for DiI stock)

2.2.3 Protocol

2.2.3.1 Sample preparation

1. Sample preparation for imaging is followed according to previously published methods for slice electrophysiology or in vivo imaging (Edwards, Konnerth, Sakmann, & Takahashi, 1989; Holtmaat et al., 2009), the discussion of which is beyond the scope of this chapter. All animal experiments were performed according to the UK Animal (Scientific Procedures) Act, 1986.

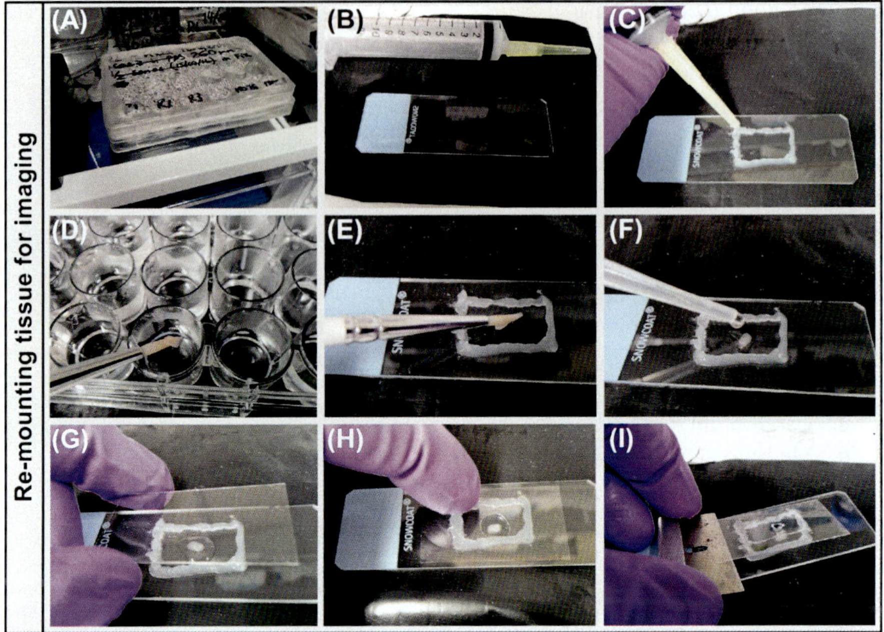

FIGURE 3

Remounting tissue after fixation. Tissue is postfixed at 4°C for at least 2 h (A). A rectangular silicone grease well is created on a microscope slide (C) using a syringe attached to a pipette tip (B). A fine paintbrush is used to handle the tissue after fixation (D) and place it in the well (E). The tissue is immersed in a small amount of buffer (PBS or cacodylate) (F) and a coverslip is placed on top (G), gentle pressure is applied on each side to spread the grease evenly (H). After imaging, the coverslip is removed using a razor blade to lift it away (I).

a. OPTIONAL (tissue not requiring live imaging): The relevant fixation steps are followed (steps **5–8**). The tissue section is washed in 0.01 M PBS and then mounted for imaging in a few drops of 0.01 M PBS inside a silicone grease well (Fig. 3).

2.2.3.2 Initial 2P imaging

2. OPTIONAL (for live tissue): Carry out functional 2P imaging (Fig. 2).
3. Low-magnification brightfield images are acquired to record the correct orientation of the tissue, identify landmarks, and indicate the relative position of the ROI (4× objective w/0.1 N.A.; Fig. 4).
4. A Z-stack of 2P images is acquired from the tissue surface to the ROI (910 nm, 1024 × 1024 pixels, 60× objective w/1.0 N.A., 1 μm steps). Afterward, another stack is acquired encompassing only the structure of interest using a higher optical zoom.

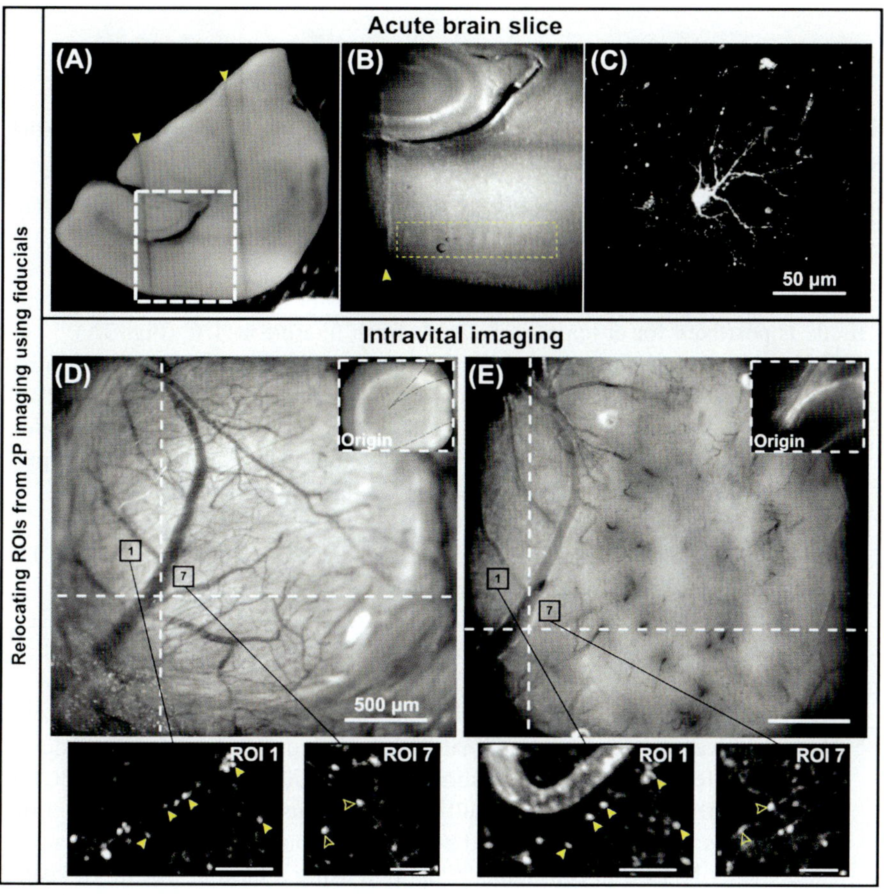

FIGURE 4

Relocation of region of interests (ROIs) after fixation using landmarks. (A) Representative low-magnification brightfield image of a fixed thalamocortical slice from a juvenile mouse (*yellow arrowheads* = anchor harp marks). (B) Brightfield image of the region outlined in A (*dashed white square*). Harp marks (*yellow arrowhead*), the hippocampus (top-left), and the barrels of the barrel cortex (*yellow, dashed rectangle*) are used to relocate a cell that was previously patched and filled during an uncaging experiment. (C) Two-photon (2P) maximum z-projection of a cell filled with Alexa 594 hydrazide. (D) Representative brightfield image of the vasculature pattern seen through a cranial window positioned over the somatosensory cortex during intravital imaging. The intersection of the *dashed lines* indicates the origin (inset; *gray dashed lines* outline vasculature) that is used to record the relative positions of ROIs during longitudinal imaging. Representative 2P images of axonal structure at ROIs 1 and 7 are shown (bottom), achieved through viral expression of cytosolic green fluorescent protein. (E) Brightfield image of a fixed tangential slice from under the cranial window in D (Note: this is the second in a series of slices, hence the vasculature has been sliced through on the right side; scale bar = 500 μm). The origin is clearly visible (inset; epifluorescence image of DiI-labeled vasculature). Representative multiphoton images of axonal structure at ROIs 1 and 7 after fixation are shown (bottom; *yellow arrowheads* = identification of same boutons before and after; scale bars = 10 μm). Note: the focal plane is not at exactly the same angle.

2.2.3.3 Fixation

CARE: Fixatives should be used in a fume hood to avoid inhalation and personal protective equipment used.

5. OPTIONAL (for acute slices): Immediately after imaging, a 400 μm brain slice is briefly washed three times in 2 mL of cold, fresh fixative (2.5% glutaraldehyde, 2% paraformaldehyde, 2 mM $CaCl_2$, 0.15 M sodium cacodylate in ddH_20, pH 7.4) using a fine paintbrush to transfer the tissue. Afterward, the tissue is postfixed for at least 2 h in 2 mL fresh fixative at 4°C (Fig. 3A).
6. OPTIONAL (for whole animal): An adult mouse is put under deep, terminal anesthesia. Cardiac perfusion is then performed first with 2–3 mL 0.01 M PBS at 5 mL/min, followed by 20–30 mL fresh, cold fixative (2.5% glutaraldehyde, 2% paraformaldehyde, 2 mM $CaCl_2$, 0.15 M sodium cacodylate in ddH_20, pH 7.4) at 5–10 mL/min. Note: Filter solutions with a 0.22 μm filter prior to perfusion. Tissue is harvested and postfixed in 5 mL fresh fixative at 4°C.
   a. OPTIONAL (to visualize vasculature): After exsanguination, 10 mL DiI (120 μg/mL) is perfused after PBS and prior to the fixative.
      i. To make DiI working solution, 200 μL of 6 mg/mL DiI in 100% EtOH is diluted in 10 mL of a 1:4 mix of 0.01 M PBS and 5% glucose (wt/vol in dH_2O) according to the protocol from Li et al. (2008).
   b. OPTIONAL (to obtain tissue under cranial window): After perfusion and before harvesting tissue, the animal is decapitated and the cranial window implant is left on during postfixation (2 h in 10 mL fresh fixative at 4°C). Subsequently, the window and skull are removed using a precision hand drill and forceps to expose the imaged brain region, but leaving the head bar intact. 0.01 M PBS is continuously applied using a Pasteur pipette to keep the area from drying. The head is then mounted in the imaging head-fix device (in-house; not shown) and a 100–400 μm slice is cut using a vibratome at the same angle as 2P imaging in 0.01 M PBS. The slice is then postfixed for a further 2 h in 2 mL fresh fixative at 4°C.
7. After postfixation, a 100–400 μm tissue section is cut from the appropriate region of the brain in 0.01 M PBS using a vibratome. Tissue sections are then left in fixative until required for branding (<24 h).

2.3 NEAR-INFRARED BRANDING

2.3.1 Principle

Branding is carried out soon after fixation (within 24 h) to avoid further changes to the ultrastructure of the tissue and quenching of fluorescence. The tissue section is washed in buffer before mounting to reduce some of the fluorescence quenching that is caused by residual fixatives. During mounting, light pressure is applied around the coverslip to spread the silicone grease and trap the tissue to prevent it from drifting during imaging (Fig. 3). The grease can spread out further during the imaging session; therefore care is taken to ensure the coverslip remains flat throughout.

The aim of branding is to make relocation of the ROI efficient and accurate during ultramicrotomy of the embedded tissue. This is achieved by creating a 3D map of NIRB marks targeting the ROI (Fig. 5A). The Ti:sapphire laser is used to create NIRB marks in the focal plane. As mentioned previously, using a higher N.A. objective lens produces more tightly confined NIRB marks in all three dimensions (greater than 1.0 N.A. is recommended).

Across different samples, the same parameters for NIRB may produce differing results, therefore branding is tested in a discrete (nonprecious) area of the tissue first. The accumulated effect of light exposure over time appears to cause branding of the tissue; a trade-off between the number of lines scanned, the pixel dwell time, and the laser power is needed to control the spread of damage. To increase the thickness of an NIRB mark, one of these parameters is increased, keeping all others the same (Fig. 5B). It is easiest to alter the number of lines scanned as this gives the greatest dynamic range. Settings are adjusted to achieve an NIRB mark with dimensions of 2–4 μm in the X/Y-axis and 5–15 μm in the Z-axis, measured by the distance between the centers of the fluorescent edges in 2P images. The fluorescent edges are not stained during heavy metal staining later in the protocol; therefore, the thickness of the line is made greater than 2 μm to make identification easier in semithin sections.

Sometimes, the NIRB mark is incomplete, not creating a fluorescent mark all the way along the defined linescan (Fig. 5C, C′, G and G′). This may be due to the heterogeneity of the tissue and certain parts being more susceptible to branding. As a rule of thumb, if the mark is not visibly fluorescent then the physical branding was unsuccessful and should be repeated. Alternatively, branding with high settings can spread damage much further than intended. Small incremental changes to settings are used to prevent this. The creation of large bubbles (10 μm+) that escape to the surface of the tissue sometimes occurs, potentially due to a weakness in the tissue and escaping microbubbles from the NIRB process. These obscure the ability to image the area below the bubble; however, they normally clear with time (Fig. 5E and F).

Note that when changing optical zoom, the pixel dwell time is manually adjusted (if it is not automatically), to account for the change in pixel size. For example, repeating a linescan that covers the same physical distance at 1× zoom and 3× zoom produces a much greater branding effect at 3× zoom if all other parameters are constant, due to the increase in pixels that make up the line. For this reason it is best to start with the branding mark closest to the ROI (at a high optical zoom), after testing in a nonprecious area first.

The NIRB mark closest to the ROI acts as a fiducial for correlation, therefore it is vital to be able to capture it in the SBF-SEM volume. The distance between the closest mark and the ROI is <10 μm in the Z-axis to avoid unnecessary imaging with SBF-SEM. This mark is also tight around the ROI in the X/Y-axis to include it in the SBF-SEM imaging FOV. Here, the final FOV is 25.7 μm and so the NIRB mark closest to the ROI is no larger than this. With a higher N.A. objective, the distance between the NIRB marks can be reduced to as little as 4 μm without damaging the

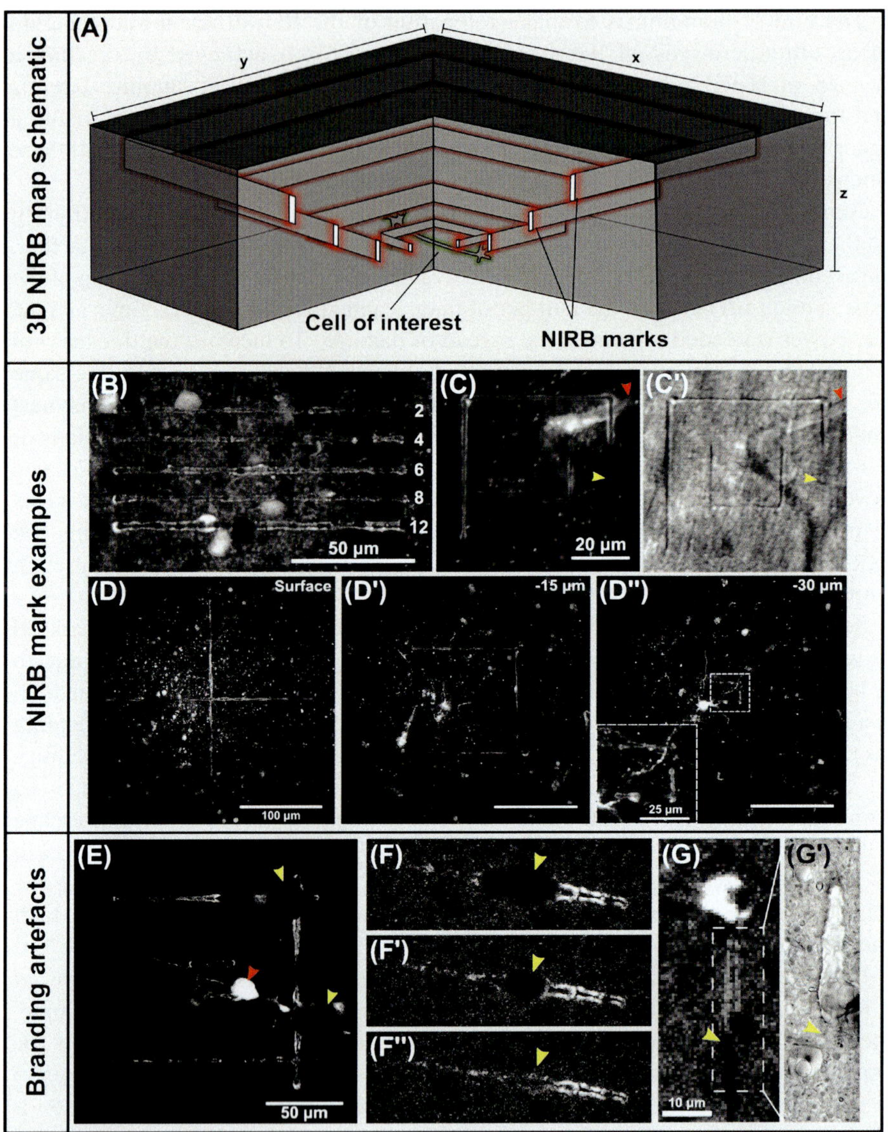

FIGURE 5

Creation of fiducial marks using a Ti:sapphire laser. (A) Schematic of a tissue cross section with a 3D NIRB map to the cell of interest; NIRB marks are made at regular intervals in the Z-axis from the surface to the ROI in a concentric pattern. (B) Representative 2P fluorescence image of NIRB marks showing successively thicker marks with increasing pixel dwell time (numbers listed are in μs/pixel). (C) 2P image of two concentric, asymmetrical NIRB marks in the same Z-plane. (C′) Brightfield image of the same NIRB marks from C (*red arrowhead* = DiI-labeled vasculature; *yellow arrowhead* = incomplete NIRB mark). (D, D′, D″) 2P images of three branding marks created at different levels in the tissue, outlining structure within the ROI (see Bishop et al., 2011). If NIRB marks are placed too close to the ROI, damage may occur. This is because the thickness of a single mark can vary considerably, due to the heterogeneity of the tissue (Fig. 5E and F).

Asymmetry of the branding mark is important to be able to accurately know the orientation of the ROI when correlating the SBF-SEM data set. A segmented linescan function is most suitable for creating branding marks such as this, because multiple straight NIRB marks can be created in a single linescan. In cases where a segmented linescan is not possible, a single straight linescan is used and repeated with rotation of the FOV to create an asymmetric shape.

The 3D map of NIRB marks is used as a reference during sectioning to keep track of how close the current block face is to the depth of the ROI (Fig. 7H). Each mark is easily differentiated by size or shape to avoid confusion during sectioning. A good guideline is to have NIRB marks every 10–15 μm in the Z-axis, so that one is always visible in each semithin section during ultramicrotomy. The use of a cross-shaped NIRB on the tissue surface (Fig. 5D) helps to relocate the ROI during sectioning if flat embedding is unsuccessful. The center of the cross will usually appear in one section, even if the tissue is embedded at an angle; this helps to identify where the center of the ROI is in the embedded block. Note that just as with MP excitation, branding requires higher power at increasing depths to combat the effects of increased light refraction.

A high-resolution map of the branding marks made from 2P Z-stack images is important to obtain the relative positions and sizes of the marks, which are referred to while sectioning the embedded tissue. To image the NIRB marks, 910 nm is used (Fig. 5C), as GFP is also excited at this wavelength. However, increasing to 1040 nm produced a stronger signal from the NIRB marks (not shown). The marks can also be seen with brightfield illumination (Fig. 5C′), which is useful to aid microdissection of large pieces of tissue (>5 mm × 5 mm).

2.3.2 Materials

2.3.2.1 Equipment

1. Power sensor (e.g., Thorlabs S120C photodiode power sensor, Thorlabs)
2. MP microscope
   a. Ti:sapphire laser (e.g., Spectra-Physics Mai Tai DeepSee; Newport Spectra-Physics, UK)

a dendrite of interest (inset of D″; zoom of *dashed box*). (E) 2P images of two asymmetrical, concentric branding marks around an axon from a dye-filled neuron (*red arrowhead* = soma), two large bubbles (*yellow arrowheads*) are visible. (F, F′, F″) 2P images showing a large bubble (*yellow arrowhead*) shrinking over time. (G) A representative 2P fluorescence image of an incomplete NIRB mark with correlated SBF-SEM data (G′) showing the lack of damage to the ultrastructure in the photobleached area. *2P*, two-photon; *NIRB*, near-infrared branding; *ROI*, region of interest; *SBF-SEM*, serial block face scanning electron microscopy.

   b. Low-magnification objective (4×, 0.1 NA. air)
   c. High-magnification objective (1.0 N.A.+)
3. Glass microscope slides
4. Coverslips (#1.0, rectangular, 22 × 40 mm)
5. Silicone grease
6. 10 mL syringe w/snipped 200 μL pipette tip attached (for silicone grease)
7. Plastic Pasteur pipettes

2.3.2.2 Reagents

1. PBS (0.01 M; 137 mM sodium chloride, 2.7 mM potassium chloride, and 10 mM phosphate buffer, pH 7.4)

2.3.3 Protocol

2.3.3.1 Remounting tissue for imaging

1. The tissue section is washed and remounted in 0.01 M PBS inside a silicone grease chamber made on a glass microscope slide and sealed with a coverslip (Fig. 3). Low-magnification brightfield reference images of the tissue are used to orient the slice correctly (correct surface facing upward; Fig. 4A and E).
2. The ROI is then relocated relative to the previously chosen fiducial marks and centerd with 2P microscopy using the high-resolution structural stack as a reference (Fig. 4C and E).
3. High-magnification 2P Z-stack images are taken of the structure of interest at this stage, before NIRB (910 nm, 1024 × 1024 pixels, 60× objective w/1.0 N.A., 1 μm steps).

2.3.3.2 Near-infrared branding

4. The segmented linescan function is selected (if possible) and an asymmetrical shape of 20 μm along the longest edge is made in the tissue at 5 μm superficial to the structure of interest (30 μm depth; Fig. 5D″), using settings previously tested in an area of the tissue away from the ROI [800 nm, 150 mW (average power as read from power meter at the sample) and 20 μs pixel dwell time]. NOTE: Some drifting may occur during imaging, be sure to recenter on the ROI before branding each mark. CAUTION: High laser power may damage the PMTs, turn them off before commencing the linescan.
5. A second, asymmetrical branding mark of 60 μm along the longest edge is created at the same depth (Fig. 5C).
6. Subsequent branding marks are made every 10–15 μm in the Z-axis from the ROI to the surface, with each mark having a progressively longer side length (Fig. 5A and D′). The final branding mark is in the shape of a cross, <2 μm below the tissue surface (Fig. 5D). NOTE: higher power is needed when branding deeper in tissue.

7. A high-resolution Z-stack is acquired (910 nm, 1024 × 1024 pixels, 60× objective w/1.0 N.A., 1 μm steps) of the 3D NIRB map with 2P microscopy. Higher optical zoom is used to acquire Z-stack images around the structure of interest, encompassing the closest NIRB mark.

2.4 SBF-SEM SAMPLE PREPARATION AND IMAGING

2.4.1 Principle

Sample preparation steps for thick tissue involve lengthy, repeated rounds of en bloc heavy metal staining to produce strong contrast and tissue conductivity for SEM imaging. This staining is sufficient to identify postsynaptic densities, synaptic vesicles, and mitochondria. Some en bloc staining steps require extra care. Uranyl acetate may precipitate in the tissue if the tissue is not washed thoroughly with ddH_2O to remove all cacodylate ions from the buffer. Further, for lead aspartate staining, the lead is removed from the solution if a white precipitate forms prior to incubation. In this case a new lead aspartate solution should be made up. Thoroughly cleaning glassware may help when making up lead aspartate solution, to prevent small impurities causing precipitation.

Resin choice is very important, because certain resins are more susceptible to charging and beam damage during SEM imaging, which effects the stability of ultrathin sectioning (Fig. 8H). Here, Durcupan ACM resin is used, as recommended by Deerinck, Bushong, Thor, and Ellisman (2010).

Care should be taken to ensure the specimen remains flat during embedding. Thicker tissue sections (400 μm) will ruffle, possibly due to greater strain on the tissue from staining and dehydration steps. This results in the plane of sectioning not matching the imaging plane. If this is the case, relocating the branding marks in the embedded tissue is difficult as they do not appear flat in the semithin sections. Therefore, it is recommended to keep slices as thin as possible before starting this part of the protocol.

After embedding and polymerization, the resin block is trimmed and semithin sections collected and imaged to relocate the NIRB marks (Fig. 7A–G). It is common for the block to be extremely brittle at this stage. Here, a cutting speed of 2.0 mm/s and section thickness of 1 μm for semithin sectioning were used. However, if major artifacts are seen in the semithin sections (tears, holes, etc.), it is recommended to reduce cutting speed to 1.5 mm/s and section thickness to 500 nm. Trimming the block face to become smaller and smoothing the sides of the pyramid also increase cutting success (Fig. 7F).

It is vital to inspect every semithin section at this stage to avoid cutting through the NIRB marks. The branding marks appear as holes in the tissue in semithin sections with no obvious bordering effect (as seen in 2P). Staining of semithin sections is required when contrast between the tissue and resin is too low to identify NIRB marks under brightfield illumination. Omitting the staining step is beneficial because it eliminates loss of sections during washing and reduces the time required to prepare each semithin section. If flat embedding of the tissue is successful, the marks are easily identified by their shape, but each mark only propagates through 5–10 μm of semithin sections (depending on size of NIRB marks). If unsuccessful, or sectioning is done at an incorrect angle (relative to the branding plane), then small parts of the NIRB marks appear in multiple semithin sections (10–20 μm), making identification harder.

On identification of an NIRB mark in semithin sections, 2P images are referred to, to calculate how far the block face is from the ROI (Fig. 7H). Once the angle and depth of the block face are calculated relative to the ROI, sectioning is sped up to reach the final NIRB mark. The tissue is checked using TEM with a small sample of ultrathin sections to confirm preservation of ultrastructure.

For SBF-SEM, the trimmed pyramid containing the ROI is detached from the main bulk of the resin. To prevent loss of the pyramid, it is covered with a small piece of parafilm, prior to pulling the razor blade through the base of the pyramid. The block is sputter-coated with 2 nm platinum to improve conductivity. After insertion into the SBF-SEM microtome, the diamond knife is aligned, and the surface polished using 100 nm cuts, taking care not to cut into the ROI. The door is closed and pumped to a pressure of either ~5–10 Pa (in a variable pressure SEM) or 10^{-3} Pa (in a high-resolution SEM). See Russell et al. (2017) for an in-depth description of this process.

Once at vacuum, the imaging and cutting conditions for the SBF-SEM run are set according to the structures to be resolved (Peddie & Collinson, 2014). The FOV, the size of the detector, and the resolution are inextricably linked. A pixel resolution of <5 nm is needed to resolve synaptic structures, which limits the FOV to 25–30 μm when using a detector with 8192 × 8192 pixels. Axial resolution is determined by the section thickness and the accelerating voltage that controls the interaction volume of the electron beam with the sample, and thus the depth from which the backscattered electrons are detected. 50 nm section thickness is routine, 25 nm section thickness can be achieved on most well-prepared samples (in our experience), and 10 nm section thickness is possible with ideal specimens and environmental conditions (Russell et al., 2017). Here, a 25.7 μm × 25.7 μm × 25 μm (XYZ) volume is used to encompass the ROI with a voxel size of 3.1 nm × 3.1 nm × 50 nm (XYZ). With this resolution it is possible to identify synaptic vesicles, mitochondria, and postsynaptic densities (Fig. 9F).

Individual .dm4 images are large, in the region of 250 MB, and a serial imaging run of 500–1000 images can easily reach 250–500 GB. The image stack is batch converted to a .tiff stack in Digital Micrography (Gatan Inc) or in ImageJ (FIJI package; Schindelin et al., 2012) using the BioFormats importer (Linkert et al., 2010) for easier handling. Large .tiff stacks are handled in FIJI and other software using a virtual stack to reduce RAM load. Contrast variation is reduced across the volume by equalizing the histogram and smoothing the images with a Gaussian filter to increase contrast (Fig. 8I).

Correlation of the SBF-SEM data set to the 2P images is not straightforward. The NIRB mark present in the EM data is used to align the rotation of the 2P images. The high-resolution 2P images of the ROI are used to measure the rough distance between the plane of NIRB and the plane of the structure of interest (Fig. 9A–C). Finally, the structure of interest is identified based on morphology (e.g., crossing processes, branching, vasculature, cell bodies, etc.).

On identification of the structure of interest, manual segmentation can be carried out using Amira (FEI) or TrakEM2 (ImageJ plugin; Cardona et al., 2012) to make a 3D model. Both have guides available online or within the software itself to cover the initial processing and segmentation followed by reconstruction, display, and measurements. For Amira, the user should refer to the software guide (https://www.fei.com/software/amira-user-guide/), alternatively the FEI YouTube channel is populated by some video guides (https://www.youtube.com/channel/UC33gA9Z-FtCccsj29Z-c2gw). For TrakEM2, the user should refer to (http://imagej.net/TrakEM2_tutorials).

2.4.2 Materials

2.4.2.1 Equipment

1. Fume hood
2. 60°C oven
3. Fine paintbrushes
4. Glass vials (7–10 mL; e.g., G100, TAAB rolled rim vials)
5. Plastic Pasteur pipettes
6. Tissue rotator (e.g., R050, TAAB rotator, 2 rpm—R050)
7. Measuring cylinders/tubes (for making solutions)
8. 10+ mL syringes and syringe filters (0.22 μm)
9. Molecular sieves (3 Å—for anhydrous ethanol)
10. Ultramicrotome (e.g., Leica EM UC7 or RMC Powertome)
11. Cocktail sticks
12. Glass microscope slides
13. ACLAR sheets (50 and 200 μm thickness; e.g., AGL4458, Agar Scientific)
14. Metal block/weight (~500 g)
15. Razor blades (single and double edged)
16. Glass (for glass knives, e.g., AGG336, Agar Scientific)
17. Glass knife maker (e.g., LKB 7800 KnifeMaker)
18. Specimen floater (for collecting semithin sections)
19. Brightfield microscope (e.g., Leica DM1000 LED)
20. Copper grids (e.g., AGG2500C, Agar Scientific)
21. TEM (e.g., FEI Tecnai T12 120 kV; JEOL JEM-1400 120 kV)
22. Parafilm
23. Aluminum pins for SBF-SEM specimen mounting (10–006002–50, EM Resolutions, UK)
24. Sputter coater
25. SEM (e.g., Zeiss Sigma VP FESEM)
26. 3View system (Gatan)
27. Amira 6.0.0 software (FEI)
28. FIJI ImageJ package (2.0.0-rc-43/1.51 g)

2.4.2.2 Reagents

1. Sodium cacodylate
2. Calcium chloride (C3306, Sigma–Aldrich)
3. ddH_2O
4. Osmium tetroxide
5. Potassium ferrocyanide
6. Thiocarbohydrazide (TCH)
7. Uranyl acetate
8. L-Aspartic acid (A9256, Sigma–Aldrich)
9. Lead nitrate
10. Potassium hydroxide
11. Ethanol
12. Propylene oxide
13. Durcupan ACM epoxy resin kit
14. Cyanoacrylate glue
15. Conductive epoxy glue for 3View sample mounting (CW2400, Farnell element14, UK)
16. Toluidine blue
17. Borax

2.4.3 Protocol

CARE: Staining and embedding steps are carried out in a fume hood using personal protective equipment (gloves, goggles, lab coat), the majority of the chemicals used are harmful or toxic.

NOTE: All incubation steps at room temperature (RT) or 4°C are done in a tissue rotator (2 rpm; Fig. 6C). All steps are carried out in the same glass vial, transferring chemicals using a plastic Pasteur pipette, unless otherwise noted. A new pipette is used for each chemical. Chemicals are not applied directly on to the tissue and the tissue is not manipulated directly at any stage, as it becomes extremely brittle after osmication.

2.4.3.1 Staining, dehydration, and flat embedding

1. After fixation, the tissue is transferred to a small glass vial (Fig. 6A) using a fine paintbrush and washed 5 × 3 min in cold 0.1 M cacodylate buffer containing 2 mM $CaCl_2$.
2. The tissue is incubated in 2 mL reduced osmium (equal parts 3% potassium ferrocyanide/0.3 M cacodylate/4 mM $CaCl_2$ and 4% aqueous osmium tetroxide) for 1 h at 4°C.
   a. During the osmium incubation, TCH solution is made. 0.1 g of TCH is added to 10 mL ddH_2O and dissolved at 60°C for 1 h. The solution is agitated gently by hand to aid dissolving. Afterward, it is filtered using a 0.22 μm syringe filter (Millipore) before use.
3. Wash 5 × 3 min w/ddH_2O at RT.
4. Filtered TCH is added and the tissue incubated for 20 min at RT.
5. Wash 5 × 3 min w/ddH_2O at RT.

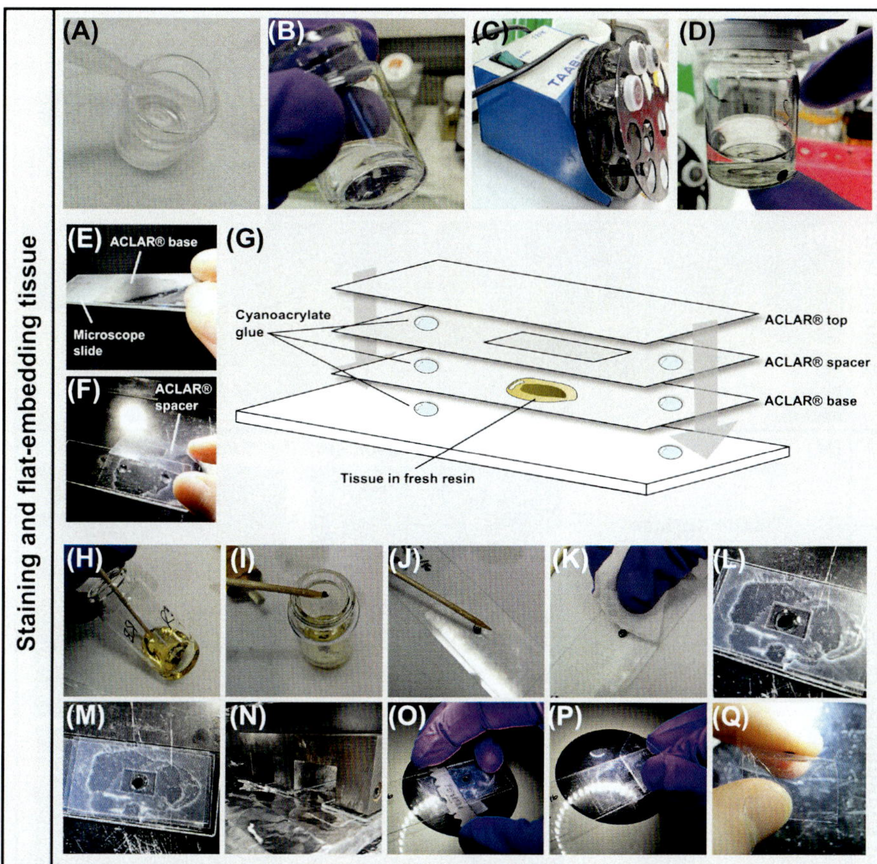

FIGURE 6

"Megametal"staining and flat-embedding fixed tissue. (A, B, D) Staining is done inside glass vials using a tissue rotator for each incubation step to provide gentle agitation (C; 2 rpm). A plastic Pasteur pipette is used for transferring chemicals in and out of the vial during incubations and washes, making sure not to touch the tissue (A). (E, F, G) A flat-embedding chamber is created from ACLAR sheets (thickness and number of spacers depends on thickness of tissue). After the final resin incubation step the tissue is gently floated to the surface using a wooden cocktail stick, taking care not to touch the tissue (H) until it can be lifted out (I). The tissue is placed in the embedding chamber (J) and excess resin is removed using tissue paper (K). A small amount of fresh resin is applied (L) and the top ACLAR sheet is stuck down using cyanoacrylate glue (M) and weighed down with a heavy metal block (N). After polymerization the ACLAR sheets are removed using a thin razor blade (O) and gentle peeling (P), finally the base ACLAR sheet is peeled away from the tissue with care (Q).

6. 2% osmium tetroxide (in ddH_2O) is then added and the tissue is incubated for 30 min at RT.
7. Wash 5 × 3 min w/ddH_2O at RT.
8. The tissue is incubated in 1% uranyl acetate (aqueous in ddH_2O; 0.22 μm filter before use) overnight (16 h) at 4°C.

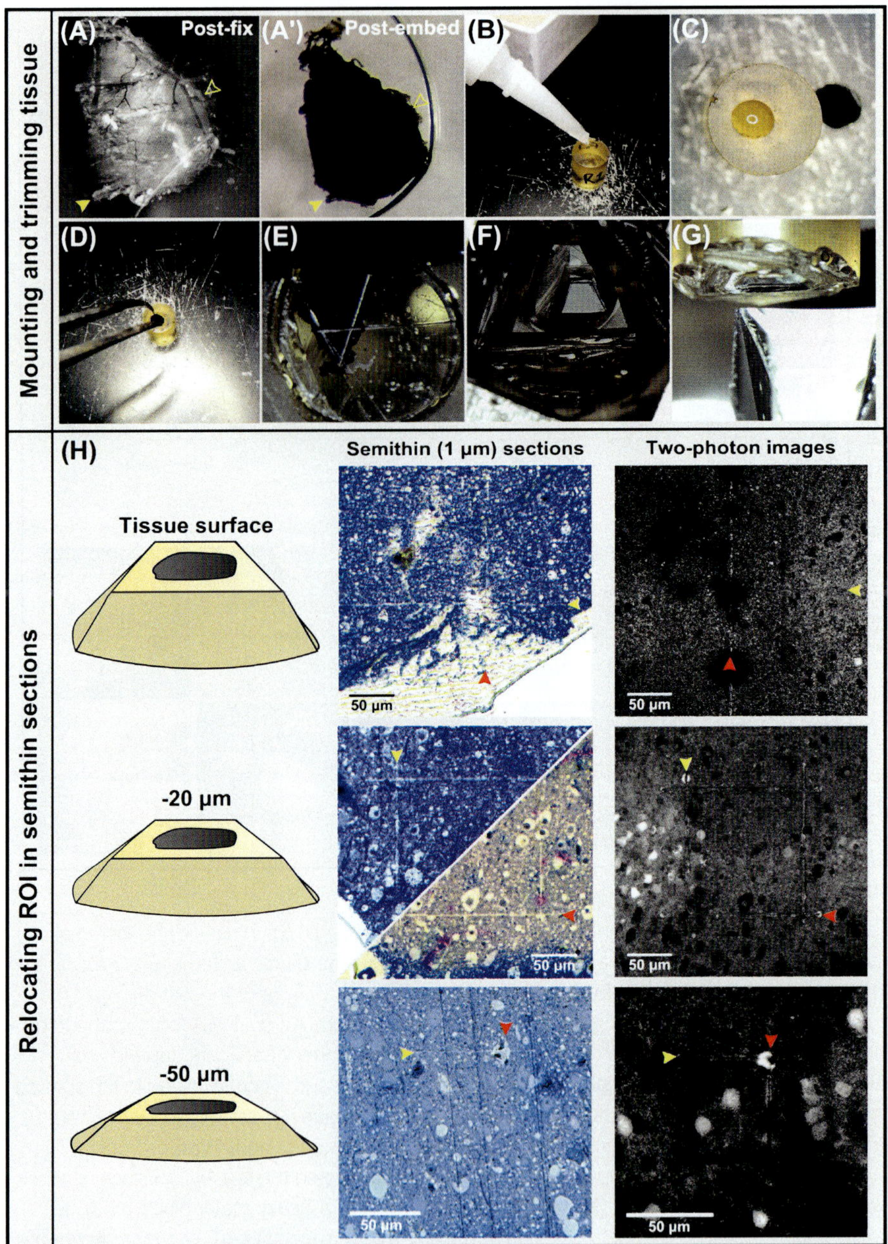

FIGURE 7

Relocating the ROI in resin-embedded tissue by correlating semithin sections and 2P images. (A) Brightfield image of a tissue slice with landmarks indicated (*yellow arrowheads*). After resin embedding, the tissue is oriented to match (A′). A thin layer of cyanoacrylate glue (B) is used to stick the flat-embedded tissue to a blank resin stub (C and D). The block face is trimmed to a trapezoid-faced pyramid that is focused roughly on the area containing the ROI

9. Wash 5 × 3 min w/ddH_2O at RT.
10. After the uranyl acetate is washed out, the tissue is incubated in Walton's lead aspartate for 30 min at 60°C.
    a. To make Walton's lead aspartate, 0.066 g of lead nitrate is added to 9 mL 0.03 M aspartic acid stock and then pH is adjusted to 5.5 using 1 M potassium hydroxide, then incubated at 60°C for 30 min. NOTE: No precipitate should form.
11. Wash 5 × 3 min w/ddH_2O at RT.
12. The tissue is then dehydrated through serial 20%, 50%, 70%, 90% (in ddH_2O), 100% EtOH, and 100% EtOH (anhydrous) incubations for 5 min each and finally fresh 100% EtOH (anhydrous) for 10 min.
13. Tissue is then incubated in propylene oxide for 10 min while Durcupan ACM resin is made. NOTE: Use a glass Pasteur pipette for propylene oxide, as plastic is corroded by it.
    a. To make Durcupan ACM resin, 11.4 g part A (epoxy resin), 10 g part B (hardener), 0.3 g part C (accelerator), 0.05–0.1 g part D (dibutyl phthalate) are added sequentially on top of each other, mixing well after each part is added.
14. Tissue is transferred through serial incubations of 25%, 50%, and 75% Durcupan in propylene oxide for 2 h each.
15. Finally, the tissue is incubated in 100% Durcupan overnight and changed to fresh 100% resin in the morning for 2 h.
16. The tissue is then carefully removed from the resin using a cocktail stick (Fig. 6H and I) and transferred to a flat-embedding chamber (Fig. 6J), excess resin is removed (Fig. 6K) and a drop of fresh resin added (Fig. 6L). The resin chamber is sealed with cyanoacrylate glue (Fig. 6M) and placed in a 60°C oven to polymerize for 72 h. A relatively heavy metal block (Fig. 6N) is placed on top to keep the specimen flat.

2.4.3.2 Relocating the region of interest in embedded tissue

17. After polymerization of resin, the flat-embedding chamber is retrieved from the oven and the layers of ACLAR are removed using a thin razor blade (Fig. 6O). Each ACLAR layer is peeled away from the resin (Fig. 6P and Q) and the tissue is imaged using a brightfield microscope to determine the correct face to begin sectioning from (Fig. 7A′).

(E, F). Semithin sections are cut from the block face using a glass knife (G) and stained with toluidine blue. (H) The 2P Z-stack images of the branding marks are used as a reference for correlating brightfield images of the stained semithin sections (*arrowheads* = indicating edges of NIRB marks). The cross is just visible on the tissue surface (top row). The second branding mark is visible in multiple section (two shown; middle row). Sectioning is stopped when the NIRB mark closest to the ROI is identified (bottom row). Note: these branding marks are not all asymmetrical, this example was shown as a good example of relocation. It is recommended to use asymmetrical marks for correlation. *2P*, two-photon; *ROI*, region of interest.

18. A thin layer of cyanoacrylate glue is used to attach the resin-embedded tissue to a flat-topped blank resin stub (Fig. 7C), making sure the tissue face containing the ROI faces upwards. The glue is allowed to polymerize for at least 2 h before attempting any sectioning.
19. The stub is mounted in an ultramicrotome and a razor blade is used to outline the rough area of the ROI on the block face (Fig. 7E). The block face is then trimmed to a trapezoid-faced pyramid (Fig. 7F).
20. Semithin sections (0.5–1.0 μm) are cut from the block face using a glass knife at 2.0 mm/s until tissue is present in the sections. Each section containing tissue is transferred to a clean glass microscope slide and dried. Every section is imaged using a brightfield microscope until the first branding mark is relocated on the tissue surface (Fig. 7H). The block face is then roughly trimmed around this region.
    a. OPTIONAL (to increase contrast of tissue): Sections are stained with 1% toluidine blue (in 1% borax).
21. Sectioning is continued until each subsequent NIRB mark is relocated (Fig. 7H). Once the final NIRB mark is relocated, the block face is trimmed tightly (<1 × 1 mm) around it (Fig. 8A).
22. OPTIONAL (to check preservation of ultrastructure): A few ultrathin sections are cut, mounted on a copper grid, and imaged using a TEM without counterstaining.

2.4.3.3 SBF-SEM preparation and imaging

23. The block is covered with parafilm and the top 2 mm of the block is trimmed off with a razor blade. The trimmed piece is mounted on an aluminum pin using conductive epoxy resin and baked overnight at 60°C. Subsequently, the block is sputter-coated with 2 nm platinum (Fig. 8B).
24. The pin is mounted in the 3View (Fig. 8C and D) and the block face polished with 100 nm cuts (Fig. 8E), before closing the chamber door and pumping to 5–10 Pa of nitrogen gas. The block is imaged during approach cuts to relocate the ROI (Fig. 8G). The ROI is centered in the FOV, and parameters are set for SBF-SEM imaging and sectioning. The SEM is operated at an accelerating voltage of 2 kV with high current mode active, a 20 μm aperture, and chamber pressure of ~5 Pa. A per pixel dwell time of 2 μs is used with a slice thickness of 50 nm. Images are acquired at 8192 × 8192 pixels (horizontal frame width of 25.7 μm, reported pixel size of 3.1 nm) and indicated magnification of 10,000×. The entire volume comprises 500 slices, totaling 16,512 μm^3.

2.4.3.4 Image processing and reconstruction

25. .dm4 images from the 3View are initially batch processed using FIJI software, reading files with the "Bioformats" plugin, and converting them to eight-bit .tiff files using the "Batch Convert" plugin.

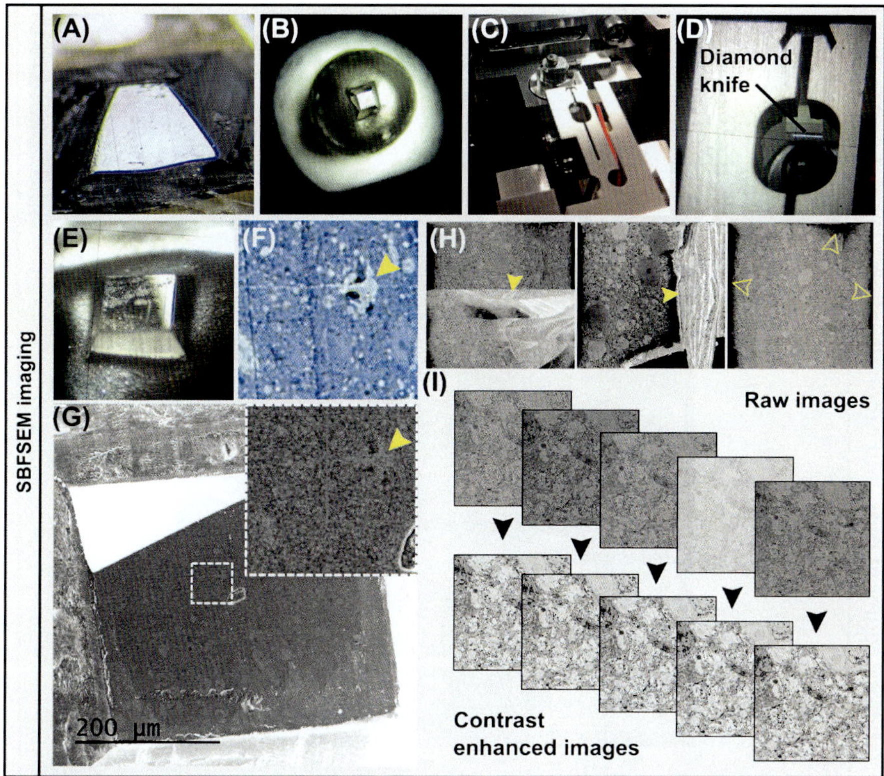

FIGURE 8

SBF-SEM sample preparation and imaging setup. The final block face (A) is accurately squared-off using a glass knife and ultramicrotome, then subsequently stuck to an aluminum pin and sputter-coated (B). The block is inserted into the SBF-SEM microtome (C and D). Initial cuts are made to polish the block face before the door is closed (E). Images are collected during approach cuts (G; inset, zoom of *dashed box*). The ROI is relocated using the final semithin section as a reference (F; *yellow arrowhead* = corner of NIRB). (H) Representative SBF-SEM images of flaking (left and middle) and charging artifacts (right). (I) Conversion of raw .dm4 files with varying relative histograms (top row) are converted to .tiff files and contrast-enhanced as well as smoothed with a Gaussian filter (bottom row). *NIRB*, near-infrared branding; *ROI*, region of interest; *SBF-SEM*, serial block face scanning electron microscopy.

a. OPTIONAL (for large files): Files are downsampled to reduce file size to make them easier to handle. This is used when doing rough correlation and tracing of structure without fine subcellular structural detail. Alternatively, large file formats (Amira) or virtual stacks (FIJI) are used to handle large data sets.

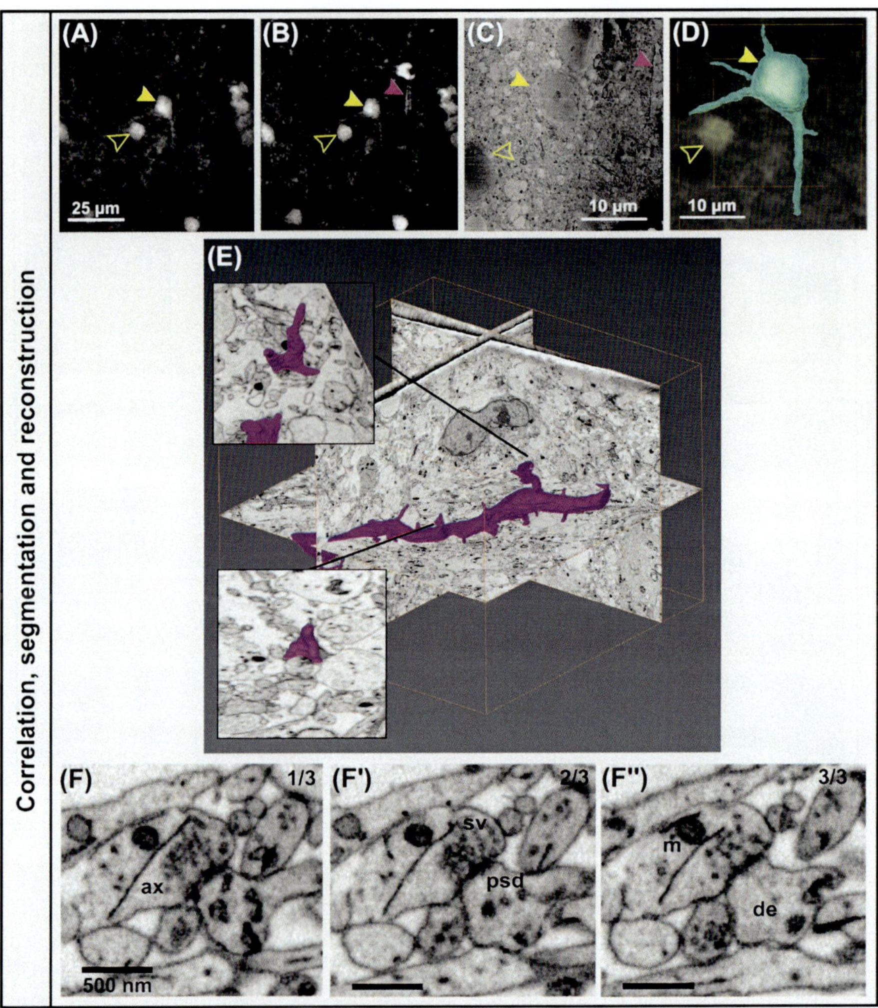

FIGURE 9

Correlation of 2P and SBF-SEM images and identification of synaptic structures. (A) A 2P image of a cell of interest (*filled arrowhead*) before branding and (B) after branding (*empty yellow arrowhead* = reference cell body, *magenta arrowhead* = section of branding mark). (C) One slice from the SBF-SEM data set for correlating to the 2P images in A and B. (D) 3D reconstruction of the cell of interest in Amira, overlaid with the 2P fluorescence image. (E) 3D reconstruction in Amira of a dendrite of interest with orthogonal planes from an SBF-SEM data set, showing filopodia-like spines at synapses (inset). (F, F′, F″) Three consecutive 50 nm sections from an SBF-SEM data set with 4 nm pixel size. *2P*, two-photon; *ax*, axon; *de*, dendrite; *m*, mitochondrion; *psd*, postsynaptic density; *SBF-SEM*, serial block face scanning electron microscopy; *sv*, synaptic vesicles.

26. To normalize the intensity across the stack of images, the .tiff files are further processed using the "Enhance Contrast" plugin in FIJI with a value of 0.3% (Fig. 8I).
27. To remove noise from the SBF-SEM images that result from nitrogen gas (used for variable pressure), a Gaussian blur filter in ImageJ with a sigma value of 2 is used.
28. The SBF-SEM images are correlated to 2P images using fiducials e.g., cell bodies, vasculature, branding marks (Fig. 9A–C).
29. Structures of interest are subsequently manually reconstructed in FIJI software using TrakEM2 or in Amira 6.0.0 software (Fig. 9D and E).

CONCLUDING REMARKS

Correlating light and electron microscopy images of intact tissue allows cell biologists to link cellular dynamics to the fine ultrastructural detail of a cell's environment. This technique has previously been unappealing for cell biologists carrying out 2P microscopy in tissue due to the difficult and laborious nature of relocating specific regions of interest. However, 3D EM techniques can make correlating tissue volumes in LM and EM a lot more accessible, allowing for the precise measurement of the geometry and organization of thousands of subcellular structures in the volume. It is clear that NIRB coupled with 3D EM (FIB-SEM or SBF-SEM) dramatically reduces the human time and skill required to achieve 3D CLEM. Here, the NIRB method is refined to go from a live sample to a correlated SBF-SEM data set in under 10 days.

There are additional ways of improving the efficiency and accuracy of this technique. The greatest technical challenge is a reliable flat-embedding technique. Sectioning is labor intensive and prone to errors, unnecessary time is added to the protocol if NIRB marks do not appear flat on sectioning, as identification is difficult. It may be possible to improve the accuracy of targeted ultramicrotomy using alternative NIRB patterning to that outlined here. Other improvements include, additional labeling for extra fiducials (e.g., cell nuclei), DAB precipitation of the NIRB marks to make them more obvious in semithin sections (Knott et al., 2009) and micro-CT imaging to highlight vasculature and potentially NIRB marks to target ultramicrotomy (Karreman et al., 2016). However, these refinements require more equipment, expertise, and time.

NIRB has mostly been limited to neuronal cell biology; marking ROIs from intravital imaging (Bishop et al., 2011; Blazquez-Llorca et al., 2017; Grillo et al., 2013; Karreman et al., 2016; Maco et al., 2014; Mostany et al., 2013). Here, some suggestions for the use of NIRB and 3D CLEM in neuronal tissue, outside of cranial window imaging, are given. NIRB has also been attempted in nonneuronal tissue, but at limited depths (Karreman et al., 2016). Therefore, it is currently open to be adapted for use in nonneuronal applications. More widespread utilization of 3D CLEM in tissue will bring powerful structural evidence to functional studies, allowing cell biologists to ask questions that are not otherwise possible.

ACKNOWLEDGMENTS

RML is funded by a Wellcome Trust PhD studentship. This work was supported by the Francis Crick Institute, which receives its core funding principally from Cancer Research UK (FC001999), the UK Medical Research Council (FC001999), and the Wellcome Trust (FC001999). In addition, this research was supported by the MRC, BBSRC, and EPSRC under grant award MR/K01580X/1 to LMC and Peter O'Toole (York University). Equipment used in MCA's laboratory was partially funded by the Medical Research Council (MR/J013188/1) and EUFP17 Marie Curie Actions (PCIG10-GA-2011-303680). We thank the Wolfson Bioimaging Facility for their support and expertise, and MRC funding of a preclinical in vivo functional imaging platform for translational regenerative medicine.

REFERENCES

Akerboom, J., Chen, T.-W., Wardill, T. J., Tian, L., Marvin, J. S., Mutlu, S., … Looger, L. L. (2012). Optimization of a GCaMP calcium indicator for neural activity imaging. *Journal of Neuroscience, 32*, 13819–13840. http://dx.doi.org/10.1523/JNEUROSCI.2601-12.2012.

Ashby, M. C., & Isaac, J. T. R. (2011). Maturation of a recurrent excitatory neocortical circuit by experience-dependent unsilencing of newly formed dendritic spines. *Neuron, 70*, 510–521. http://dx.doi.org/10.1016/j.neuron.2011.02.057.

Bishop, D., Nikić, I., Brinkoetter, M., Knecht, S., Potz, S., Kerschensteiner, M., & Misgeld, T. (2011). Near-infrared branding efficiently correlates light and electron microscopy. *Nature Methods, 8*, 568–570. http://dx.doi.org/10.1038/nmeth.1622.

Bishop, D., Nikić, I., Kerschensteiner, M., & Misgeld, T. (2014). The use of a laser for correlating light and electron microscopic images in thick tissue specimens. In T. Muller-Reichert, & P. Verkade (Eds.), *Correlative light and electron microscopy: Vol. 124. Methods in cell biology* (pp. 328–338). Academic Press.

Blazquez-Llorca, L., Valero-Freitag, S., Rodrigues, E. F., Merchán-Pérez, Á., Rodríguez, J. R., Dorostkar, M. M., … Herms, J. (2017). High plasticity of axonal pathology in Alzheimer's disease mouse models. *Acta Neuropathologica Communications, 5*, 14. http://dx.doi.org/10.1186/s40478-017-0415-y.

Cardona, A., Saalfeld, S., Schindelin, J., Arganda-Carreras, I., Preibisch, S., … Douglas, R. J. (2012). TrakEM2 software for neural circuit reconstruction. *PLoS One, 7*, e38011. http://dx.doi.org/10.1371/journal.pone.0038011.

Deerinck, T. J., Bushong, E. A., Thor, A., & Ellisman, M. H. (2010). *NCMIR methods for 3D EM: A new protocol for preparation of biological specimens for serial block face scanning electron microscopy* (Online). https://ncmir.ucsd.edu/sbem-protocol

Edwards, F. A., Konnerth, A., Sakmann, B., & Takahashi, T. (1989). A thin slice preparation for patch clamp recordings from neurones of the mammalian central nervous system. *Pflugers Archive, 414*, 600–612. http://dx.doi.org/10.1007/BF00580998.

Grillo, F. W., Song, S., Ruivo, L. M. T.-G., Huang, L., Gao, G., Knott, G. W., … De Paola, V. (2013). Increased axonal bouton dynamics in the aging mouse cortex. *Proceedings of the National Academy of Sciences of the United States of America, 110*, E1514–E1523. http://dx.doi.org/10.1073/pnas.1218731110.

Helmstaedter, M. (2013). Cellular-resolution connectomics: Challenges of dense neural circuit reconstruction. *Nature Methods, 10*, 501–507. http://dx.doi.org/10.1038/nmeth.2476.

Hodgson, L., Nam, D., Mantell, J., Achim, A., & Verkade, P. (2014). Retracing in correlative light electron microscopy: Where is my object of interest? In T. Muller-Reichert, & P. Verkade (Eds.), *Correlative light and electron microscopy: Vol. 124. Methods in cell biology* (pp. 1−21). Academic Press.

Holtmaat, A., Bonhoeffer, T., Chow, D. K., Chuckowree, J., De Paola, V., Hofer, S. B., … Wilbrecht, L. (2009). Long-term, high-resolution imaging in the mouse neocortex through a chronic cranial window. *Nature Protocols, 4*, 1128−1144. http://dx.doi.org/10.1038/nprot.2009.89.

Karreman, M. A., Mercier, L., Schieber, N. L., Solecki, G., Allio, G., Winkler, F., … Schwab, Y. (2016). Fast and precise targeting of single tumor cells in vivo by multimodal correlative microscopy. *Journal of Cell Science, 129*, 444−456. http://dx.doi.org/10.1242/jcs.181842.

Kasthuri, N., Hayworth, K. J., Berger, D. R., Schalek, R. L., Conchello, J. A., Knowles-Barley, S., … Lichtman, J. W. (2015). Saturated reconstruction of a volume of neocortex. *Cell, 162*, 648−661. http://dx.doi.org/10.1016/j.cell.2015.06.054.

Knott, G. W., Holtmaat, A., Trachtenberg, J. T., Svoboda, K., & Welker, E. (2009). A protocol for preparing GFP-labeled neurons previously imaged in vivo and in slice preparations for light and electron microscopic analysis. *Nature Protocols, 4*, 1145−1156. http://dx.doi.org/10.1038/nprot.2009.114.

Li, Y., Song, Y., Zhao, L., Gaidosh, G., Laties, A. M., & Wen, R. (2008). Direct labeling and visualization of blood vessels with lipophilic carbocyanine dye DiI. *Nature Protocols, 3*, 1703−1708. http://dx.doi.org/10.1038/nprot.2008.172.

Linkert, M., Rueden, C. T., Allan, C., Burel, J.-M., Moore, W., Patterson, A., … Swedlow, J. R. (2010). Metadata matters: Access to image data in the real world. *The Journal of Cell Biology, 189*, 777−782. http://dx.doi.org/10.1083/jcb.201004104.

Maco, B., Cantoni, M., Holtmaat, A., Kreshuk, A., Hamprecht, F. A., & Knott, G. W. (2014). Semiautomated correlative 3D electron microscopy of in vivo−imaged axons and dendrites. *Nature Protocols, 9*, 1354−1366. http://dx.doi.org/10.1038/nprot.2014.101.

Maco, B., Holtmaat, A., Jorstad, A., Fua, P., & Knott, G. (2014). Correlative in vivo 2-photon imaging and focused ion beam scanning electron microscopy: 3D analysis of neuronal ultrastructure. In T. Muller-Reichert, & P. Verkade (Eds.), *Correlative light and electron microscopy: Vol. 124. Methods in cell biology* (pp. 339−361). Academic Press.

Micheva, K. D., O'Rourke, N., Busse, B., & Smith, S. J. (2010). Array tomography: High-resolution three-dimensional immunofluorescence. *Cold Spring Harbor Protocols, 2010*. http://dx.doi.org/10.1101/pdb.top89.

Mostany, R., Anstey, J. E., Crump, K. L., Maco, B., Knott, G., & Portera-Cailliau, C. (2013). Altered synaptic dynamics during normal brain aging. *Journal of Neuroscience, 33*, 4094−4104. http://dx.doi.org/10.1523/JNEUROSCI.4825-12.2013.

Peddie, C. J., & Collinson, L. M. (2014). Exploring the third dimension: Volume electron microscopy comes of age. *Micron, 61*, 9−19. http://dx.doi.org/10.1016/j.micron.2014.01.009.

Renier, N., Wu, Z., Simon, D. J., Yang, J., Ariel, P., & Tessier-Lavigne, M. (2014). iDISCO: A simple, rapid method to immunolabel large tissue samples for volume imaging. *Cell, 159*, 896−910. http://dx.doi.org/10.1016/j.cell.2014.10.010.

Russell, M. R. G., Lerner, T. R., Burden, J. J., Nkwe, D. O., Pelchen-Matthews, A., Domart, M.-C., … Collinson, L. M. (2017). 3D correlative light and electron microscopy of cultured cells using serial blockface scanning electron microscopy. *Journal of Cell Science, 130*, 278−291. http://dx.doi.org/10.1242/jcs.188433.

Schindelin, J., Arganda-Carreras, I., Frise, E., Kaynig, V., Longair, M., Pietzsch, T., ... Cardona, A. (2012). Fiji: An open-source platform for biological-image analysis. *Nature Methods, 9*, 676–682. http://dx.doi.org/10.1038/nmeth.2019.

Shoham, S., O'Connor, D. H., Sarkisov, D. V., & Wang, S. S.-H. (2005). Rapid neurotransmitter uncaging in spatially defined patterns. *Nature Methods, 2*, 837–843. http://dx.doi.org/10.1038/nmeth793.

Svoboda, K., & Yasuda, R. (2006). Principles of two-photon excitation microscopy and its applications to neuroscience. *Neuron, 50*, 823–839. http://dx.doi.org/10.1016/j.neuron.2006.05.019.

Yang, G., Pan, F., Parkhurst, C. N., Grutzendler, J., & Gan, W.-B. (2010). Thinned-skull cranial window technique for long-term imaging of the cortex in live mice. *Nature Protocols, 5*, 201–208. http://dx.doi.org/10.1038/nprot.2009.222.

Zipfel, W. R., Williams, R. M., & Webb, W. W. (2003). Nonlinear magic: Multiphoton microscopy in the biosciences. *Nature Biotechnology, 21*, 1369–1377. http://dx.doi.org/10.1038/nbt899.

CHAPTER

Find your way with X-Ray: using microCT to correlate in vivo imaging with 3D electron microscopy

13

Matthia A. Karreman*[,1], Bernhard Ruthensteiner[§], Luc Mercier[¶,||,#,], Nicole L. Schieber*, Gergely Solecki[§§,¶¶], Frank Winkler[§§,¶¶], Jacky G. Goetz[¶,||,#,**], Yannick Schwab***

**European Molecular Biology Laboratory, Heidelberg, Germany*

[§]Zoologische Staatssammlung München, Munich, Germany

[¶]MN3T, Inserm U1109, Strasbourg, France

[||]Université de Strasbourg, Strasbourg, France

[#]LabEx Medalis, Université de Strasbourg, Strasbourg, France

***Fédération de Médecine Translationnelle de Strasbourg (FMTS), Université de Strasbourg, Strasbourg, France*

[§§]University Hospital Heidelberg, Heidelberg, Germany

[¶¶]German Cancer Research Center (DKFZ), Heidelberg, Germany

[1]Corresponding author: E-mail: karreman@embl.de

CHAPTER OUTLINE

Methods in Cell Biology, Volume 140, ISSN 0091-679X, http://dx.doi.org/10.1016/bs.mcb.2017.03.006

Abstract

Combining in vivo imaging with electron microscopy (EM) uniquely allows monitoring rare and critical events in living tissue, followed by their high-resolution visualization in their native context. A major hurdle, however, is to keep track of the region of interest (ROI) when moving from intravital microscopy (IVM) to EM. Here, we present a workflow that relies on correlating IVM and microscopic X-ray computed tomography to predict the position of the ROI inside the EM-processed sample. The ROI can then be accurately and quickly targeted using ultramicrotomy and imaged using EM. We outline how this procedure is used to retrieve and image tumor cells arrested in the vasculature of the mouse brain.

INTRODUCTION

Intravital correlative light and electron microscopy (intravital CLEM) is a powerful approach to study developmental and pathological processes in animal models (Durdu et al., 2014; Karreman, Hyenne, Schwab, & Goetz, 2016; Maco et al., 2013). While in vitro model systems never fully recapitulate the complexity of living tissue, intravital microscopy (IVM) enables to monitor processes over time and in their native environment (Ellenbroek & van Rheenen, 2014; Follain, Mercier, Osmani, Harlepp, & Goetz, 2016). Extending IVM with electron microscopy (EM) reveals the process of interest at high resolution within its ultrastructural context.

The main challenge in intravital CLEM is to keep track of the region of interest (ROI) when moving from in vivo imaging to EM. For small organisms or embryos, it is possible to process and image the full sample in EM and retrieve the area of interest by browsing through the resulting images (Müller-Reichert, Srayko, Hyman, O'Toole, & McDonald, 2007; Zito, Parnas, Fetter, Isacoff, & Goodman, 1999). However, for larger model systems, a biopsy (e.g., a small piece of tissue or a vibratome section) containing the ROI needs to be selected and subsequently processed for EM. Retrieval of the ROI in the electron microscope is particularly difficult, due to the difference in image formation between IVM and EM, the small field of view (FOV) of EM and the deformation of the tissue that results from the EM sample preparation approaches.

In this chapter, we outline an approach that enables correlating IVM to three-dimensional (3D) EM of voluminous samples. As an example, we demonstrate how to retrieve single tumor cells inside mouse brain biopsies. Here, we aim to study how metastatic JIMT1 breast cancer cells cross the blood−brain barrier as part of the metastatic process (Karreman et al., 2016). We have developed a multimodal correlative microscopy workflow that relies on microscopic X-ray computed tomography (microCT) imaging as a guide to retrieve the area(s) of interest inside the EM-processed sample (Karreman et al., 2016) (Fig. 1). The workflow starts out with IVM of fluorescent JIMT1 tumor cells inside the mouse brain, through a cranial window (Kienast et al., 2010). Following perfusion fixation, the position of the ROI is marked onto the mouse brain surface by near-infrared branding (NIRB) (Bishop et al., 2011). A small biopsy, which contains the area of interest, is then dissected from the fixed mouse brain. Next, the biopsy is prepared for EM by microwave-assisted processing, followed by embedding in resin. The sample is then imaged by microCT, enabling to get a 3D volume showing the resin block, the biopsy, and the structural features therein. The microCT data set of the processed sample is subsequently correlated to the IVM data set, which is the critical step in this correlative workflow because it translates the position of the target cell inside the volume of the opaque resin-embedded piece of tissue.

Using Amira software, 3D models of both the IVM and microCT volumes are generated by segmentation, and the corresponding points that can be found in each model are marked using the *Landmarks* module. The IVM volume, showing the vasculature and the tumor cell(s), is then registered into the microCT volume, which also shows the outlines of the resin block, the global topology of the biopsy and, importantly, the vasculature. Even though the fluorescent signal is lost during the sample preparation for EM, this warping procedure precisely positions the tumor cell inside the microCT data set, enabling to measure its location with respect to the

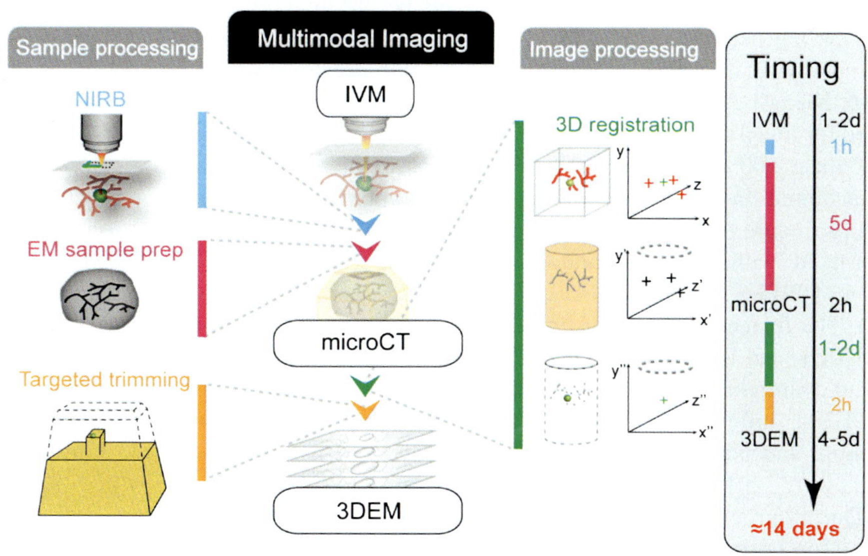

FIGURE 1

Overview of the multimodal correlative microscopy workflow. For a detailed description of the different steps involved, please refer to the main text. This figure appeared earlier in the original publication of the approach (Karreman et al., 2016). *EM*, electron microscopy.

resin block surface in x, y, and z. Based on these measurements, the resin block is accurately trimmed in an ultramicrotome to approach the position of the tumor cell. 3DEM is then performed, e.g., by focused ion beam—scanning electron microscopy (FIB-SEM), serial block face imaging or by serial section TEM (and serial electron tomography).

In this chapter, we outline step-by-step the protocol from taking the biopsy to accurately approaching the cell of interest using 3D-targeted trimming. In particular, we detail how both the IVM and the microCT data sets can be segmented and registered in 3D using Amira software and how this subsequently allows to sculpt the resin block to expose the target to 3DEM. The workflow is demonstrated here on mouse brain biopsies in which one or more tumor cells are retrieved. Importantly, this method can also be applied to other model systems, such as mouse skin tissue (Karreman et al., 2016) and starfish oocytes. The high success rate and improved throughput of this approach makes it highly suitable and promising for intravital CLEM. Moreover, its versatility allows its use in different fields, as it can be applied to retrieve any rare event from in vivo imaging to EM.

1. METHODS

1.1 PROCESSING FOR ELECTRON MICROSCOPY

A cranial window is grafted into 8- to 10-week-old nude mice and is followed by a 3-week healing period. Three or seven days before imaging, the mice are injected into the left heart ventricle with cytoplasmic GFP—expressing JIMT1 tumor cells. During IVM, described in more detail in our previous work (Karreman et al., 2016; Kienast et al., 2010), arrested and potentially extravasated tumor cells are targeted for imaging. On finding a cell of interest, a large-FOV z-stack (e.g., 600 μm × 600 μm × 500 μm) and a small-FOV z-stack (e.g., 200 μm × 200 μm × 250 μm) are acquired. In a later stage, both these IVM volumes are required for the correlation procedure (see Section 1.3). Following IVM, the anesthetized mouse is perfusion-fixed with 2.5% glutaraldehyde and 2% formaldehyde in PHEM buffer (60 mM PIPES, 25 mM HEPES, 10 mM EGTA, and 2 mM MgCl, pH adjusted to 6.9). The imaged volumes of interest are marked on the level of the brain surface by NIRB, and the brain is removed from the skull, immersed in the fixative, and stored overnight at 4°C. The next day, the fixative is replaced by 1% formaldehyde in PHEM buffer, for prolonged storage. For more details about IVM, perfusion fixation, and NIRB, please refer to earlier work (Bishop et al., 2011; Karreman et al., 2014, 2016).

During processing samples for EM, the effective infiltration of chemicals into the tissue is limited. For this reason, it is critical to dissect a small biopsy. The NIRB markings indicate the x-y position of the area of interest, and the IVM data set will roughly indicate the depth of the ROI inside the brain. Using this information as a guide, dissect a <600 μm × 600 μm × 900 μm biopsy (see Fig. 2).

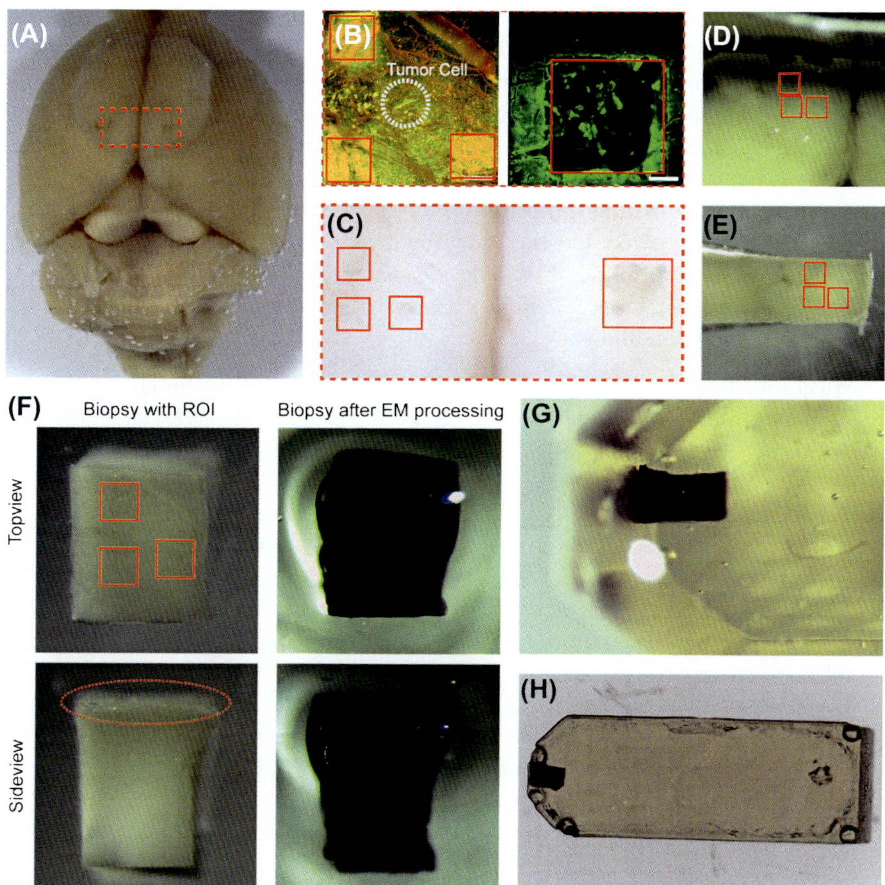

FIGURE 2

Dissecting and electron microscopy (EM) processing of the mouse brain biopsy containing the region of interest (ROI). (A) The position of the in vivo imaged tumor cell inside the mouse brain is marked by near-infrared branding (NIRB), visible inside the boxed area (see panel C). (B) Maximum intensity z-projections of the intravital microscopy data sets, showing in the left panel the position of the tumor cell (*dotted circle*) with respect to small NIRB markings (*boxes*). The right panel depicts a larger NIRB square that is generated on the opposing hemisphere to facilitate retrieval of the smaller markings around the tumor cell (see panel A and C). Scale bars: 100 μm. (C) On the brain surface, the small and large NIRB markings are visible as small scars (*boxes*). (D and E) A small biopsy is cut around the NIRB markings (*boxes*), using a razor blade. (F) Before EM processing (left panels, "Biopsy with ROI"), the NIRB markings are still visible (*boxes*, top of biopsy indicated in bottom left panel with *dotted oval*). During EM processing, the biopsy turns black due to osmification and the NIRB markings are no longer visible. (G and H) Due to the asymmetric shape of the biopsy, it can be positioned in the resin mold so that the NIRB markings are closest to the future block face.

Note: The NIRB markings are not always clearly recognizable on the brain surface, but their visibility is critical for this step of the workflow (Fig. 2C–F). Therefore, first test and optimize the NIRB conditions before embarking in the correlative studies. It is helpful to create an asymmetrically shaped biopsy, so that it is unambiguous which side of the sample is the brain surface with the NIRB marks. Following EM processing, the marks will no longer be visible (Fig. 2F–H) and the asymmetric shape will help position the biopsy during embedding.

The biopsy is prepared for EM imaging by microwave-assisted processing, which improves the infiltration of the solutions and greatly speeds up the procedure. The PELCO BioWave Pro (Ted Pella) can be preprogrammed, and we use the following steps and conditions:

1. Wash four times for 5 min in 0.1 M cacodylate buffer in the hood.
2. Primary postfixation: 1% OsO_4 + 1.5 $K_3Fe(CN)_6$ in 0.1 M cacodylate buffer, in the microwave: vacuum on, wash seven times for 2 min consecutive steps, 100 W cycling on–off.
3. Wash two to ten times briefly in 0.1 M cacodylate buffer; in the hood, and then in the microwave: wash two times for 40 s, 250 W, change buffer in between washes.
4. Secondary postfixation: 1% OsO_4 in 0.1 M cacodylate buffer, in the microwave: vacuum on, wash seven times for 2 min consecutive steps, 100 W cycling on–off.
5. Wash as described in step 3, but using water.
6. Staining with uranyl acetate (UA): 1% UA in water, in the microwave: vacuum on, wash seven times for 1 min consecutive steps, 100 W cycling on–off.
7. Wash in water as described in step 3.
8. Dehydration: Use a graded series of ethanol dilutions in water: 25%, 50%, 75%, 90%, 95%, in the microwave: 40 s of each step, 250 W. Finally, wash the sample two times for 40 s in 100% ethanol at 250 W.
9. Resin infiltration: Use a graded series of resin (Durcupan or Epon extra hard mix) in ethanol: 25%, 50%, 75%, 90%, in the microwave: vacuum on, wash seven times for 3 min, 250 W, cycling on–off.
   Here, we choose hard types of resin because they provide the required stability for FIB-SEM imaging. Our preferred choice is Durcupan. For serial sectioning TEM, other medium-hard Epon mixes would also be suitable.
10. Place the biopsy in the resin block mold, so that the surface with the NIRB mark will be close to the future block surface (Fig. 2G and H).
11. Polymerize the sample at 60°C for at least 48 h.

1.2 TRIMMING THE RESIN BLOCK AND microCT IMAGING

For microCT imaging, the sample should be trimmed to a small size (less the 2–5 mm in x, y, z). The microCT employed in our work (Borrego-Pinto, Somogyi, & Karreman, 2016; Karreman et al., 2016) uses a cone-shaped beam, and the effective voxel size thus depends on the distance between the detector and the sample. Removing the excess of resin around the tissue will allow to position the object closer to the X-ray source, which will result in an improved voxel size and resolution. Mount the sample in a microtome and trim the sample using a razor blade (Fig. 3A and B) and a trimming diamond knife (or a glass knife). Trim the block as close as possible to the biopsy. Note: It is critical to create a flat block surface during this first trimming step. Trimming the sides with a knife will also be helpful for the 3D registration (see part 5), but it is not required to trim along the full depth of

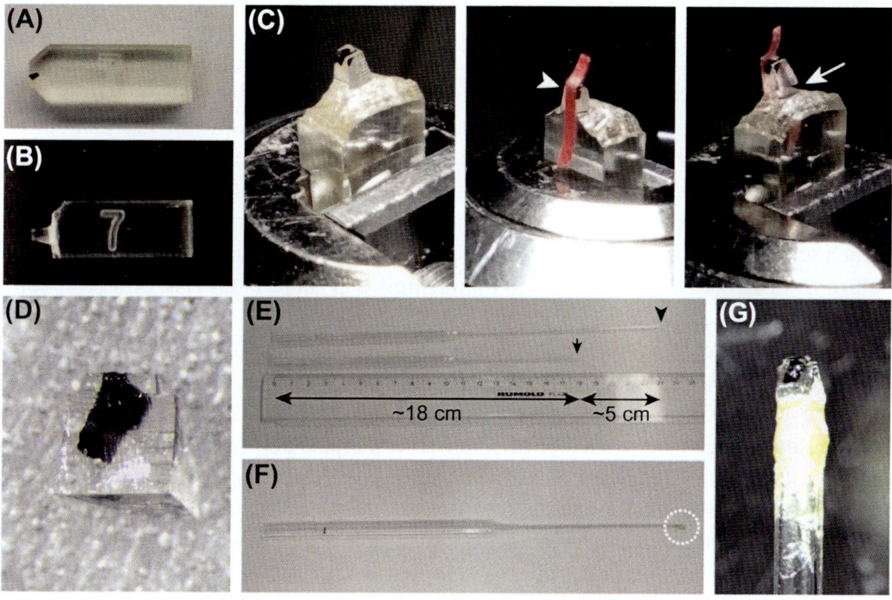

FIGURE 3

Trimming the sample for microCT imaging. (A) The untrimmed resin block, the brain biopsy is visible on the left side. (B) The resin block is roughly trimmed, using a razor blade, around the position of the biopsy. (C) The block surface and three out of four sides are trimmed using a 90-degree diamond knife (left panel). To ensure that the samples is not lost while cutting it off the large block, a small strip of tape (middle panel, red tape indicated with *white arrowhead*) is attached to the block that keeps the samples in place after separating it from the large block using a razor blade (*arrow* in right panel). (D) The small, chopped-off block can be placed on a strip of double-sided tape stuck to a glass slide. This allows to, if necessary, further trim the block with a razor blade. (E) For microCT imaging in the nanotom m, the sample is mounted on a glass rod. Hereto, take a long (approximately 23 cm) glass pipette and break off the tip (*arrowhead*) to remove approximately 5 cm of length (small *arrow*). Then, carefully round off the tip in a flame. (F) Mount the resin-embedded sample on the tip of the glass rod (*dotted circle*) using glue. (G) The sample is mounted on the tip of the glass rod.

the biopsy. The trimmed part with the biopsy can then be chopped off the larger block using a fresh razor blade (Fig. 3C and D).

For microCT imaging, the general steps include (1) mounting the sample on a holder that fits into the microCT setup, (2) aligning the sample to the center of the FOV in the microCT, (3) setting the imaging conditions (beam voltage and current, voxel size, exposure time, image averaging, etc.) and, finally, (4) reconstructing the data set. In our work, we have used the nanotom m (GE sciences), which operation is described here in more detail. First, the small sample is mounted on a holder for microCT imaging. For the nanotom m, rods can be used that are made from long glass Pasteur pipettes. Select a straight pipette; remove ~5 cm from the tip; and round it off in a flame (Fig. 3E). The sample can be mounted on the tip using dental wax or glue (e.g., all-purpose glue, as long as it is removable after imaging) (Fig. 3F and G). Install the sample in the microCT setup and move it 10–15 cm away from the source. Start the X-ray beam at a low current (e.g., 45 μA) and 60 kV. In the microCT *xs control* software, position the sample in the center of the FOV by adjusting its vertical position. The sample will be fully rotated during acquisition and should remain centered during imaging. To achieve this, rotate the sample to 0, 90, 180, and 270 degrees, and position it to the middle of the FOV for each angle. Turn off the X-ray beam, open the machine, and position the sample as close as possible to the tube, while confirming that there is no risk of a collision with the tube while the sample rotates. The Phoenix *datos|x* software will indicate the voxel size, which should be in between 0.5 and 1.5 μm. Adjust the X-ray settings to the acquisition values: typically, we use 430 μA/60 kV [focus 3 (smallest) for a voxel size up to 0.7 μm, focus 2 (medium) for 0.7–1.3 μm]. To reduce the voxel size, the detector can be moved further backward (max 500 mm), but beware that this also reduces the signal count. It is recommended to perform the "centering 1" and the "Adjust Filament" procedure in the *xs|control* software before each scan. Next, store the position in *datos|x*, move the sample fully downward, and perform a detector calibration. After moving the sample back to its acquiring position, indicate the acquisition parameters: number of images (e.g., 1440), exposure time (e.g., 1000 ms), averaging (4), binning (e.g., 1), image size, etc. Indicate the filter (MolyB) and activate *Auto|Sco* (which allows to evaluate and compensate for sample movement during the scan) and *Shift* (which moves the detector slightly during the scan to prevent image artefacts generated by damaged pixels). Start the scan, which will take 1–2 h, depending on the image parameters. After completion, reconstruct the volume in the Phoenix *datos|x* reconstruction software.

1.3 SEGMENTATION AND 3D REGISTRATION IN AMIRA

The microCT imaging provides an image stack, which visualizes the structural features of the resin-embedded sample in 3D. In the current step, the IVM image volumes are correlated to this data set. In general, this requires the following steps:

1. segmentation of the biopsy and vasculature in the microCT data set,
2. segmentation of the vasculature and tumor cell(s) in the IVM data sets,

3. registration of the large-FOV IVM data set into the full microCT volume, cropping the microCT volume around the expected position of the tumor cell, and
4. finally, registration of the small-FOV IVM data set into the cropped microCT volume.

Since cropping the microCT data set preserves its coordinates with respect to the uncropped volume, it is then possible to measure the position of the tumor cell (from the registered small-FOV IVM data set) with respect to the resin block surface, visible in the uncropped microCT data set.

In this section, these steps are described in detail for the correlation of IVM and microCT imaging of mouse brain biopsies. Importantly, a similar approach can be applied for the correlation of different samples or tissues. However, this may rely on using different landmarks for the correlation between the two data sets, and the segmentation parameters should be adjusted accordingly.

1.3.1 Segmenting the microCT data set in Amira

1. Open Amira and start a blank project (Fig. 4)
2. Import the microCT volume in Amira (select "read full volume into memory") by dragging the file directly into Amira's Project View (indicate the voxel size in x, y, and z in the pop-up window, if applicable). Visualize the data set (Fig. 5) by creating a *Slice* or *Orthoslice*: left mouse click on the icon in the Project View, *Display* > *Slice* or *Orthoslice*. *Slice* can be rotated freely (in version 5 and older versions of Amira, this is called *Obliqueslice*).
3. The microCT can be resampled to speed up further processing steps. Left click on the data set, and select *Compute* > *Volume Operations* > *Resample*. In the Properties Area, change the voxel size to a higher value (between 1 and 2 μm). A novel file is generated and shown in the Project View with the extension "resampled" (Fig. 5A). Remove the original file from the project.
4. To facilitate the segmentation, filter the microCT data set. Apply a *Non-Local Means Filter* (*Image Processing* > *Smoothing and Denoising* > *Non-Local Means Filter*, Similarity Value: 0.5) (Fig. 5B).
5. Move to the "Segmentation" tab and select the filtered data set in "Image Data." A label data file is automatically generated with two "Materials": "Exterior" and "Inside" (Fig. 5D).
6. Select the pixels that depict biopsy inside the resin block, using the "Threshold" option. In "Display and Masking," select part of the histogram, so that the biopsy is highlighted in blue. In "Options" tick "All slices" to highlight the biopsy throughout the full data set. In "Action," click "Select."
7. Create a Material "Biopsy" in the label file. Add the selected pixels to this Material by selecting it, and click the red-circled plus button in "Selection" (Fig. 5D).

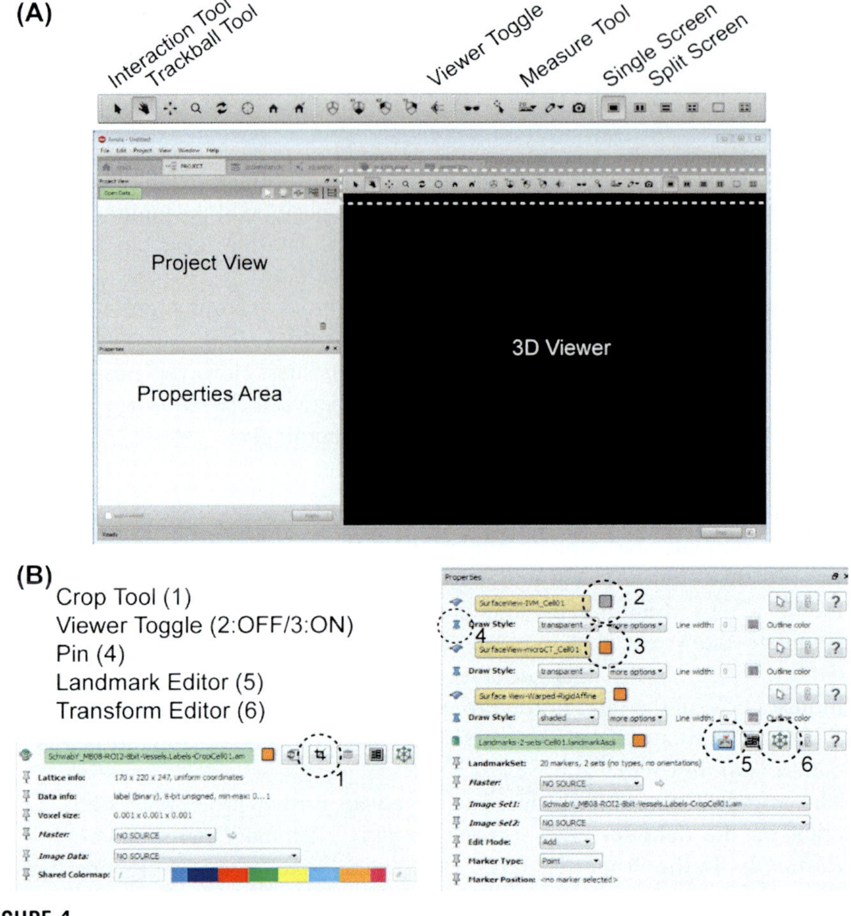

FIGURE 4

An overview of Amira's Project screen and the tools used for the workflow. (A) Amira's main Project screen shows the Project View, the Properties Area and the 3D viewer. The toolbar above the 3D viewer (boxed with a *dotted line*) has several useful tools to visualize and interact with the data. (B) When files or modules are selected in the Project View window, their properties and parameters are shown in the Properties Area (A). Here, also the conditions for specific modules can be set. Some tools that are used during the workflow are highlighted with *dotted circles*.

8. The vessels and nuclei inside the biopsy are not selected via thresholding since they appear darker than the surrounding tissue. Add these to the Material in the menu *Segmentation > Fill holes > All slices.*
9. Depending on the gray levels chosen for the thresholding, some parts outside the biopsy may be selected. These can be removed using the

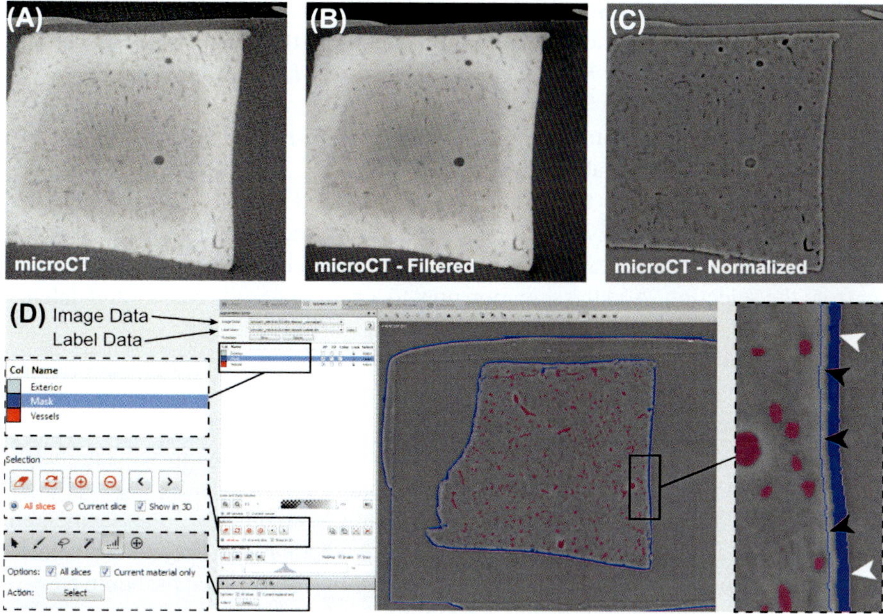

FIGURE 5

Segmentation of the microCT data set in Amira. (A) Visualization of a virtual section through the microCT data set. The biopsy appears in light/gray white, the vessels and nuclei are dark, and the outline of the resin block is dark gray. The outside of the sample (above) is shown in black/very dark gray. (B) A virtual section through the filtered microCT data set, at the same position as in A. (C) A virtual section through the filtered and normalized microCT data set, at same position as in A and B. (D) Segmentation of the filtered and normalized microCT data set. Commonly used tools are highlighted (left panels). The right panel shows the outline of the segmented biopsy (Mask, *thin line, black arrowheads*). The thresholded pixels that fall outside of the mask (*thick line, white arrowheads*) are not selected, but the thresholded pixels inside of the mask (left side, rounded structures that represent nuclei and vessel cross sections) are selected.

Segmentation > *Remove Islands*. In the pop-up window "Remove Islands," choose size 1000–5000, tick "All Slices" and press "Select." The selection can be excluded from the Material by pressing "Remove." The brush tool, the magic wand, and *Selection* > *Interpolate* can be used to improve the segmentation.

10. To segment the vasculature, the Biopsy label file will be used as a mask in which the darker gray values of the emptied vessels can be selected by thresholding. Duplicate the Biopsy label file, which will function as the basis for the mask.
11. In case the infiltration of the heavy metals is not optimal, there might be a density gradient visible from the inside of the biopsy toward the outside.

However, the thresholding used to segment the vessels requires a homogeneous contrast—difference between the vessels and the surrounding tissue. To achieve this, use the *Normalize Image (Background Detection Correction)* module (Fig. 5C).

12. Go to the segmentation tab and select the normalized data set (Image Data) and the duplicated Biopsy label file (Fig. 5D).
13. Select the pixels in the Biopsy Material and shrink the volume by *Selection > Shrink > Volume* (repeat two to four times), so that the dark edges on the outside of the biopsy, visible in the normalized data set, are excluded from the selection. Add the selection now to a new Material "Mask" and delete "Biopsy" (Fig. 5D).
14. To select the vessels, choose a range of gray values, so that the vasculature is highlighted. Note: Since the nuclei are low in density, they will inadvertently also be selected during this procedure. Select the "Mask" material and tick "All Slices" and "Current Material Only" in the segmentation panel (Fig. 5D). Select these pixels and add these to a new material "Vessels."
15. Delete the "Mask" material, the label file now contains only the segmented vasculature and nuclei. To remove the nuclei, go back to the Project tab, select the label file in Project View and create a *Remove Small Spots* module (in *Image Segmentation*, Extension: XImagePAQ). Select in Interpretation "3D" and size 350 ($\sim$1 μm voxel size). This creates a new filtered label file. Alternatively, and much more time-consuming, is to go to the menu *Segmentation > Remove Islands* (Segmentation tab, select size 350 in "3D volume").
16. To visualize the segmentation, in Project View, create a *Generate Surface* module from the label file. The 3D surface visualization can be smoothed to different extends in the "Smoothing type" drop-down menu and is visualized using *Surface View.*

1.3.2 Segmenting the intravital microscopy z-stack

1. Import the IVM imaging stack (as an RGB tiff file) into Amira. In the pop-up window, select "Channel Conversion: All Channels" and provide the voxel size in x, y, and z. The file and the different channels are shown separately in the Project View.
2. Move to the Segmentation tab, and select channel 1 (red) in Image Data. Using the histogram tool, select the blood vessels and add these pixels to a new material "Vessels." Note: Due to light scattering and absorption deep into the tissue, there will be an intensity gradient along the z-axis of the data set. Manual selection of vessels might be required.
3. Get rid of noise by smoothing the labels (*Segmentation > Smooth Labels* 3, 3D Volume) and selecting and removing islands in 3D as described before.
4. Create a new Material for the tumor cell(s) in the same label file that contains the vessel segmentation. Select the green channel in "Image

Data," and segment the fluorescent tumor cell(s) using local thresholding with the magic wand tool.

5. The label file with the segmented vessels and the tumor cell(s) can be visualized by creating *Generate Surface* and *Surface View* in Project View.

1.3.3 Registration of the large-field of view intravital microscopy z-stack into the microCT volume

To register both data sets in 3D, the IVM surface file [vessels and tumor cell(s)] will be warped into the microCT surface file (vessels). Hereto, the *Landmark Surface Warp* module is used. Landmarks, shown as yellow and blue spheres, are manually placed in corresponding points in the two surface files, e.g., vessel forks or branches. To achieve higher accuracy, the landmark should be placed in the center of the vessel branch. For this, an Auto Skeleton view of the surface needs to be created, on which the landmarks can be placed (Fig. 6). To facilitate the docking, it is easiest to first perform a manual docking of the IVM surface into the microCT surface.

1. Open a new project for the 3D registration and load the label files of the microCT vessels segmentation and the IVM vessels and tumor cell(s) segmentation. Note: Display the vessels from the different imaging modalities in different colors (Fig. 7A).
2. Duplicate the IVM surface file, add "-PreWarp" to its name, and visualize it. Deactivate the display of the original IVM surface file using the viewer toggle (Fig. 4B).
3. Select the "-PreWarp" surface, and, in the Properties Area, click the Transform Editor (Fig. 4B). In the 3D Viewer, the IVM surface is now surrounded by a "transformation cube" that enables scaling, translation, and rotation, using the interact tool.

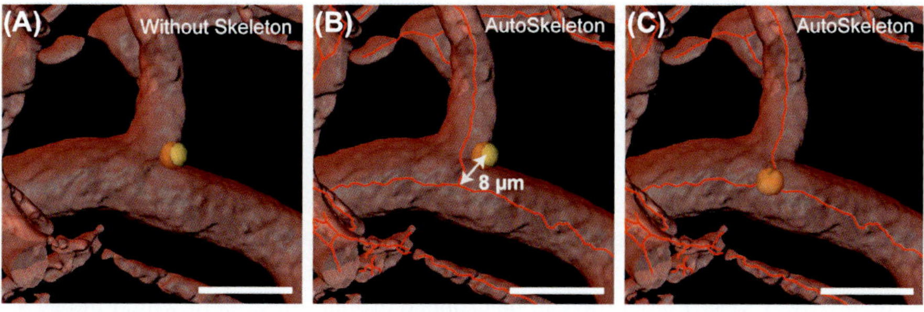

FIGURE 6

Placing the landmarks. (A) A view of the intravital microscopy surface file, showing a vessel bifurcation. Without the use of an *Auto Skeleton*, the landmark is placed on the surface. (B and C) In presence of an *Auto Skeleton*, the landmark can be positioned in the center of the bifurcation, improving the accuracy of the registration. The distance between the position of the landmark on the surface and on the *Auto Skeleton* is 8 μm. Scale bars: 20 μm.

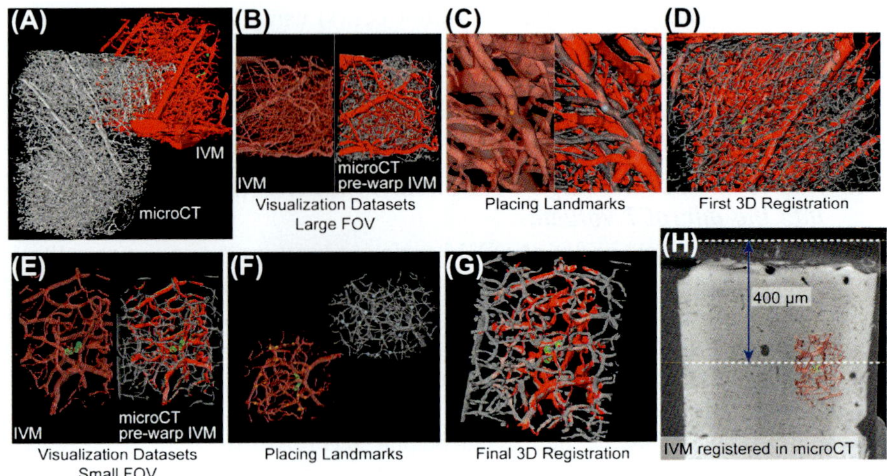

FIGURE 7

3D registration of the intravital microscopy (IVM) data sets into the microCT volume. (A) The IVM large-field of view (FOV) and microCT surface files are visualized in Amira's 3D viewer window. (B) Using a split screen, the IVM surface (left) and the prewarped IVM surface and cropped microCT surface (right) can be visualized simultaneously. (C) Landmarks (shown as small *spheres*) are placed in corresponding positions in the IVM and microCT data set. To improve accuracy of the registration, the landmarks are placed on the *Auto Skeleton* (see Fig. 6) that represents the core of the surface. (D) Based on the initial placement of 5–10 landmarks, a first 3D registration can be performed. (E) The first registration enables to crop the microCT label file to a small area around the prewarped position of the cell of interest (right panel). In a split screen, show the small-FOV Cell01 IVM surface (left), the prewarped IVM surface, and the microCT-CropCell01 surface. (F) Place the landmarks (*spheres*) so that these are evenly distributed around the tumor cell. (G) Perform the final registration of the small-FOV IVM surface into the microCT-CropCell01 surface. Since Amira preserves the coordinates of the data set during cropping, this registration also applies to the full, uncropped microCT data set. (H) Registration of the small-FOV Cell01 IVM surface in the microCT data set allows to measure the depth of the tumor cell with respect to the block surface (*top dotted line*).

4. By adjusting the position, rotation, and scale of the IVM data set, manually fit it inside the microCT data set. Then, inactivate Transform Editor and save the surface file (Fig. 7B).
5. The microCT vessel label file can be cropped around the area in which the IVM PreWarp surface is manually positioned. Activating the Crop Editor (Fig. 4B), a dialogue window appears and the outlines of the volume are shown in the 3D Viewer. Using the interaction tool, the volume can now be reduced around the area of interest by reducing the size of the box. Save the cropped label file under a new name, and visualize the segmentation.

1.3.3.1 Creating Auto Skeletons of the label files

6. The *Auto Skeleton* module (Extension: XSkeleton) shows the centerline and nodes of the label file. A simplified label file will generate less noisy centerlines; it is thus recommended to create a smoothened label file especially for the Auto Skeleton. Duplicate the label files, and add "-forSkeleton" to the name, in the Segmentation tab, select the new label file (in Label Data), and go to *Segmentation > Smooth Labels* (size 3, 3D volume). Do this for both the cropped microCT vessel label file and the IVM label file.
7. Select the "-forSkeleton" label files, and create an Auto Skeleton (*Image Processing > Skeletonization > Auto Skeleton*). A new *Spatial Graph View* appears, untick the box "Nodes" (Properties Window).
8. The *Surface Views* and *Spatial Graph Views* can be visualized simultaneously in Amira's 3D Viewer window. In the Properties Area, make the *Surface View* semitransparent by selecting *Draw Style: Transparent.* Change *Base Trans* to 0.6.

1.3.3.2 Placing the landmarks

9. Change the 3D viewer to a split screen (two viewers, vertical, Fig. 4A) to be able to visualize both data sets independently (Fig. 7B).
10. In the left screen, visualize the IVM vessels surface and Auto Skeleton. In the right screen, show the microCT vessels surface and Auto Skeleton and the prewarped IVM surface. The latter will function as a reference to identify corresponding points in the data sets (Fig. 7B).
11. Right-click into the Project View area, and select "Create Object…" in the menu and select *Points and Lines > Landmarks (2 sets).* Create two *Landmark View* modules (*Display > Landmark View*). Select the *Landmark View* and change *Point Set: Point Set 1* in the Properties Area. Limit its visualization to the left panel, using the viewer toggle. In the second *Landmark View2*, select *Point Set: Point Set 1* and show it only in the right panel.
12. To facilitate placing the landmarks, it is easier to pin a couple of useful objects in the Properties Area. Hereto, select the object and use the "pin" icon (Fig. 4B) to keep it visible in the Properties Area even though it is not selected. Pin the *Surface Views* of the IVM, the microCT (cropped), the prewarped IVM and the *Landmark-2-sets.*
13. To start placing the landmarks, select the Landmark Editor in the *Landmark-2-sets* (Properties Area). Select *Edit Mode: Add.*
14. Zoom into the right panel and identify a feature, i.e., vessel branches and bifurcations, which is visible in both the data sets. When a position is found, center the same area in the IVM surface in the left panel.
15. With *Landmark-2-sets* selected, place the first Landmark with the interact tool. Hereto, disable the visualization of the IVM surface and click on the appropriate position on the underlying Auto Skeleton. A yellow sphere appears in that position. Next, disable the visualization of the IVM prewarp surface and the microCT vessel surface, and click on the corresponding position on the

microCT vessel Auto Skeleton. A blue sphere appears. Change to Trackball mode, enable the visualization of the surfaces, and confirm if the Landmark spheres are placed correctly (Fig. 7C). Incorrectly placed landmarks can be moved or removed using the corresponding settings in *Edit Mode* (Properties *Landmark-2-sets,* Landmark Editor).

16. Seed five to seven well-spread Landmarks throughout the data set.
17. At this point, it makes sense to perform a first Landmark surface warp, to generate a reference prewarped surface, which is better than the manual fit (Fig. 7D). Hereto, select *Landmarks-2-sets* and create a *Landmark Surface Warp* (*Compute > Landmark Surface Warp*). Select in *Surface Data* the original IVM surface file, and choose *Direction 1>2.* These settings will result in the warping of the IVM surface into the microCT surface, by calculating the transformation of Point Set 1 into 2. Select *Method*: *Rigid* and choose *Affine.* This will enable translation, rotation of the surface file, and independent scaling of its x, y, and z dimensions. Visualize the new warped surface.
18. This warped surface provides an improved reference data set, enabling to place more Landmarks. If, however, the registration already looks satisfactory, it is possible to proceed from here with warping the smaller FOV IVM image volume(s) into the microCT data set (Fig. 7D).

1.3.4 Registration: small field of view around the tumor cells(s)

1. Start a new Amira Project, and import the small-FOV IVM imaging stack ("small-FOV Cell01"), which shows the ROI at a higher magnification.
2. Segment the vessels and the tumor cell(s) in the volume, and visualize the small-FOV Cell01 IVM surface.
3. Load the warped large-FOV IVM surface (Section 1.3.3) and the cropped microCT vessel surface file and label file.
4. Duplicate the cropped microCT vessel label file and further crop it around the tumor cell, to a similar volume as shown in the small-FOV Cell01 IVM surface. Save this microCT label file without overwriting the full-size file, e.g., by adding "-CropCell01" to its name.
5. Remove the warped IVM large-FOV surface, the tiff stack, the older microCT vessel surface file and label file from the network, but keep the new microCT-CropCell01 label file.
6. Create Auto Skeletons from both the IVM and the microCT label files (Figs. 6 and 7E).
7. Place Landmarks into the data sets, and warp the small-FOV Cell01 IVM surface into the microCT-CropCell01 surface. While placing the 7–20 Landmarks, ensure to find common points close to the tumor cell, and well dispersed in the space around it (Fig. 7F). Following warping, carefully check the fit of the IVM Cell01 surface into the microCT-CropCell01 surface, particularly around the tumor cell. The quality of the fit can be judged by the level of overlap between the segmented vessels visible in both the surfaces (Fig. 7G).

8. In a new project, load the resampled microCT z-stack (the full volume) and the warped IVM Cell01 surface file. The warped small-FOV Cell01 IVM surface is shown in its registered position within the microCT z-stack. By scrolling through the images, confirm that, indeed the vessels from the IVM surface, align with the vessels visible in the microCT. We emphasize here that cropping and reloading the label files and data sets *do not* influence where these are loaded in Amira's virtual coordinate system. This means that 3D registration of the small-FOV IVM volume in a cropped microCT volume also registers the IVM volume in the corresponding position within the full-sized microCT data set.
9. Align a *Slice* of the microCT z-stack with the top surface of the resin block. This can be done manually by *Options: Rotate* in the Properties Area. Use the Interact tool on the "Trackball" that appears on the *Slice* to rotate it to the desired orientation. Alternatively, use the *Options: Fit to points.* Using the Interact tool, click on three different positions on the top surface of the resin block (easily visible using a *Slice* that shows the side of the block in cross section). The *Slice* will then automatically align to the block face. In the Properties Area, select *Sampling: Finest* to achieve the highest resolution.
10. Duplicate the block face *Slice* and move it to a position 3–5 μm above the predicted position of the tumor cell. Note: Make sure that the 3D Viewer (Fig. 4A) is set to "Orthographic" and not "Perspective."
11. Measure the distance between the block face slice and the ROI (Fig. 7H). Hereto, use the 2D measurement tool.
    Note: In Amira 6.1 and later versions, the measurement tool is by default set to "3D measurements." This can lead to incorrect measurements of the distance between two slices. To change the measurement tool to 2D, type "measure useNewMeasureTools 0" into the console window and restart the software.

We have created a data package that contains Amira files, which allows the interested reader to interact with representative IVM, microCT, and registered data sets. A zipped folder, containing the data and a pdf file describing each of the files, can be downloaded at this link: https://www.embl.de/download/schwab/2017-Karreman_etal-MCB.zip.

1.4 TARGETED TRIMMING

Following registration of the IVM volume into the microCT data set, the sample can be trimmed in an ultramicrotome to approach the ROI (Fig. 8). Adhere to the small sample, imaged with microCT, to a blank resin block. When using resin to attach the sample, let it polymerize for 24–48 h at 60°C.

Mount the sample in an ultramicrotome, and ensure to keep track of its orientation, e.g., by noting which side of the resin block is facing up when it sits in the microtome. Here, it helps to have trimmed the sample asymmetrically. For targeted trimming, it is critical to be able to correlate the position of the sample in microtome to the 3D visualization of the microCT data set in Amira.

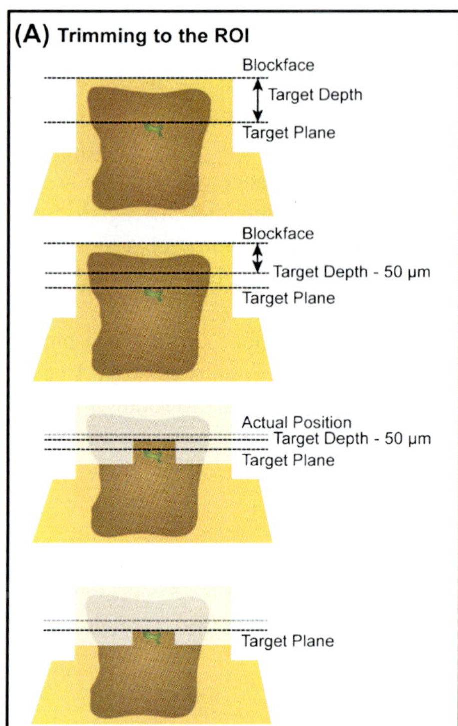

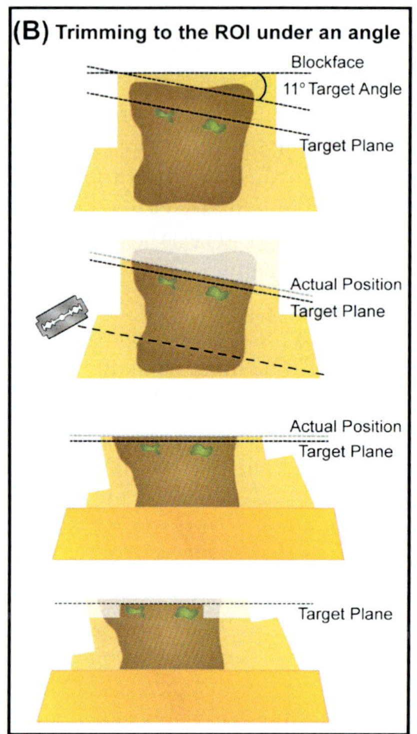

FIGURE 8

Trimming the resin block to expose the tumor cell for 3DEM. (A) Based on the 3D registration of the intravital microscopy data sets into the microCT volume, the distance between the block surface (*top dotted line*) and the tumor cell (*bottom dotted line*, Target Plane) can be measured: This is the Target Depth. Trim the front of the block to a depth of 50 μm above the Target Plane ("Target Depth—50 μm"), to stay at a safe distance of the tumor cell. Check the progression into the block based on an LM section, as described in the text and Fig. 9). Based on the prior knowledge of the position of the tumor cell with respect to the sides of the block, trim sides to reduce the block surface to ~300 × 500 μm. Finally, approach the tumor cell in z. (B) In case multiple cells are targeted simultaneously, or the tumor cell is under a specific angle, it could be worthwhile to change the angle of the approach to the region of interest (ROI) (*bottom dotted line*). In this case, measure in Amira the angle between the block face and the desired angle (here 11-degree Target Angle). This angle can be created by tilting the trimming diamond knife (see Fig. 8). If the angle is too high, subsequent trimming and sectioning will be affected. In that case, cut the sample from the resin block (as shown in Fig. 2C) and trim its base with a razor blade, so that it is parallel to the new surface angle. Then, remount the sample on a new resin block and trim to the ROI as described in A.

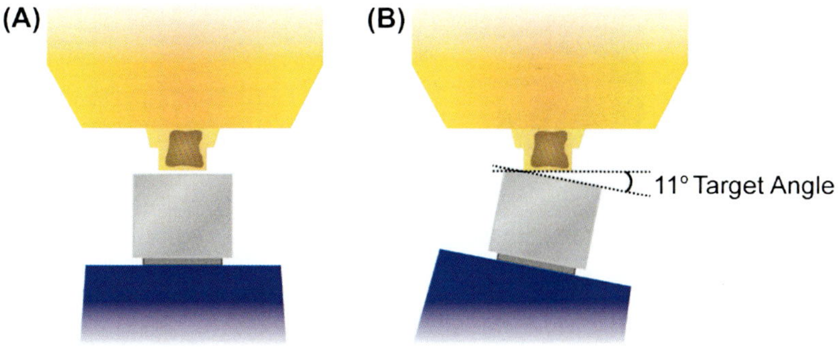

FIGURE 9

Creating an angled block surface. The sample should be mounted in the microtome, so that the angle can be introduced horizontally by rotating the trimming diamond. (A) To create a new, angled block surface, align the trimming diamond knife accurately to the resin block. Take particular care to align also the bottom and top side of the sample very well parallel to the cutting edge of the knife. (B) Change the rotation of the knife, so that the desired angle is achieved. Note: The knife could already be under an angle due to the alignment. Next, the knife will start cutting only one side of the block, creating an angled block surface.

Align a trimming diamond accurately to the block face. If required, an angle can be introduced to the block surface by rotating the knife following alignment (Fig. 8B: Target Angle, Fig. 9).

Trim to a depth that is 20–50 μm above the predicted ROI (Fig. 8, Target Plane), and check the progression into the resin block. Hereto, obtain a thick 300- to 500-nm section, and place it on a (Superfrost++) glass slide. Dry the section on a hot plate, stain it with toluene blue (or a comparable histological stain), and image it with a light microscope (LM, magnification 10–20×). Correlate the LM image to a virtual section (*Slice*) of the microCT data set, which is at the expected distance and angle with respect to the block surface (Fig. 8). Generally, the error margin of the microtome is around 5% (we experienced a similar offset with two different Leica UC7 microtomes); it removes less material than is indicated on the counter. When the knife is trimming under an angle (Fig. 9B), this error margin is even larger. By moving the *Slice* through the microCT data set, the best match to the LM image can be found. This enables to measure the actual progression toward the Target Plane (Figs. 8 and 10).

In Amira, determine how much material may be removed from the sides of the resin block to generate a smaller block face (Figs. 8 and 10C). Trim the desired amount of material using the straight side of a 90-degree diamond knife. To be on the safe side, the trimming can be performed to a limited depth, which is still above the ROI. Then, the dimensions of the new block face can be confirmed after trimming by correlating again an LM image of a thick section from the trimmed block to a virtual section in Amira. If this trim test is satisfactory, trim the sides further to below the volume of interest.

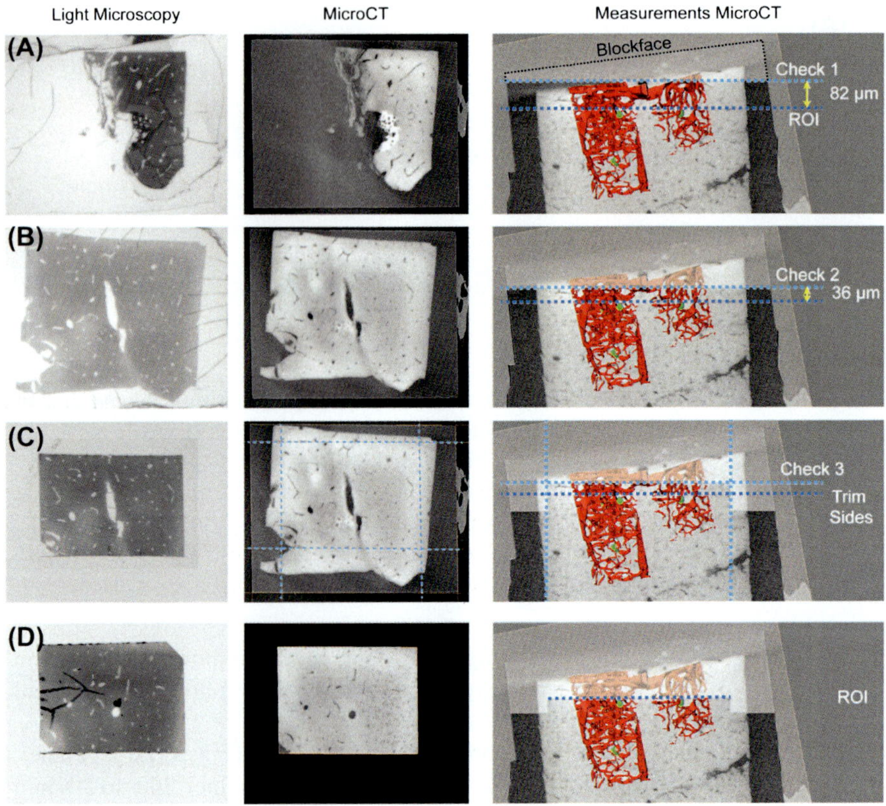

FIGURE 10

Checking the progression to the region of interest (ROI) as part of the targeted trimming procedure. (A) An angle is introduced to the block face, and a 300-nm section of the new surface is stained and imaged with LM (left panel). The matching section is found in the 3D volume of the microCT data set (middle panel) and is shown in the right panel as the top *thick dotted line* (Check 1). This confirms that the angle is correctly introduced (with respect to the original block face, *thin dotted outline*) and that the distance to the ROI (*bottom dotted line*, ROI) is 82 μm. (B) Following a second approach, the procedure described in A allows to determine that the current position is 36 μm from the ROI. This is a suitable position to start trimming the sides of the future block face. (C) Measure, based on the Amira 3D registration, how much material can be removed from each side and trim this using the straight sides of a 90-degree trimming diamond or glass knife. (D) After trimming the sides, carefully approach the ROI. Obtain a 300-nm section from the resin block before starting serial sectioning or mounting the block for SBF-SEM or FIB-SEM. *FIB-SEM*, focused ion beam–scanning electron microscopy; *SBF-SEM*, serial block face scanning electron microscopy.

Trim the block face up to approximately 3−5 μm above the predicted position of the ROI (Fig. 8, Target Plane). The approach to the ROI can be monitored as described before. It is helpful to obtain an LM image of the final 300- to 500-nm section that is taken from the block face, to correlate with the subsequent EM imaging (Fig. 10D). The resin block is now ready for serial sectioning or to be mounted on an SEM stub for FIB-SEM imaging or serial block face SEM (SBF-SEM).

2. INSTRUMENTATION AND MATERIALS

2.1 PROCESSING FOR ELECTRON MICROSCOPY

Instrumentation and Materials:

1. PELCO BioWave Microwave with Coldspot (Ted Pella)
2. Resin-embedding mold
3. Oven, set to 60°C

Solutions and Reagents:

1. Glutaraldehyde (EM grade EMS cat#16220)
2. Formaldehyde (EM grade EMS cat#15710)
3. Cacodylate buffer (pH 7.2)
4. Osmium tetroxide in water (EM grade EMS cat#19150)
5. $K_3Fe(CN)_6$ (Merck, art. 4973)
6. UA (research grade, Serva lot 150126)
7. Ethanol
8. Resin: Durcupan (Sigma Cat# 44610) or Epon (Serva)

2.2 TRIMMING THE RESIN BLOCK AND microCT IMAGING

1. Razor blades
2. Ultramicrotome UC7 (Leica Microsystems)
3. Trimming diamond (TrimTool 90 degree, DiATOME) or glass knife
4. Long glass Pasteur pipettes
5. Wax or glue
6. MicroCT system (phoenix nanotom m, GE Sciences)
7. MicroCT data reconstruction and processing software

2.3 SEGMENTATION AND THREE-DIMENSIONAL REGISTRATION IN AMIRA

1. Amira v.6 (FEI visualization group, Thermo Fisher Scientific), with extensions XImagePAQ (optional) and XSkeleton (required).
2. 64-bit computer with Intel Xeon processor CPU 3.50 GHz, 32-bit RAM, and an NVIDIA Quadro K4000 graphics card, running Windows 7 professional. Minimal technical requirements to run Amira can be found on www.fei.com/software/amira-3d-for-life-sciences/.

2.4 TARGETED TRIMMING

1. "Dummy" blocks: blank resin block to remount the sample on
2. Ultramicrotome UC7 (Leica Microsystems)
3. Trimming diamond knife (TrimTool 90 degree, DiATOME) or glass knife
4. Histo diamond knife (DiATOME) or glass knife with attached boat
5. Superfrost++ glass slides and pick-up loops
6. Toluene blue stain
7. Hot plate, set to 100°C
8. Light microscope with 10× and 20× air objectives and digital camera

3. DISCUSSION

The method described here enables to reliably and easily correlate between IVM and 3DEM through an intermediate step of X-ray microCT imaging of the EM-processed sample. The power of the method can be found in its throughput and versatility; we are now routinely using this approach to study various model systems and biological questions that focus on monitoring and retrieving rare events in large voluminous samples. Moreover, it is applicable to prepare the sample for serial section TEM, FIB-SEM, and serial block face imaging.

Others demonstrated the use of microCT in the prescreening of a resin-embedded sample (Burnett et al., 2014; Bushong et al., 2014; Handschuh, Baeumler, Schwaha, & Ruthensteiner, 2013; Sengle, Tufa, Sakai, Zulliger, & Keene, 2013) and the correlation between fluorescence microscopy and EM (Bushong et al., 2014; Shami et al., 2016). MicroCT imaging provides a unique insight into the organization and orientation of the sample following EM processing. This information assists in retrieving the ROI postprocessing, as described in our work (Borrego-Pinto et al., 2016; Karreman et al., 2016) and that of others (Bushong et al., 2014; Shami et al., 2016). Moreover, it also enables adjusting the imaging orientation for 3DEM, which can be critical to study the organization of tissues and small organisms.

Multimodal intravital correlative microscopy as presented here is an approach that in principle can be relatively easily implemented since the different procedures described here are quickly learned. However, the method does require a range of high-end equipment and techniques, including IVM, a microCT setup, an ultramicrotome, and access to appropriate software (Amira) and to an electron microscope (TEM, FIB-SEM, or serial block face SEM). The application of this workflow thus may rely on establishing collaborations with different laboratories and/or companies that can offer access to, and experience with, one or more of these instruments. In our specific example, the IVM, microCT, and EM imaging were each performed at different research institutes.

Although this approach already offers a satisfying throughput and reliability (~14 days from IVM to EM), it is still possible to facilitate and speed up certain parts of the procedure. Most importantly, registration of both data sets could be automated (in part) to gain time and improve the ease of use of the workflow. Currently, it takes on average 1−2 days to perform the segmentation and registration in Amira (see Section 1.3), depending also on the experience of the operator and the number of target areas. Further advancements of the software or the development of specialized plug-ins may allow to minimize the input of the operator and further simplify the protocol. In addition, the targeted trimming to approach the ROI may be subject to improved accuracy or even automation. The microCT data set and 3D registration of the IVM data set provide the exact dimensions of the resin block and the coordinates of the ROI inside (see Sections 1.3 and 1.4). Based on this information, an automated "sculpting" of the resin block to expose the ROI could thus be envisioned. Finally, the most time-consuming step of the full protocol is the EM imaging and subsequent processing of the data. The latter is a general problem for 3DEM, and software packages are currently developed to help and facilitate these steps (Belevich, Joensuu, Kumar, Vihinen, & Jokitalo, 2016; Schindelin et al., 2012; Sommer, Straehle, Kothe, & Hamprecht, 2011).

In summary, this chapter outlines the most critical and specific steps involved in the multimodal correlative microscopy workflow. This approach allows to quickly move from in vivo imaging of temporary, rare events in pathological or development models, to imaging these at high resolution using 3DEM. Although we demonstrate the approach here on a specific example, the study of tumor cells arrest in the vasculature of the mouse brain, the method is versatile and can be applied to different model systems and to answer a diversity of biological research questions.

ACKNOWLEDGMENTS

We would like to thank Robert Brandt and Peter Westenberger from FEI visualization sciences group (Thermo Fischer Scientific) for their help and suggestions on the use of Amira software. Furthermore, we would like to thank Dr. Heinz Schwarz for his advice on the use of buffers during EM sample preparation.

REFERENCES

Belevich, I., Joensuu, M., Kumar, D., Vihinen, H., & Jokitalo, E. (2016). Microscopy image browser: A platform for segmentation and analysis of multidimensional datasets. *PLoS Biology, 14*(1), e1002340. http://dx.doi.org/10.1371/journal.pbio.1002340.

Bishop, D., Nikić, I., Brinkoetter, M., Knecht, S., Potz, S., Kerschensteiner, M., & Misgeld, T. (2011). Near-infrared branding efficiently correlates light and electron microscopy. *Nature Methods, 8*(7), 568−570. http://dx.doi.org/10.1038/nmeth.1622.

Borrego-Pinto, J., Somogyi, K., & Karreman, M. A. (2016). Distinct mechanisms eliminate mother and daughter centrioles in meiosis of starfish oocytes. *The Journal of Cell.*

Burnett, T. L., McDonald, S. A., Gholinia, A., Geurts, R., Janus, M., Slater, T., … Withers, P. J. (2014). Correlative tomography. *Scientific Reports, 4.* http://dx.doi.org/10.1038/srep04711.

Bushong, E. A., Johnson, D. D., Kim, K.-Y., Terada, M., Hatori, M., Peltier, S. T., … Ellisman, M. H. (2014). X-ray microscopy as an approach to increasing accuracy and efficiency of serial block-face imaging for correlated light and electron microscopy of biological specimens. *Microscopy and Microanalysis, 21*(01), 231–238. http://dx.doi.org/10.1017/S1431927614013579.

Durdu, S., Iskar, M., Revenu, C., Schieber, N., Kunze, A., Bork, P., … Gilmour, D. (2014). Luminal signalling links cell communication to tissue architecture during organogenesis. *Nature, 515*(7525), 120–124. http://dx.doi.org/10.1038/nature13852.

Ellenbroek, S. I. J., & van Rheenen, J. (2014). Imaging hallmarks of cancer in living mice. *Nature Reviews Cancer, 14*(6), 406–418. http://dx.doi.org/10.1038/nrc3742.

Follain, G., Mercier, L., Osmani, N., Harlepp, S., & Goetz, J. G. (2016). Seeing is believing: Multi-scale spatio-temporal imaging towards in vivo cell biology. *Journal of Cell Science*. http://dx.doi.org/10.1242/jcs.189001. jcs.189001.

Handschuh, S., Baeumler, N., Schwaha, T., & Ruthensteiner, B. (2013). A correlative approach for combining microCT, light and transmission electron microscopy in a single 3D scenario. *Frontiers in Zoology, 10*(1), 1–16. http://dx.doi.org/10.1186/1742-9994-10-44.

Karreman, M. A., Hyenne, V., Schwab, Y., & Goetz, J. G. (2016). Intravital correlative microscopy: Imaging life at the nanoscale. *Trends in Cell Biology*. http://dx.doi.org/10.1016/j.tcb.2016.07.003.

Karreman, M. A., Mercier, L., Schieber, N. L., Shibue, T., Schwab, Y., & Goetz, J. G. (2014). Correlating intravital multi-photon microscopy to 3D electron microscopy of invading tumor cells using anatomical reference points. *PLoS One, 9*(12), e114448. http://dx.doi.org/10.1371/journal.pone.0114448.s006.

Karreman, M. A., Mercier, L., Schieber, N. L., Solecki, G., Allio, G., Winkler, F., … Schwab, Y. (2016). Fast and precise targeting of single tumor cells in vivo by multimodal correlative microscopy. *Journal of Cell Science, 129*(2), 444–456. http://dx.doi.org/10.1242/jcs.181842.

Kienast, Y., Baumgarten, von, L., Fuhrmann, M., Klinkert, W. E. F., Goldbrunner, R., Herms, J., & Winkler, F. (2010). Real-time imaging reveals the single steps of brain metastasis formation. *Nature Medicine, 16*(1), 116–122. http://dx.doi.org/10.1038/nm.2072.

Maco, B., Holtmaat, A., Cantoni, M., Kreshuk, A., Straehle, C. N., Hamprecht, F. A., & Knott, G. W. (2013). Correlative in vivo 2 photon and focused ion beam scanning electron microscopy of cortical neurons. *PLoS One, 8*(2), e57405. http://dx.doi.org/10.1371/journal.pone.0057405.g003.

Müller-Reichert, T., Srayko, M., Hyman, A., O'Toole, E. T., & McDonald, K. (2007). Correlative light and electron microscopy of early *Caenorhabditis elegans* embryos in mitosis. *Methods in Cell Biology, 79*, 101–119. http://dx.doi.org/10.1016/S0091-679X(06)79004-5.

Schindelin, J., Arganda-Carreras, I., Frise, E., Kaynig, V., Longair, M., Pietzsch, T., … Cardona, A. (2012). Fiji: An open-source platform for biological-image analysis. *Nature Methods, 9*(7), 676–682. http://dx.doi.org/10.1038/nmeth.2019.

Sengle, G., Tufa, S. F., Sakai, L. Y., Zulliger, M. A., & Keene, D. R. (2013). A correlative method for imaging identical regions of samples by micro-CT, light microscopy, and electron microscopy: Imaging adipose tissue in a model system. *The Journal of Histochemistry and Cytochemistry, 61*(4), 263–271. http://dx.doi.org/10.1369/0022155412473757.

Shami, G. J., Cheng, D., Huynh, M., Vreuls, C., Wisse, E., & Braet, F. (2016). 3-D EM exploration of the hepatic microarchitecture – lessons learned from large-volume in situ serial sectioning. *Scientific Reports, 6*(36744). http://dx.doi.org/10.1038/srep36744.

Sommer, C., Straehle, C., Kothe, U., & Hamprecht, F. A. (2011). Ilastik: Interactive learning and segmentation toolkit. In *Presented at the eighth IEEE international symposium on biomedical imaging (ISBI)* (pp. 230–233). http://dx.doi.org/10.1109/ISBI.2011.5872394.

Zito, K., Parnas, D., Fetter, R. D., Isacoff, E. Y., & Goodman, C. S. (1999). Watching a synapse grow: Noninvasive confocal imaging of synaptic growth in *Drosophila*. *Neuron, 22*, 719–729.

CHAPTER

14

triCLEM: combining high-precision, room temperature CLEM with cryo-fluorescence microscopy to identify very rare events

Nicholas R. Ader[*,§], **Wanda Kukulski**[*,1]

*MRC Laboratory of Molecular Biology, Cambridge, United Kingdom

§National Institutes of Health, Bethesda, MD, United States

[1]Corresponding author: E-mail: kukulski@mrc-lmb.cam.ac.uk

CHAPTER OUTLINE

Methods in Cell Biology, Volume 140, ISSN 0091-679X, http://dx.doi.org/10.1016/bs.mcb.2017.03.009

Abstract

Fiducial-based correlation of fluorescence and electron microscopy data from high-pressure frozen and resin-embedded samples allows for high-precision localization of fluorescent signals to subcellular ultrastructure. Here we introduce the triCLEM procedure to facilitate the identification of very rare events for high-precision correlation. We present a detailed protocol to screen high-pressure frozen cell monolayers on sapphire disks for very rare signals by cryo-fluorescence microscopy, relocate the cells of interest after freeze substitution and Lowicryl embedding, and perform fiducial-based correlation of the identified fluorescent signals to high-magnification electron tomograms. We show the applicability of the protocol to localize and image damaged mitochondria marked by the presence of Parkin, a protein involved in initiating mitophagy. We discuss how this extension to previously published fiducial-based correlation procedures has potential to both allow identifying very rare events and assess the quality of preservation in high-pressure frozen samples.

INTRODUCTION

In recent years, the application of correlative light and electron microscopy (CLEM) to biological samples has benefited from many technical advances (de Boer, Hoogenboom, & Giepmans, 2015). In particular, preservation of fluorescence following high-pressure freezing (HPF), freeze substitution (FS), and Lowicryl embedding (Nixon et al., 2009) has allowed for the expansion of correlative techniques that capture information about the same biological time point by fluorescence microscopy (FM) and electron microscopy (EM) or electron tomography (ET). Furthermore, the use of fluorescent fiducial markers—visible in both FM and EM images—has allowed high-precision (<100 nm) spatial correlation of fluorescent protein (FP)-labeled proteins to discrete localities of cellular ultrastructure (Kukulski et al., 2011).

While preserving fluorescent signals in resin sections facilitates imaging transient biological time points, it remains challenging to acquire significant data sets of very rare structures. For instance, when fluorescent signals are present in a low percentage of a heterogeneous population of cells, due to low transient transfection of a reporter plasmid, or when the signal represents a particularly ephemeral event. Assuming a transient transfection system with low transfection efficiency (~20% of cells are transfected with a plasmid encoding the FP-tagged protein of interest), low cellular expression (one fluorescent spot per cell), and a resin section thickness of 300 nm [capturing about ~10%–20% of volume of a HeLa cell (Zhao et al., 2008)], only about 2%–4% of cell profiles in the sections would be expected to display fluorescence signals. In such cases, it would be helpful to include an additional step to identify cells containing fluorescent signals of interest prior to sectioning. This can be done before HPF; various CLEM protocols are available for live imaging prior to HPF, FS, and resin embedding. For example, specialized assemblies of carrier, spacers, and sapphire disks facilitate live-cell FM immediately followed by HPF for imaging of cell monolayers (Brown, Van Weering, Sharp, Mantell, & Verkade, 2012) or for controlling preparations of fragile cellular extracts (Tranfield, Heiligenstein, Peristere, & Antony, 2014). To localize the cell or region of interest (ROI), targeted ultramicrotomy procedures have been developed that are based on a grid pattern in the carbon film evaporated on the sapphire disk (McDonald et al., 2010). These approaches can be combined with in-resin fluorescence imaging and high-precision correlation (Heiligenstein et al., 2014).

Even though the delay between live imaging and HPF can be minimized to approximately 5 s (Verkade, 2008), it remains an issue for very dynamic structures. Moreover, cells can change location, and fragile ultrastructures can disintegrate when handling the sample during HPF (Tranfield et al., 2014). Therefore, in many cases, it would be desirable to screen for cells with fluorescent signals after they have been high-pressure frozen, prior to selecting the region of the resin block that will be sectioned. This could be achieved by imaging the whole resin block by laser scanning microscopy (Hohn et al., 2015). However, faint signals from events marked by low FP copy numbers may be bleached during this first FM step, and may therefore not be detectable in the second step of resin-section imaging, which is critical for high-precision CLEM.

Here, we describe a modification to the CLEM procedure described in detail previously in this series (Kukulski et al., 2012), incorporating screening of high-pressure frozen cells by cryogenic FM (cryo-FM). As cryo-FM allows for FM in fully vitrified samples, it is becoming increasingly employed to identify structures of interest that will subsequently be imaged by cryo-EM (Schorb & Briggs, 2014) or cryo-ET (Koning et al., 2014; Mahamid et al., 2015; Rigort, Villa, Bauerlein, Engel, & Plitzko, 2012). While most cryo-FM has been performed on cells grown on EM grids or purified and reconstituted protein assemblies, Strnad et al. (2015) recently demonstrated imaging of mammalian cells cryo-immobilized on sapphire disks by cryo-FM combined with subsequent scanning EM analysis. Similarly, Peddie et al. (2014) used cryo-FM to image cell pellets in HPF carriers. Here, we employ cryo-FM to image mammalian cells high-pressure frozen on sapphire disks. These samples are then subjected to the previously described FS and resin-embedding protocol that allows retaining FP signals (Avinoam, Schorb, Beese, Briggs, & Kaksonen, 2015; Nixon et al., 2009). The cryo-FM images are used for guidance in targeted ultramicrotomy. Furthermore, we use the cryo-FM images to roughly correlate fluorescent signals of interest to room temperature FM (RT-FM) images of resin sections on grids. The RT-FM data set is then precisely correlated to ET data, following the procedure described in Kukulski et al. (2012). Importantly, the fluorescent signals remain preserved until RT-FM of resin sections is performed because photobleaching of FPs is strongly reduced at cryogenic temperatures (Schwartz, Sarbash, Ataullakhanov, McIntosh, & Nicastro, 2007). This three-microscope correlation procedure (triCLEM) allows for the identification of very rare events via cryo-FM and then achieves high-precision, fiducial-based correlation between RT-FM and ET data of the identified fluorescent signals, all at the same biological time point. An overview of the workflow and the timeline is provided in Fig. 1.

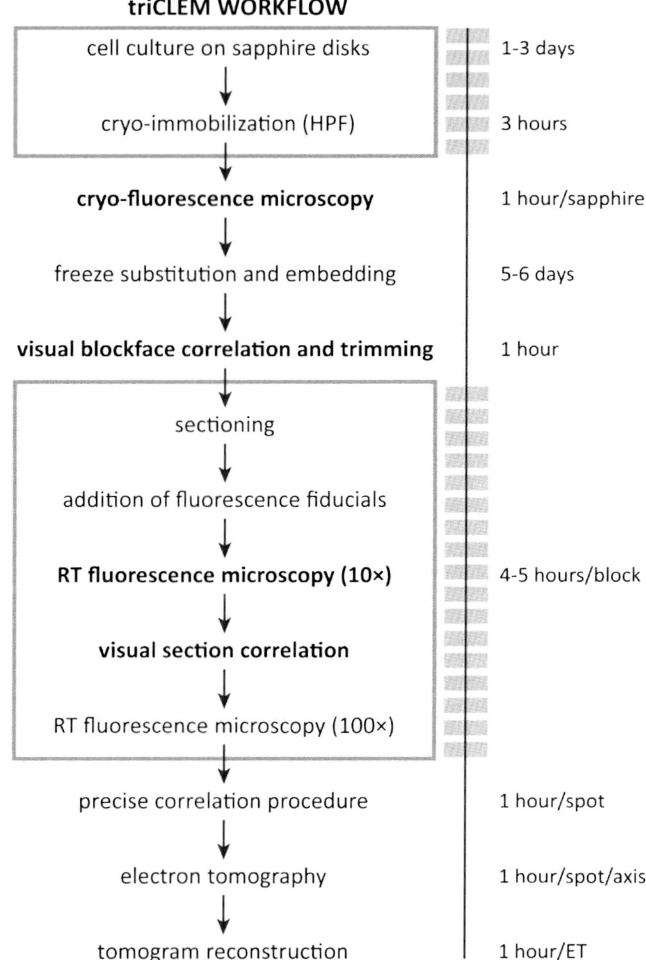

FIGURE 1 triCLEM textual overview and time schedule.

Additional steps compared to Kukulski et al. (2012) are in bold text; approximate labor timeline is on the right. Steps to be carried out without significant interruptions (i.e., on the same day) are in *boxes*. Following cell culture on sapphire disks, samples are high-pressure frozen. Cryo-immobilized cells on sapphires are screened using cryo-fluorescence microscopy prior to freeze substitution and Lowicryl HM20 embedding. Visual correlation between cryo-fluorescence images and resin block face allows for trimming of trapezoid around cells of interest. Sections, collected on grids, are imaged by room temperature fluorescence microscopy at low magnification (10×) and visually correlated with cryo-fluorescence images. High-magnification (100×) images are collected in identified grid squares of interest. Following precise, fiducial-based correlation between high-magnification fluorescence and a defocused electron micrograph, high-magnification tilt-series are collected and reconstructed.

We use this triCLEM procedure to identify and image damaged mitochondria marked by the presence of the ubiquitin ligase Parkin in human cervical cancer cells (HeLa). Parkin ubiquitinates mitochondrial outer membrane proteins, leading to recruitment of autophagy machinery and eventually to degradation of the damaged mitochondrion (Youle & van der Bliek, 2012). The HeLa cell line we use here stably expresses YFP-Parkin, which is normally cytosolic, but localizes to mitochondria upon mitochondrial insult (Narendra, Tanaka, Suen, & Youle, 2008). We achieve mitochondrial insult with a doxycycline (DOX)-inducible system to express a deletion mutant of a mitochondrial matrix-targeted protein that readily misfolds and leads to protein aggregation in the matrix (Jin & Youle, 2013). After DOX treatment, single YFP-Parkin foci form on mitochondria, but only in approximately 20% of cells. Despite low occurrence of fluorescent signals of interest in this system, we demonstrate that triCLEM allows for robust identification of these rare events in cryo-immobilized, whole cells and subsequent high-precision localization to the ultrastructure using RT-FM and ET.

This method description is accompanied by figures that follow the very same sample through the complete procedure, illustrating its robustness and reproducibility (See Fig. 5). Besides a detailed protocol for triCLEM, we describe several subtle modifications to our high-precision CLEM procedure to simplify and vary its applicability.

1. METHODS

1.1 CULTURE AND HIGH-PRESSURE FREEZING OF MAMMALIAN CELLS

We prepare 3 mm sapphires for cell culture by cleaning with detergent, carbon coating, and sterilization, a procedure modified from Walther, Wang, Liessem, and Frascaroli (2010). Following carbon coating, we scratch a "2" onto the carbon-coated surface, as this allows for orientation during subsequent handling (McDonald et al., 2010). Immediately prior to use in culture, we sterilize sapphires through baking at 120°C for at least 3 h or 5 min in an 800 W microwave.

HeLa cells are grown in the presence of 1 μg/mL DOX in supplemented DMEM to about 80% confluency on sapphire disks in 6-well plates. Prior to freezing, cells are incubated in 10 nM MitoTracker Deep Red, washed, and returned to fresh media.

HPF is accomplished using a Leica HPM100 in the provided temperature- and humidity-controlled chamber. To high-pressure freeze cells, we use a carrier method described in the manual for the Leica EM HPM100 CLEM 3 mm system for HPF of sapphire disks. In brief, carriers are assembled as follows between two plastic half cylinders: (1) 6 mm copper gold-plated support ring in a 6 mm CLEM middle plate, (2) a 3 mm sapphire (cells up), (3) a 3 mm spacer ring, (4) a clean sapphire, and (5) a 6 mm cover ring. Materials in direct contact with cells, including the cell-coated sapphire, are washed with FluoroBrite DMEM, as we find this enhances signal-to-noise ratio in cryo-FM.

1.2 CRYOGENIC-FLUORESCENCE MICROSCOPY

To image cryo-immobilized cells on sapphire disks, we use the Leica cryo-CLEM system (Schorb et al., 2016) in a humidity-controlled room. Thus, the humidity in the room is typically 20%–25%, which we find to be essential for imaging as well as sample handling with minimal contamination. In particular, imaging multiple samples over a longer time period is facilitated, as the stage openings and the objective remain frost-free.

Kept immersed in liquid nitrogen (LN_2), the sapphire is transferred to the LN_2-cooled cryo-FM transfer shuttle using a metal bottle cap. The sapphire (cell-side up) is mounted into the cartridge as described for EM grids (Schorb et al., 2016). We then use the shuttle to load the sapphire into the precooled (−195°C) cryo-stage. Once the cartridge with the sapphire is placed in the cryo-stage, we first use bright field (BF) imaging to find the z-height at which the surface of the sapphire is in focus. The cells are not visible beneath the ice in BF. However, the "2" etched into the carbon coat is visible in BF, helping to set an initial focus position. To map the cell distribution pattern on the entirety of the sapphire, we use the tile scan function in the Leica LAS X software to generate a stitched 2.22 × 2.22 mm image with a 50 μm (2.5 μm step) autofocus in the z-direction using the green fluorescent protein (GFP) channel, taking advantage of the visibility provided by cytosolic fluorescence background. BF and MitoTracker signal are captured at the same z-height (Fig. 2A). This area is sufficient to image all cells on the sapphire that were inside the 2 mm hole of the spacer ring during HPF. Using this map, smaller ROI (750 × 750 μm) can be identified and subsequently imaged at multiple focal planes over twenty 1 μm intervals (automatic z-stack) (Fig. 2B), allowing identification of cells that contain fluorescent signals of interest. We do not collect a z-stack for the entire sapphire, as trimming of the resin block at a later stage will reduce the final imaged area.

Once all images are acquired, the sapphire is removed from the cryo-stage using the transfer shuttle, transferred to a standard cryo-EM grid box, and stored in LN_2 until FS.

1.3 FREEZE SUBSTITUTION AND EMBEDDING

We follow the FS and Lowicryl HM20 embedding protocols that allow for retention of fluorescence (Kukulski et al., 2011; Nixon et al., 2009), with modifications catering to mammalian cells (Avinoam et al., 2015) and other minor adjustments. We use a temperature-controlling Leica AFS2 with FSP. FS is performed at −90°C for 24–36 h in 0.01%–0.04% (w/v) uranyl acetate in glass-distilled acetone. The temperature is then increased to −45°C (5°C/h). Next, the samples are washed three times with acetone and infiltrated with increasing concentrations (10%, 25%, 50%, 75%, 2 h each) of Lowicryl HM20 in acetone. During the final mix, the temperature is raised to −35°C (2.5°C/h). The temperature is then raised further to −25°C (2.5°C/h), while 100% Lowicryl is exchanged three times in 4 h steps with agitation. Then, UV light is applied for 24 h to initialize Lowicryl polymerization. The

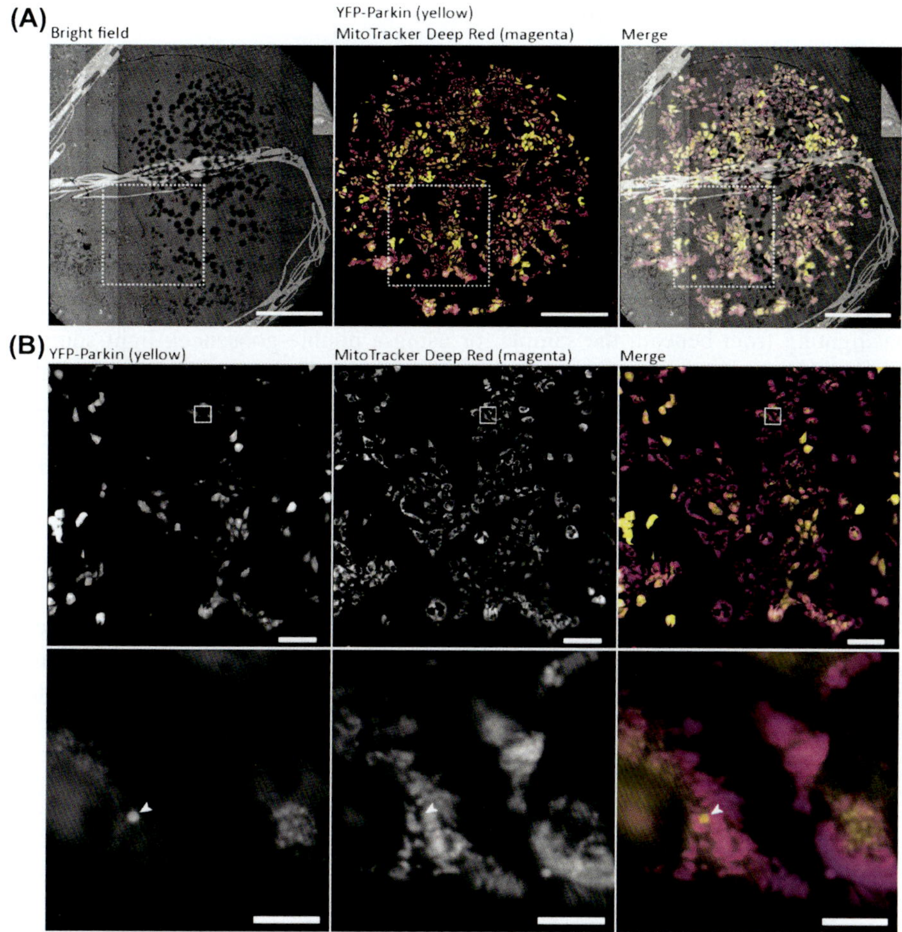

FIGURE 2 Cryo-fluorescence microscopy mapping of sapphire and region of interest.

High-pressure frozen cells on 3 mm sapphire disks imaged by cryo-fluorescence microscopy. (A) Tile scan of entire sapphire, acquiring bright field, YFP-Parkin and MitoTracker Deep Red signals. (B) Slice of z-stack (20 μm range) at region of interest (*dashed box*) in (A). In the *zoom box*, a YFP-Parkin signal is indicated by an *arrowhead*. Scale bars are 500 μm (A), 100 μm [(B), upper panel], and 10 μm [(B), lower panel].

temperature is then raised to 20°C (5°C/h). At this point, samples can be taken out of the AFS2, but we wait at least 2 days before removing blocks from the plastic wheel to ensure complete polymerization. Previously, we used 0.1% uranyl acetate (Kukulski et al., 2011). However, we find that in mammalian cells, the cellular ultrastructure can be equally well preserved with uranyl acetate concentrations as low as 0.01%. To maximize preservation of fluorescent signals, we therefore lowered the uranyl acetate concentrations. In our hands, 0.04% uranyl acetate provides an ideal balance of contrast and signal preservation, though this may vary depending on cell type.

1.4 VISUAL CORRELATION AND TRIMMING

After embedding, the sapphire is carefully removed from the polymerized Lowicryl block using a razor blade. An inverted "2" should now be visible on the block face without magnification. Cells are visible under a phase-contrast stereomicroscope (Loussert, Forestier, & Humbel, 2012; van Weering et al., 2010). Due to the low contrast of the cells in resin, it can be difficult to initially identify cells. We find that lighting from beneath the sample, or using a double-gooseneck light source, greatly aids in enhancing contrast, though rotation of the sample to highlight different areas of the block face is often necessary (Fig. 3A). Under the stereomicroscope of the microtome, illumination solely with a double-gooseneck light source is sufficient to identify cells. By comparing the distribution pattern of cells with the cryo-FM map and the cryo-BF image of the "2," we locate the previously identified ROI and trim the block accordingly by cutting a trapezoid around this location with a fine razor blade (Fig. 3B). We then section off roughly 10 sections of 300 nm thickness and collect each individually on an EM grid.

For trimming the block, we find a combination of the "2" and the observed cell distribution pattern provides sufficient accuracy to include the ROI (compare Fig. 2A with Fig. 3). The "2" allows for a rough localization of the ROI, and the visible cell distribution pattern is enough to trim away unwanted areas. While the use of gridded sapphires or an evaporated finder grid pattern would also greatly aid localization of the ROI, we find they are not necessary for our cell system.

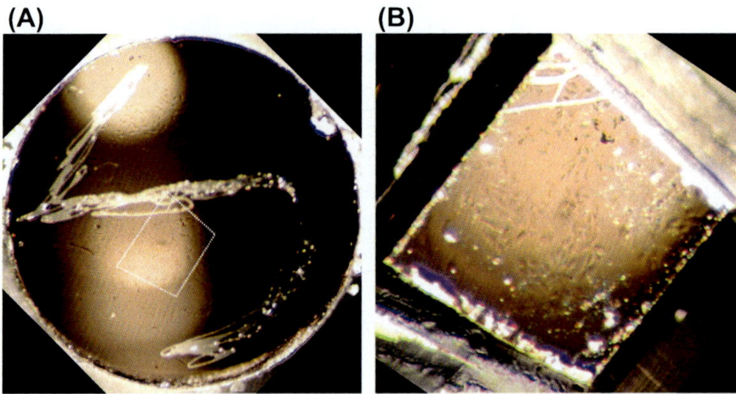

FIGURE 3 Resin-embedded cells pre- and post-trimming.

Resin-embedded cells from the same sapphire disk as in Fig. 2 visualized using a stereomicroscope with carefully adjusted illumination, both pre- (A) and post-trimming (B). Trimmed area shown in (B) is outlined in (A).

1.5 SECTIONING AND FLUORESCENT FIDUCIAL APPLICATION

For sectioning and application of fluorescent fiducial markers, we again follow a protocol modified from Kukulski et al. (2012). After sectioning at 300 nm thickness, sections are picked up on carbon-coated 200 mesh copper grids. For later precise correlation between RT-FM and ET, we apply 50 nm TetraSpeck microspheres as fluorescent fiducials, at a 1:100−1:300 dilution in phosphate buffered saline (PBS) pH 8.4 through adsorption (Suresh et al., 2015).

1.6 ROOM TEMPERATURE FLUORESCENCE MICROSCOPY

For RT-FM, we place the grid onto a drop of PBS on a coverslip, section side facing the coverslip. A second drop of PBS is placed on a glass slide. The coverslip is then turned and placed onto the glass slide, such that the grid is sandwiched between the two and immersed into PBS. Residual PBS is blotted from the side using filter paper.

Previously, we imaged resin sections on EM grids by sandwiching them in a drop of water between two round coverslips held together by a ringlike sample holder (Kukulski et al., 2011). We now recommend using PBS rather than water to ensure imaging at basic pH, which is crucial for optimal GFP fluorescence. It has been shown that quenched GFP fluorescence in resin-embedded samples can be restored using alkaline buffers of pH 8−11 (Xiong et al., 2014). It remains to be systematically explored which buffers at the same time preserve section quality for EM. In our hands, PBS at pH 8.4 allows good fluorescence imaging without compromising the ultrastructure.

We also find it faster to sandwich the grid between a coverslip and glass slide without sealing with vacuum grease, as we had previously done (Kukulski et al., 2011). Although the vacuum grease efficiently prevents drying and therefore sticking of the sections to the coverslip during long imaging periods, we found that, generally, the time it takes to image a grid does not lead to drying. However, it is important to not use too much PBS because that can increase the distance between the section and the objective, and thereby impair image quality. Furthermore, too much liquid can make the grid move during imaging. Thus, residual buffer should be removed by blotting with filter paper from the edge of the coverslip.

Low Magnification (10× Objective). To visually correlate sections to cryo-FM images, we image the entire section in BF and as a short (∼500 ms) single exposure in the GFP channel using a 10× air objective. These channels can be merged in acquisition software to visualize grid square layout and cytosolic background fluorescence, respectively (Fig. 4A). The cytosolic background fluorescence of the cell cross sections allows identifying the ROI by visual correlation to the cryo-FM images. This is easiest in earlier sections, as a larger cross section of the cytosol is visible.

High Magnification (100× Objective). Once visual correlation is established, a 100× TIRF objective is used to acquire high-magnification images of the grid squares that contain cells of interest. Due to the often uneven profile of the section,

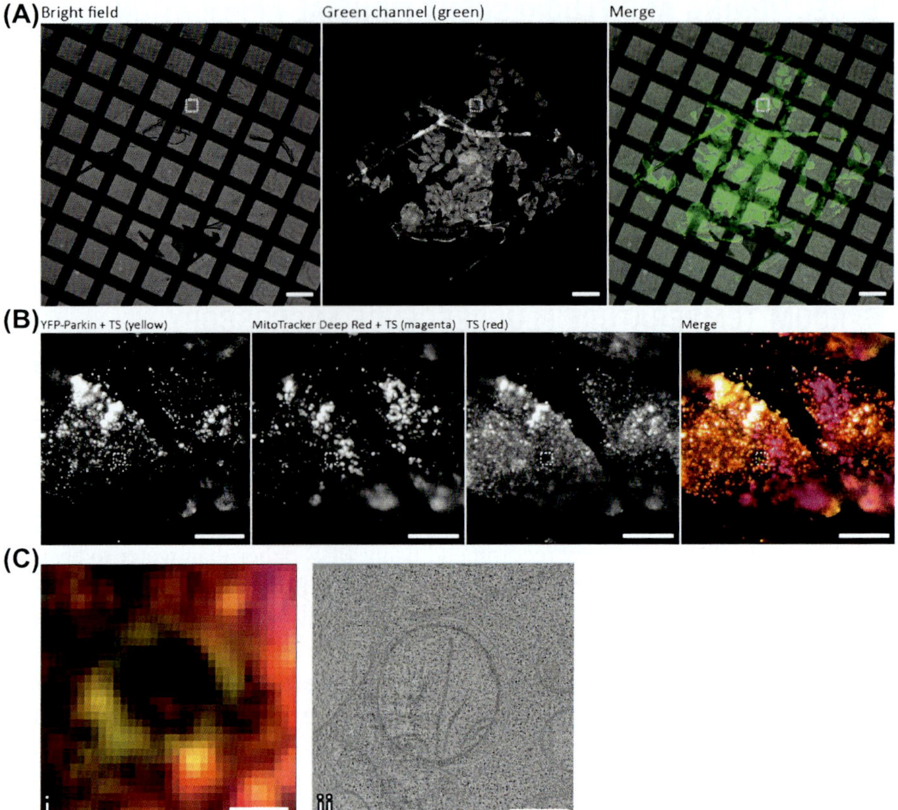

FIGURE 4 Room temperature fluorescence microscopy correlation to electron tomogram.

Section from resin-embedded cells shown in Figs. 2 and 3 on electron microscopy grid, imaged at room temperature by widefield fluorescence microscopy, first at low magnification (10×; A) and then at high magnification (100×; B). (A) Low-magnification imaging of cytoplasmic background in the green channel allows for visual correlation of section fluorescence with cryo-fluorescence data (compare Fig. 4A with Fig. 2B). (B) High-magnification imaging of area indicated in (A) in three fluorescence channels to visualize YFP-Parkin, MitoTracker Deep Red, and TetraSpecks. (C, i) Zoom onto spot of interest marked by *dashed box* in (B) and virtual slice through an electron tomogram collected following precise, fiducial-based correlation (ii), showing the corresponding cellular ultrastructure. In this example, the YFP-Parkin spot marks a mitochondrion engulfed by an autophagosome. Scale bars are 100 μm (A), 10 μm (B), and 500 nm (C, i and C, ii).

we find it best to take approximately 10 images (100–200 nm steps in z-direction) in each channel. To ensure signals of interest originate truly from FP-tagged protein or fluorescent dye rather than fiducials, acquisition of TetraSpeck fluorescence in an otherwise empty channel is necessary (Fig. 4B).

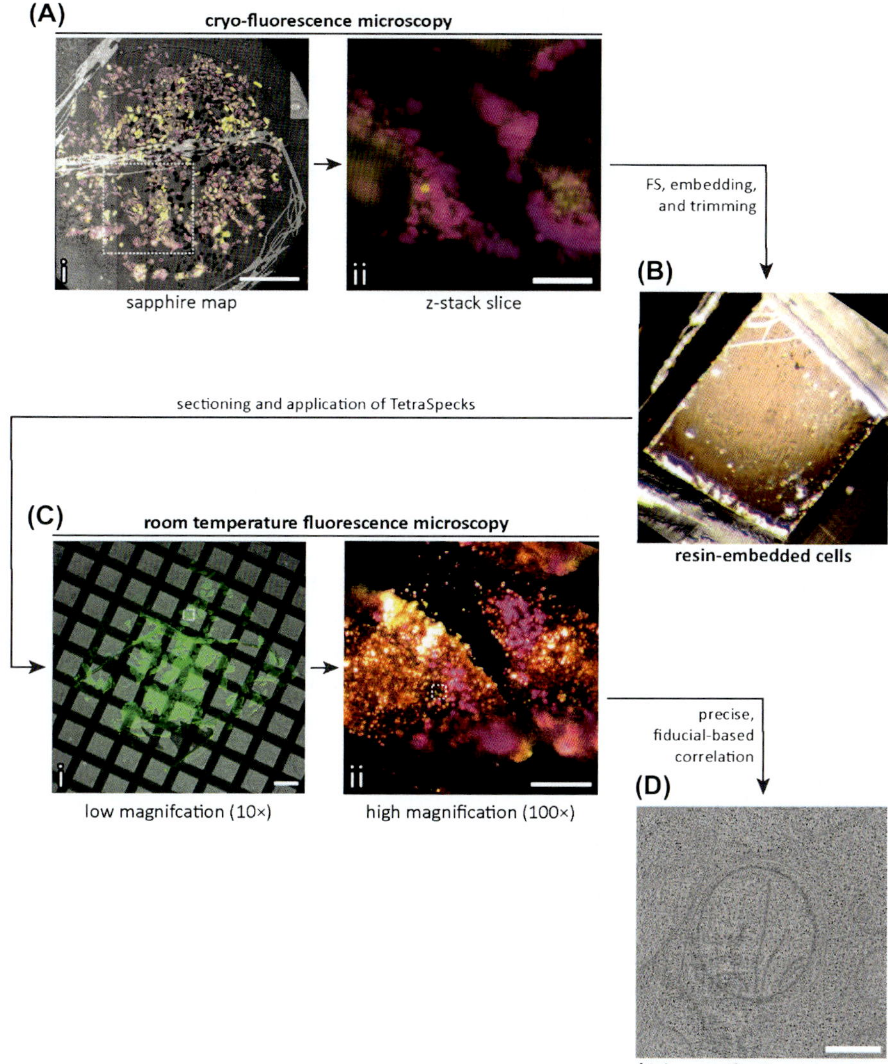

FIGURE 5 triCLEM diagrammatic overview.

Using triCLEM, rare, fluorescentlyly labeled events are tracked throughout the preparation for high-precision correlative light and electron microscopy. Cells containing signals of interest are first identified through a tile scan of whole, high-pressure frozen cells on a sapphire disk (A, i) and imaged for rough correlation as a z-stack (A, ii). Following freeze-substitution and resin embedding, visual correlation of the block face to cryo-fluorescence images allows for accurate trimming (B). Visual correlation of room temperature fluorescence microscopy at low magnification (C, i) with cryo-fluorescence images facilitates collection of high-magnification images at areas of interest (C, ii). Following precise, fiducial-based correlation of high-magnification room temperature fluorescence image to a defocused electron micrograph, a high-magnification (1.1 nm/pixel) tilt-series is collected and reconstructed (D). Scale bars are 500 μm (A, i), 10 μm (A, ii), 100 μm (C, i), 10 μm (C, ii), and 500 nm (D).

The cryo-FM focus stacks only provide a rough estimate at which z-position the fluorescent spots of interest are. Furthermore, cells of interest on a particular section can sometimes be obscured by the grid bars. For these reasons, multiple sections need to be imaged before all the fluorescent spots of interest are identified. By the 10th serial section, most of the cytosolic volume has been removed and only the nucleus remains. Thus, in the cell model used here, we generally image only the first 10 sections from a block. A careful estimation of what range of sections is expected to contain the signals of interest is very helpful to reduce the labor-intensive screening of sections on EM grids for signals by FM.

1.7 ELECTRON TOMOGRAPHY AND FIDUCIAL-BASED CORRELATION PROCEDURE

Tomographic fiducial markers (10 nm, 20 nm colloidal gold, or 15 nm antibody-coupled gold) are adsorbed on both sides of the grids. For EM and ET, we use a Tecnai F20 electron microscope. To image the TetraSpeck fiducials for the correlation procedure to FM, transmission EM (TEM) images are taken at approximately 100 μm defocus using SerialEM. At this high defocus value, the TetraSpecks display enough contrast to be unambiguously distinguished from the cellular content in single projection images (Cecilia Bebeacua, unpublished). If TetraSpecks are present at low density, we use the montaging function in SerialEM (Mastronarde, 2005) to image the field of view required to include enough fiducial markers for a significant correlation procedure. Correlation is carried out using the scripts written at EMBL in Matlab as described in Kukulski et al. (2012). A high-magnification tilt-series is then collected in scanning transmission EM (STEM) mode, typically as a dual-axis tilt-series from −60 to 60 degrees (1 degree increment) at 1.1 nm pixel size over a 2048 × 2048 pixel image. We use an axial BF detector for imaging, as recommended by Hohmann-Marriott et al. (2009), typically a C2 aperture of 50 μm and a camera length of 200 mm. Using STEM for tomography of resin-embedded samples has several advantages over TEM, as described by Hohmann-Marriott et al. (2009), Sousa, Azari, Zhang, and Leapman (2011), and Villinger et al. (2014). For instance, because no image forming lenses are used, inelastically scattered electrons do not induce chromatic aberration, which particularly aids imaging thick sections. Furthermore, dynamic focusing allows adjusting the focus during image acquisition, resulting in in-focus images over the whole field of view even at high tilt angles. We find that implementation in the current SerialEM setup makes applicability of STEM tomography very user-friendly. The tomograms are reconstructed using IMOD (Kremer, Mastronarde, & McIntosh, 1996), just like TEM tomograms, either manually or, using a more recent IMOD release, automatically (Mastronarde & Held, 2017). Correlation between low-magnification TEM images and high-magnification tomograms can be performed using the scripts written at EMBL in Matlab as described in Kukulski et al. (2012).

2. INSTRUMENTATION AND MATERIALS

2.1 CULTURE AND HIGH-PRESSURE FREEZING OF MAMMALIAN CELLS

Instrumentation: HPM100 equipped with a humidity- and heat-controlled chamber (Leica Microsystems).

Materials: High glucose, GlutaMAX, DMEM (Thermo) supplemented with 10% FBS (Labtech), 1× MEM NEAA (Thermo), and 10 mM HEPES; MitoTracker Deep Red FM (Thermo); 3 mm flat sapphire discs, 50 μm thick (Engineering Office M. Wohlwend, Switzerland), carbon-coated; 6 mm plastic half cylinders (Leica Microsystems); 6 mm CLEM middle plate (Leica Microsystems); 6 mm support ring, copper gold-plated (Leica Microsystems); 3 mm nickel spacer ring with 2 mm hole, 50 μm thick (Agar Scientific Ltd.); 6 mm cover ring (Leica Microsystems); FluoroBrite DMEM (Thermo) supplemented with 10% FBS (Labtech), 1× MEM NEAA (Thermo), and 10 mM HEPES.

2.2 CRYOGENIC-FLUORESCENCE MICROSCOPY OF SAPPHIRE DISKS

Instrumentation: DM6 FS microscope controlled by LAS X software, LN_2-cooled stage, HCX PL APO 50× cryo objective with NA = 0.9 (all: Leica Microsystems), Orca Flash 4.0 V2 SCMOS camera (Hamamatsu Photonics), Sola Light Engine (Lumencor) and the following filters: L5 excitation 480/40, dichroic 505, emission 527/30 for YFP-Parkin and Y5 excitation 620/60, dichroic 660, emission 700/75 for MitoTracker Deep Red (all filters: Leica Microsystems). The microscope is used in a humidity-controlled room.

Materials: Cryotransfer shuttle for loading of sapphire disk, cryo CLEM cartridges (both: Leica Microsystems), cryo grid boxes (made by in-house workshop; similar devices available from Agar Scientific).

2.3 FREEZE-SUBSTITUTION/LOWICRYL EMBEDDING

Instrumentation: AFS2 with FSP robot for automated reagent handling (Leica Microsystems).

Materials: AFS2 consumables (reagent containers, flow-through rings, reagent baths, dispenser syringes) (Leica Microsystems).

Reagents: Glass-distilled acetone, 20% uranyl acetate in dried methanol, Lowicryl HM20 (Polysciences, Inc.).

2.4 ULTRAMICROTOMY, ELECTRON MICROSCOPY GRIDS, AND FIDUCIALS

Instrumentation: Stereomicroscope (Zeiss, product 475022-9902), Ultracut E Microtome (Reichert), diamond knife (Diatome), and double-gooseneck KL 1500 lamp (Schott).

Materials: 200 mesh copper grids with carbon support film (Agar Scientific Ltd., product S160) and filter paper (Whatman No. 1).

Reagents: TetraSpeck microspheres, 50 nm (custom made, Invitrogen).

2.5 ROOM TEMPERATURE FLUORESCENCE MICROSCOPY OF SECTIONS

Instrumentation: TE2000-E widefield microscope (Nikon UK Ltd.) controlled by Nikon NIS Elements 4.4, 10× air objective with NA = 0.45, 100× oil-immersion TIRF objective with NA = 1.49, an NEO sCMOS DC-152Q-C00-FI camera (Andor), a Lambda DG-4 lamp (Sutter Instruments), filters: 89006ET CFP/YFP/mCherry (Chroma), excitation 560/20, dichroic 89008bs, emission 535/30 for YFP-Parkin; 49005ET DSRED (Chroma), excitation 545/30, dichroic T570lp, emission 620/60, with additional emission filter 605/70 for TetraSpecks to reduce bleed through of MitoTracker Deep Red signal; and 49006ET CY5 (Chroma), excitation 520/60, dichroic T660lpxr, emission 700/75 for MitoTracker Deep Red.

Materials: Glass slides (Menzel-Glaser, number 50), square coverslips (VWR, 22 × 22 mm, thickness no. 1), and filter paper (Whatman No. 1).

Reagents: PBS, pH 8.4.

2.6 ELECTRON TOMOGRAPHY

Instrumentation: FEI Tecnai F20 operated at 200 kV in STEM mode equipped with an axial BF detector and a 50 μm C2 aperture, Orius CCD camera (Gatan) for TEM imaging, and high-tilt tomography holder (Model 2020; Fischione Instruments).

Software: SerialEM for automated STEM tilt-series acquisition with dynamic focusing (Mastronarde, 2005). IMOD software package for tomogram reconstruction (Kremer et al., 1996; Mastronarde & Held, 2017).

Reagents: Tomographic fiducial markers (10–20 nm gold colloids or gold conjugates, Agar Scientific Ltd.).

2.7 FLUORESCENT FIDUCIAL-BASED CORRELATION

Software: Matlab 9.0 (The MathWorks, Inc.) with the Image Processing Toolbox installed. Scripts for correlation as described in Schorb et al. (2016), available from the EMBL. ImageJ 1.51c (National Institutes of Health, USA).

3. DISCUSSION

In this book chapter, we present triCLEM, a protocol for incorporating cryo-FM screening of high-pressure frozen cells on sapphire disks, with additional modifications to the CLEM protocol published in Kukulski et al. (2011).

The following assessment gives an estimate of the efficiency of our protocol: For a typical experiment, we start with cells grown on 12–14 sapphires. Of these, we typically high-pressure freeze 10–12 successfully without cracking the sapphire. This can be accomplished in one afternoon. Cryo-FM screening then identifies 8–10 of the best sapphires for FS and Lowicryl embedding and is typically done over 2 days or 1 long day. Approximately 50%–75% of sapphires preserve cells during HPF at a confluency level comparable to what we observe via the stereomicroscope attached to the HPM100 immediately prefreezing. The remainder of sapphires display preservation of only a fraction of the cells observed prefreezing, often with a distribution pattern that suggests they were subjected to extreme shearing during HPF. For the 8–10 sapphires judged to be of adequate quality for FS and Lowicryl embedding, we take z-stacks of two ROIs of approximate size of the trapezoid to be cut (750 × 750 μm). While only one ROI will be sectioned, we image two possible ROIs in case one is damaged during the embedding process. Though this rarely occurs, the time necessary to image an ROI (~2 min with stitching) is marginal compared to mapping the entire sapphire (~20 min with stitching).

During RT-FM, we typically find back about 50% of signals from a cryo-FM ROI. In addition to significant portions of the section being obscured by grid bars, fluorescent signals seen in cryo-FM are usually not seen in all sections analyzed by RT-FM. We attribute this to two factors: (1) signals observed in whole cells were not captured in a particular section or, in the case of proteins with very weak signals; (2) the fluorescent signal observed in cryo-FM was a false positive. In total, one ROI from one sapphire with 4–5 cells of interest with one signal each will usually lead to 2–3 confirmed signals in RT-FM and an equal number of tomograms. In total, the process from culturing cells to collecting tomograms for one experiment could be done in 2–3 weeks. Significant labor is required for HPF (~3 h), cryo-FM screening (~1 full day), RT-FM (~4–5 h/block), and correlation/tilt-series acquisition (~1–2 h/spot) (Fig. 1).

triCLEM currently can be applied to any sample high-pressure frozen on 3 mm sapphire disks. While 6 mm sapphires would increase the field of view for searching for signals of interest, these cannot be imaged using the current cartridge and transfer system of the Leica cryo-FM microscope. Further, to take full advantage of the larger field of view and more ROIs on 6 mm sapphires, one would need to take additional steps of splitting and attaching the block face to new resin blocks. In addition to facilitating the localization of rare fluorescence, we believe cryo-FM of sapphire disks is also most useful to screen the quality of HPF, particularly for delicate samples or cells.

ACKNOWLEDGMENTS

We would like to thank Richard Youle and Jonathon Burman for cell lines, continuous support, and discussions; Gillian Howard for advice on sample preparation and for sharing reagents; Ori Avinoam for advice on sample preparation; Neil Grant for assistance in digitally capturing representative images of embedded cells; and Cveta Tomova for input on high-pressure freezing. We are grateful to David Mastronarde for help with SerialEM for STEM tomography and Christos Savva for help setting it up on the electron microscope. This work was supported by the Medical Research Council (MC_UP_1201/8).

REFERENCES

Avinoam, O., Schorb, M., Beese, C. J., Briggs, J. A., & Kaksonen, M. (2015). ENDOCYTOSIS. Endocytic sites mature by continuous bending and remodeling of the clathrin coat. *Science, 348*(6241), 1369–1372. http://dx.doi.org/10.1126/science.aaa9555.

de Boer, P., Hoogenboom, J. P., & Giepmans, B. N. (2015). Correlated light and electron microscopy: Ultrastructure lights up! *Nature Methods, 12*(6), 503–513. http://dx.doi.org/10.1038/nmeth.3400.

Brown, E., Van Weering, J., Sharp, T., Mantell, J., & Verkade, P. (2012). Capturing endocytic segregation events with HPF-CLEM. *Methods in Cell Biology, 111*, 175–201. http://dx.doi.org/10.1016/B978-0-12-416026-2.00010-8.

Heiligenstein, X., Heiligenstein, J., Delevoye, C., Hurbain, I., Bardin, S., Paul-Gilloteaux, P., … Raposo, G. (2014). The CryoCapsule: Simplifying correlative light to electron microscopy. *Traffic, 15*(6), 700–716. http://dx.doi.org/10.1111/tra.12164.

Hohmann-Marriott, M. F., Sousa, A. A., Azari, A. A., Glushakova, S., Zhang, G., Zimmerberg, J., & Leapman, R. D. (2009). Nanoscale 3D cellular imaging by axial scanning transmission electron tomography. *Nature Methods, 6*(10), 729–731. http://dx.doi.org/10.1038/nmeth.1367.

Hohn, K., Fuchs, J., Frober, A., Kirmse, R., Glass, B., Anders-Osswein, M., … Dietrich, C. (2015). Preservation of protein fluorescence in embedded human dendritic cells for targeted 3D light and electron microscopy. *Journal of Microscopy, 259*(2), 121–128. http://dx.doi.org/10.1111/jmi.12230.

Jin, S. M., & Youle, R. J. (2013). The accumulation of misfolded proteins in the mitochondrial matrix is sensed by PINK1 to induce PARK2/Parkin-mediated mitophagy of polarized mitochondria. *Autophagy, 9*(11), 1750–1757. http://dx.doi.org/10.4161/auto.26122.

Koning, R. I., Celler, K., Willemse, J., Bos, E., van Wezel, G. P., & Koster, A. J. (2014). Correlative cryo-fluorescence light microscopy and cryo-electron tomography of *Streptomyces*. *Methods in Cell Biology, 124*, 217–239. http://dx.doi.org/10.1016/B978-0-12-801075-4.00010-0.

Kremer, J. R., Mastronarde, D. N., & McIntosh, J. R. (1996). Computer visualization of three-dimensional image data using IMOD. *Journal of Structural Biology, 116*(1), 71–76. http://dx.doi.org/10.1006/jsbi.1996.0013.

Kukulski, W., Schorb, M., Welsch, S., Picco, A., Kaksonen, M., & Briggs, J. A. (2011). Correlated fluorescence and 3D electron microscopy with high sensitivity and spatial precision. *The Journal of Cell Biology, 192*(1), 111–119. http://dx.doi.org/10.1083/jcb.201009037.

Kukulski, W., Schorb, M., Welsch, S., Picco, A., Kaksonen, M., & Briggs, J. A. (2012). Precise, correlated fluorescence microscopy and electron tomography of lowicryl sections using fluorescent fiducial markers. *Methods in Cell Biology, 111*, 235–257. http://dx.doi.org/10.1016/B978-0-12-416026-2.00013-3.

Loussert, C., Forestier, C. L., & Humbel, B. M. (2012). Correlative light and electron microscopy in parasite research. *Methods in Cell Biology, 111*, 59–73. http://dx.doi.org/10.1016/B978-0-12-416026-2.00004-2.

Mahamid, J., Schampers, R., Persoon, H., Hyman, A. A., Baumeister, W., & Plitzko, J. M. (2015). A focused ion beam milling and lift-out approach for site-specific preparation of frozen-hydrated lamellas from multicellular organisms. *Journal of Structural Biology, 192*(2), 262–269. http://dx.doi.org/10.1016/j.jsb.2015.07.012.

Mastronarde, D. N. (2005). Automated electron microscope tomography using robust prediction of specimen movements. *Journal of Structural Biology, 152*(1), 36–51. http://dx.doi.org/10.1016/j.jsb.2005.07.007.

Mastronarde, D. N., & Held, S. R. (2017). Automated tilt series alignment and tomographic reconstruction in IMOD. *Journal of Structural Biology, 197*(2), 102–113. http://dx.doi.org/10.1016/j.jsb.2016.07.011.

McDonald, K., Schwarz, H., Muller-Reichert, T., Webb, R., Buser, C., & Morphew, M. (2010). "Tips and tricks" for high-pressure freezing of model systems. *Methods in Cell Biology, 96*, 671–693. http://dx.doi.org/10.1016/S0091-679X(10)96028-7.

Narendra, D., Tanaka, A., Suen, D. F., & Youle, R. J. (2008). Parkin is recruited selectively to impaired mitochondria and promotes their autophagy. *The Journal of Cell Biology, 183*(5), 795–803. http://dx.doi.org/10.1083/jcb.200809125.

Nixon, S. J., Webb, R. I., Floetenmeyer, M., Schieber, N., Lo, H. P., & Parton, R. G. (2009). A single method for cryofixation and correlative light, electron microscopy and tomography of zebrafish embryos. *Traffic, 10*(2), 131–136. http://dx.doi.org/10.1111/j.1600-0854.2008.00859.x.

Peddie, C. J., Blight, K., Wilson, E., Melia, C., Marrison, J., Carzaniga, R., … Collinson, L. M. (2014). Correlative and integrated light and electron microscopy of in-resin GFP fluorescence, used to localise diacylglycerol in mammalian cells. *Ultramicroscopy, 143*, 3–14. http://dx.doi.org/10.1016/j.ultramic.2014.02.001.

Rigort, A., Villa, E., Bauerlein, F. J., Engel, B. D., & Plitzko, J. M. (2012). Integrative approaches for cellular cryo-electron tomography: Correlative imaging and focused ion beam micromachining. *Methods in Cell Biology, 111*, 259–281. http://dx.doi.org/10.1016/B978-0-12-416026-2.00014-5.

Schorb, M., & Briggs, J. A. (2014). Correlated cryo-fluorescence and cryo-electron microscopy with high spatial precision and improved sensitivity. *Ultramicroscopy, 143*, 24–32. http://dx.doi.org/10.1016/j.ultramic.2013.10.015.

Schorb, M., Gaechter, L., Avinoam, O., Sieckmann, F., Clarke, M., Bebeacua, C., … Briggs, J. A. (2016). New hardware and workflows for semi-automated correlative cryo-fluorescence and cryo-electron microscopy/tomography. *Journal of Structural Biology*. http://dx.doi.org/10.1016/j.jsb.2016.06.020.

Schwartz, C. L., Sarbash, V. I., Ataullakhanov, F. I., McIntosh, J. R., & Nicastro, D. (2007). Cryo-fluorescence microscopy facilitates correlations between light and cryo-electron microscopy and reduces the rate of photobleaching. *Journal of Microscopy, 227*(2), 98–109. http://dx.doi.org/10.1111/j.1365-2818.2007.01794.x.

Sousa, A. A., Azari, A. A., Zhang, G., & Leapman, R. D. (2011). Dual-axis electron tomography of biological specimens: Extending the limits of specimen thickness with bright-field STEM imaging. *Journal of Structural Biology, 174*(1), 107–114. http://dx.doi.org/10.1016/j.jsb.2010.10.017.

Strnad, M., Elsterova, J., Schrenkova, J., Vancova, M., Rego, R. O., Grubhoffer, L., & Nebesarova, J. (2015). Correlative cryo-fluorescence and cryo-scanning electron microscopy as a straightforward tool to study host-pathogen interactions. *Scientific Reports, 5*, 18029. http://dx.doi.org/10.1038/srep18029.

Suresh, H. G., da Silveira Dos Santos, A. X., Kukulski, W., Tyedmers, J., Riezman, H., Bukau, B., & Mogk, A. (2015). Prolonged starvation drives reversible sequestration of lipid biosynthetic enzymes and organelle reorganization in *Saccharomyces cerevisiae*. *Molecular Biology of the Cell, 26*(9), 1601–1615. http://dx.doi.org/10.1091/mbc.E14-11-1559.

Tranfield, E. M., Heiligenstein, X., Peristere, I., & Antony, C. (2014). Correlative light and electron microscopy for a free-floating spindle in *Xenopus laevis* egg extracts. *Methods in Cell Biology, 124*, 111–128. http://dx.doi.org/10.1016/B978-0-12-801075-4.00006-9.

Verkade, P. (2008). Moving EM: The rapid transfer system as a new tool for correlative light and electron microscopy and high throughput for high-pressure freezing. *Journal of Microscopy, 230*(Pt 2), 317–328. http://dx.doi.org/10.1111/j.1365-2818.2008.01989.x.

Villinger, C., Schauflinger, M., Gregorius, H., Kranz, C., Hohn, K., Nafeey, S., & Walther, P. (2014). Three-dimensional imaging of adherent cells using FIB/SEM and STEM. *Methods in Molecular Biology, 1117*, 617–638. http://dx.doi.org/10.1007/978-1-62703-776-1_27.

Walther, P., Wang, L., Liessem, S., & Frascaroli, G. (2010). Viral infection of cells in culture—approaches for electron microscopy. *Methods in Cell Biology, 96*, 603–618. http://dx.doi.org/10.1016/S0091-679X(10)96025-1.

van Weering, J. R., Brown, E., Sharp, T. H., Mantell, J., Cullen, P. J., & Verkade, P. (2010). Intracellular membrane traffic at high resolution. *Methods in Cell Biology, 96*, 619–648. http://dx.doi.org/10.1016/S0091-679X(10)96026-3.

Xiong, H., Zhou, Z., Zhu, M., Lv, X., Li, A., Li, S., … Zeng, S. (2014). Chemical reactivation of quenched fluorescent protein molecules enables resin-embedded fluorescence microimaging. *Nature Communications, 5*, 3992. http://dx.doi.org/10.1038/ncomms4992.

Youle, R. J., & van der Bliek, A. M. (2012). Mitochondrial fission, fusion, and stress. *Science, 337*(6098), 1062–1065. http://dx.doi.org/10.1126/science.1219855.

Zhao, L., Kroenke, C. D., Song, J., Piwnica-Worms, D., Ackerman, J. J., & Neil, J. J. (2008). Intracellular water-specific MR of microbead-adherent cells: The HeLa cell intracellular water exchange lifetime. *NMR in Biomedicine, 21*(2), 159–164. http://dx.doi.org/10.1002/nbm.1173.

CHAPTER

15

Matrix MAPS—an intuitive software to acquire, analyze, and annotate light microscopy data for CLEM

Martin Schorb*, Frank Sieckmann§,[1]

*European Molecular Biology Laboratory, Heidelberg, Germany

§Leica Microsystems GmbH, Mannheim, Germany

[1]Corresponding author: E-mail: Frank.Sieckmann@leica-microsystems.com

CHAPTER OUTLINE

Abstract

Matrix MAPS provides an intuitive interface for acquiring light microscopy data during a correlative light and electron microscopy experiment either at room or cryogenic temperatures. First, the user graphically defines the geometry of the acquisition region on top of preview images. Multiple independent regions can then be imaged in an automated way, each with individual settings. The same interface allows the user to mark and select points or regions of interest whose coordinates can subsequently be transferred directly to the electron microscope.

Methods in Cell Biology, Volume 140, ISSN 0091-679X, http://dx.doi.org/10.1016/bs.mcb.2017.03.012

INTRODUCTION

Software plays an important role during various steps of a typical correlative light and electron microscopy (CLEM) experiment. Most crucially, it is used to register the independent coordinate frames from the different imaging modalities. The second role lies in providing an intuitive interface for acquiring microscopy data, while aiming for maximum automation of the imaging process.

The registration of the imaging coordinate frames can happen at various points in the CLEM workflow (de Boer, Hoogenboom, & Giepmans, 2015; Bykov, Cortese, Briggs, & Bartenschlager, 2016). The highest possible correlation accuracy can be achieved when performing a registration of the final acquired images (Ader & Kukulski, 2017; Kukulski et al., 2011; Schellenberger et al., 2014; Schorb & Briggs, 2014). This postacquisition correlation relies on matching pairs of landmarks that are present in both the data from light microscopy (LM) and electron microscopy (EM); specialized software is available for this purpose (Fukuda et al., 2014; Heiligenstein, Paul-Gilloteaux, Raposo, & Salamero, 2017; Keene et al., 2014; Kukulski et al., 2012; Paul-Gilloteaux et al., 2017; Schorb et al., 2017; Thévenaz, Ruttimann, & Unser, 1998).

In a similar fashion, image landmarks can also be used to guide the acquisition in the second imaging modality—in the CLEM case this is EM. To enable targeted acquisition, the EM imaging software needs to provide the means of importing images and/or coordinate data from previous imaging modalities. It also needs to be able to register the imported coordinates with the coordinate frame of the EM acquisition. Manufacturers that offer both light and scanning electron microscopy (SEM) solutions provide software packages that are able to control both microscopes and transfer images and coordinate data internally (Loussert Fonta & Humbel, 2015; Wacker et al., 2015). The coordinate registration is facilitated by the use of identical specimen carriers with built-in landmarks that are detected in both microscopy modalities. In the case of transmission electron microscopy (TEM), approaches that integrate LM require a significant modification of the expensive equipment (Agronskaia et al., 2008; Iijima et al., 2014). For unmodified TEM systems there are commercial software packages available ("Pop software," JEOL) or from independent developers (Mastronarde, 2005) that can import and register LM data (Briegel et al., 2010).

For the dedicated cryo-light microscopy system Leica Cryo CLEM, we have added the CLEM option to the MatrixScreener HCSA module (Schorb et al., 2017).

The MatrixScreener HCSA module is part of the Leica LAS X software suite. As a platform, it allows the definition of complex microscopy experiments for biology. The development of MatrixScreener HCSA began in 2004 as a part of the European MitoCheck project to research the regulation of mitosis by phosphorylation (Neumann et al., 2010). This project, as well as follow-up high-content microscopy screens (Conrad et al., 2011), was a close collaboration between Leica Microsystems and EMBL Heidelberg. MatrixScreener, in its current version 5, supports a wide range of microscopy platforms and experiment types. Since 2005, the software incorporates an easy-to-learn interface, called Computer-aided microscopy (CAM), that can be used to control the experiment remotely from all modern computing languages (Tischer, Hilsenstein, Hanson, & Pepperkok, 2014). This feature is often referred to as feedback microscopy or intelligent microscopy.

1. THE SOFTWARE

The Matrix MAPS module within Leica's LAS X microscope control software is version 5 of the MatrixScreener HCSA module. Matrix MAPS allows the acquisition of tiled mosaic image stacks of any geometrical shape using various assigned acquisition jobs. The philosophy of the new CLEM module is to convert the complexity of a CLEM experiment into intuitive, comprehensive workflows while providing all necessary tools for direct interaction with the sample at your fingertips.

The user can analyze the acquired image mosaics and select the points or regions of interest to be further investigated by EM within the same user interface (UI). The specified coordinates can then be exported in file formats suitable for EM acquisition. The currently implemented CLEM export provides data formats compatible with SerialEM (Mastronarde, 2005)—for TEM acquisition—as well as Leica file formats to enable further LAS X integrated application modules, such as the 3-D viewer or image analysis modules. The generic XML export can also be used by software packages for postacquisition image registration.

The software is compatible with Leica's portfolio of fluorescence microscopes for life-science applications. In particular, it can be used with both wide field as well as confocal scanning microscopes and therefore for any application of CLEM at room temperature. It was developed to control the Leica Cryo CLEM system and the workflow was specifically adapted to cryo-CLEM experiments.

1.1 GENERAL LAYOUT

The window composition within the software divides the main UI components in different columns. This partition is coherent between the different tabs that serve setting up the several levels of an experiment. The leftmost area of the screen shows the controls to edit the experimental parameters. The last acquired image (series) or live view is displayed in the viewer on the right side of the screen. The main and central part of the window shows either the *Stage Overview* of the specimen to be acquired—in the *Start* tab (Figs. 1 and 4)—or offers to adjust the acquisition settings in the *Adj. Experiment* tab (Fig. 2).

Start (Fig. 1)—in this tab you define the scanning properties and the experimental workflow. It is composed of a control panel on the left-hand side where the experiment is defined and individual steps are configured. We will describe the individual controls as part of the experimental workflow. The main display area in the center shows the *Stage Overview*. Here, the software displays the underlying scan geometry based on the chosen experimental preset and overlays any image acquired to set up the experiment. This is where you define the acquisition

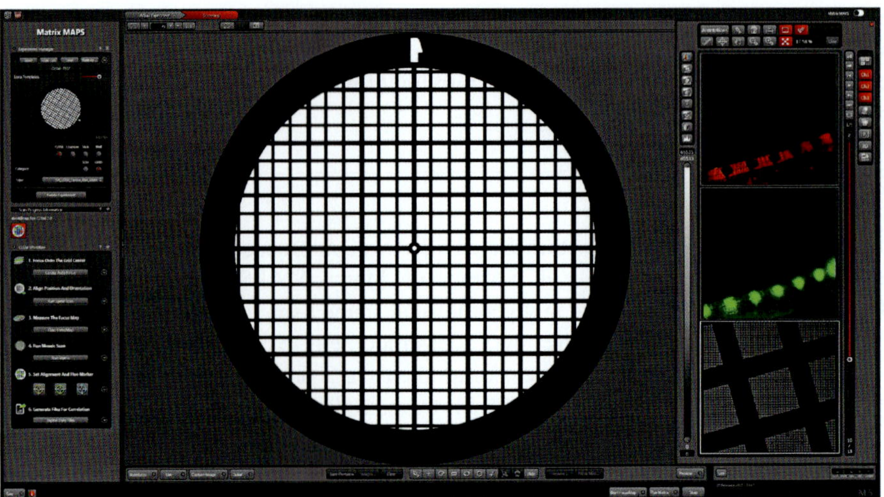

FIGURE 1

Screenshot of the Matrix MAPS correlative light and electron microscopy module—*Start Screen* tab with *Experiment Manager* active in the left column. Here you can select a predefined experiment template or load presets from a previous experiment. The selected specimen support type will be drawn as an overlay in the central *Stage Overview*. The controls at the top of the *Stage Overview* adjust its current zoom. The controls at the bottom (left to right) start several imaging actions at the current stage position (indicated by a *white rectangle* in the overview) or select the images that should be displayed. The right-hand controls provide the tools to define the geometry and properties of the acquisition region(s), place the AF points and adjust the focus scans at these positions. The right-hand column of the window is occupied by the image viewer showing the last acquired image(s). Its display can be toggled inactive using the slider at its top. In the given case, the viewer shows data from a previous experiment on cryosections (Kolovou et al., 2017).

region(s) as well as the position of the autofocus (AF) points. You can find the tools for defining points and any geometrical shapes underneath the *Stage Overview*. The controls at the top of the *Stage Overview* allow the adjustment of the display region as well as changing the display and contrast of the visible image tiles. If a multi-z acquisition is shown inside the *Stage Overview*, a slider at its right-hand edge allows browsing through the slices. The image viewer on the right side can be toggled on or off using the slider in the top right corner.

Adj. Experiment (Fig. 2)—all imaging parameters are set in this tab. Here, you define the acquisition settings for the individual jobs (in the *collecting pattern* area at the top). Jobs can either be defined for image acquisition or for focusing. The exact imaging parameters are then defined for each of the channels within this job. The available options go as far as acquiring a multichannel multi-z-slice time series for an acquisition job or can simply define a single channel to be scanned during an autofocus job.

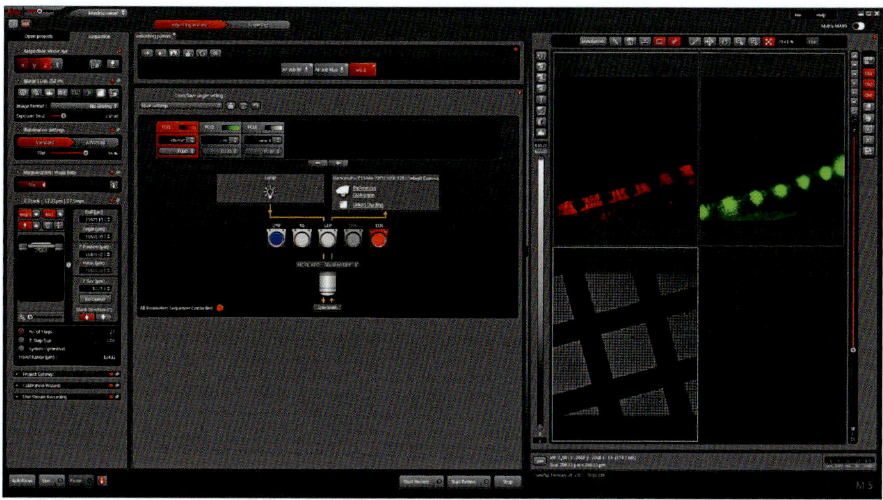

FIGURE 2

Screenshot of the *Adj. Experiment* Tab. The left-hand column contains the specific acquisition parameters of the currently selected channel. These range from illumination and detection settings to specifying the scan range if a multi-*z* image is desired. The central part of the window shows the job definitions at the top—here two AF jobs and one acquisition job. Below, you specify the optics settings for each channel of the currently active job (both *highlighted in red*). The start button(s) at the bottom of the window will perform the chosen action at the current position of the microscope stage. The right part of the window is the identical image viewer as described in Fig. 1.

The controls adjust the imaging parameters for each channel. These depend on the installed microscopy hardware and include the settings of the confocal scanner, or filter cube selection and camera exposure times in the wide-field case. Each acquisition job can also be set up as an image stack over a range of z values. In the image viewer on the right a region of interest (ROI) can be chosen within the detection field of view (FOV).

A typical CLEM experiment would have the following job arrangement:

- a single acquisition job composed of multiple channels, fluorescence plus transmitted light;
- one autofocus job with a single bright-field (BF)/transmitted light channel;
- one AF job with a single fluorescence channel for refinement.

2. WORKFLOW

The acquisition of LM data in the Matrix MAPS CLEM module follows the procedure illustrated in Fig. 3. The basic idea is to first obtain a quick overview about

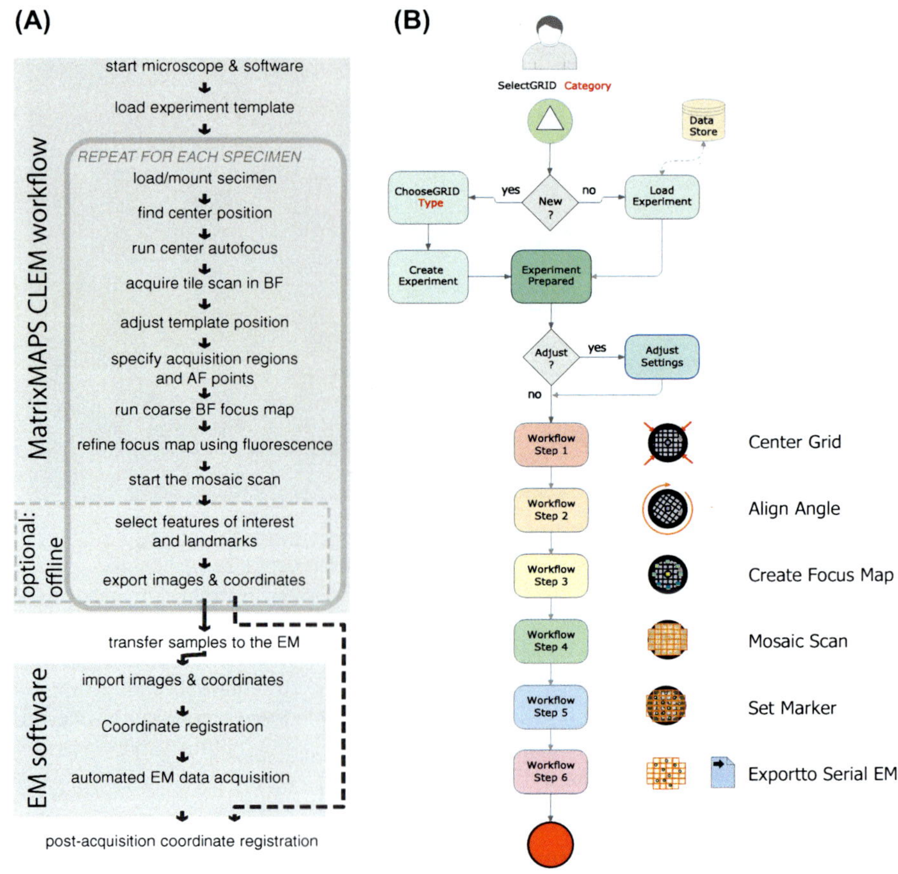

FIGURE 3

Workflow of a correlative light and electron microscopy (CLEM) experiment. (A) Workflow diagram describing the individual steps of the procedure. The steps that are performed with Matrix MAPS are grouped in the upper *gray box*. The feature selection and data export can also be performed at a separate PC to ensure optimal use of microscopy time.
(B) Schematic chart of the different elements of the workflow on the software level. *AF*, autofocus; *BF*, bright-field; *EM*, electron microscopy.

the positioning and orientation of the specimen and then mark the regions for acquisition on the preview images displayed in the *Stage Overview*. The user then defines positions where the focus is measured. From these AF measurements the software will define the focus position for the entire specimen upon starting the final acquisition. In a final step the coordinates and images are exported in suitable file formats.

2.1 STARTUP AND INITIAL EXPERIMENT SETUP

- Start the LASX software and select the appropriate microscope configuration.
- Start MatrixScreener by selecting the module drop-down list in the top-left corner.

In the *Start* tab that is displayed now (Fig. 1), the left-hand column shows various managers where you configure the experiment. Advanced properties are hidden by default but can be accessed by expanding the arrows. The spatial order of the managers and associated property editors follows the experiment workflow from top to bottom.

Experiment Manager (Fig. 1): The selection of the template in this property editor defines the presets for all experiment properties. You can load a saved preset from a previous experiment and thereby restore all properties of the performed scan(s), from specimen geometry to AF settings and acquisition parameters or create a new experiment. Besides the CLEM case described here, scans of a similar kind can be performed using well plates, culture dishes, or other specimen supports and geometries. In the following selection, you choose the category of specimen support. For CLEM there is the distinction between sample carriers of fixed geometry for to be correlated with SEM data and standard TEM grids. You then select the specific specimen support type from the drop-down list. By clicking *Create Experiment* the presets for all experiment properties are applied.

If the *CLEM Workflow* button is enabled—indicated by red highlighting—all other advanced manager options will be hidden (Fig. 4). If it is deactivated, the following options will be available for configuration.

The *Attribute Manager* shows certain options for the scan. *DEX* (Data Exporter) should be enabled, so that the image data is automatically exported during the scan. The *DRIFT* correction enables AF scans repeated at defined intervals to maintain the focus during a long scan.

The *Settings Manager* offers several buttons that enable advanced configurations for certain aspects of the scan. For example the property editor hidden under the JOB button selects which acquisition job (if multiple are desired) is assigned to which scan region. Most relevant for CLEM would be *AF*, that toggles the manager for the autofocus channels and is especially important when using multiple focus channels.

2.2 CORRELATIVE LIGHT AND ELECTRON MICROSCOPY ACQUISITION WORKFLOW

- Activate the *CLEM workflow*. This will show each step of the procedure as a button to call the action plus optional advanced settings (Fig. 4).
- In the BF channel, activate the live view and navigate to the grid's center mark using the stage's *XY* movement.
- Click the *Center Auto Focus* button to set the initial global focus position. Check that the *Use Grid Center Position* is disabled.

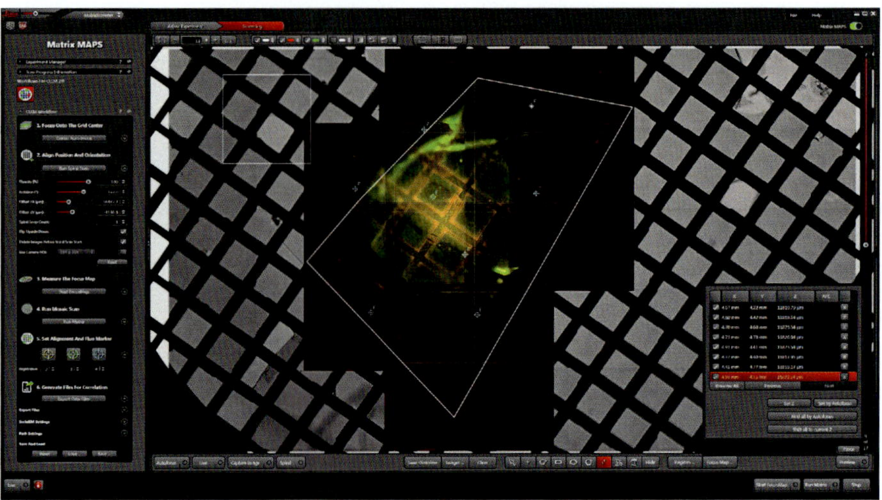

FIGURE 4

Screenshot of the *Start Screen* tab with the *CLEM workflow manager* active. The second step of the procedure is flapped open to display the advanced properties. Here is where the positioning of the grid template is adjusted. In the main display—the *Stage Overview*—the typical set up of a complete scan is shown. The transmitted light image displayed in the back is composed of preview tiles that were generated using the spiral scan. The grid template visible at the very edge of the overview is aligned to it. On top of it, the completed mosaic tiles are placed. The positions of the individual overlapping tiles are indicated with *thin white lines*. These positions were determined from the acquisition area defined by the user, in this case the *white polygon* surrounding the microtome section. The *blue positions* indicate desired autofocus points. The *white rectangle* at the top-left indicates the current position of the camera field of view given the actual sage position. The table shown at the bottom-right provides the coordinates of the autofocus points and allows modifying the *z* height of each point. The additional controls at the top of the *Stage Overview* allow adjusting the properties (contrast, lookup table) of the separate channels of all image tiles that are displayed on top of the grid template. The slider at the right edge browses through the different *z* slices. The specimen displayed here is a 300-nm Lowicryl section as described in Hampoelz et al. (2016).

- Set the spiral loop count to 4 (depends on the camera FOV, in our case this corresponds to about 2 × 2 mm).
- Click *Run Spiral Scan*. The microscope will now—starting from the current stage position (if *Use Grid Center Position* is disabled)—move outward in spirals and acquire a tiled image. You can always stop the scan and maintain the display of the already acquired image tiles on screen.
- Align the position of the grid overlay to the imaged mosaic. You can move the sliders to adjust the displacement in x and y and rotate the overlay.

- Select the region(s) you wish to acquire using one of the geometry tools provided below the display. You have the ability to mark single individual images or place a tiled mosaic in various shapes from rectangles to ellipses or variable polygons.
- In the selected regions, place *Focus Map Points* (the blue cross symbol). This is where AF measurement is performed to calculate the global focus map of the specimen.
- Run the focus scan using the BF AF job. We use the following scan settings: scan range 55 μm, 15 AF slices (the other parameters in the AF manager do not have an influence).
- Change the active AF job to the fluorescence AF job in the AF manager (our settings: scan range 12 μm, 13 AF slices).
- If needed, move to a representative area of your specimen and adjust the acquisition parameters of the fluorescence AF job in the *Adj. Experiment* tab. The settings here will depend on the fluorescence signal present in the specimen. Usually, we use the channel with the brightest signal—typically generated by the beads used for the coordinate registration—where bleaching is of the least concern. It might be beneficial to restrict the FOV of the camera to the center while running an AF job to accommodate for focus gradients. Applying camera binning will reduce the time required to collect the AF scan.
- Run the *Focus Map Scan* again using the fluorescence AF job.
- You now want to move to a representative area of your specimen and check and if necessary adjust the acquisition parameters of the various acquisition channels using the *Adj. Experiment* tab.

We typically acquire a multi-z-slice acquisition (a z stack) at each tile to compensate for gradients in the sample height. This ensures that all parts of the camera FOV are acquired properly in focus. Typical parameters for the z stack are: 10–15 μm thickness at 1 μm increment.

- Run the full acquisition by clicking *Mosaic Grid Scan.*
- The software will now acquire a tiled mosaic of the selected regions. At each tile the defined job is performed and images in all desired channels and at different z heights are acquired.
- Once finished, Matrix MAPS will automatically stitch the acquired tiles to single global images for each z layer. You can browse through the layers using the slider at the very right of the *Stage Overview* (Fig. 4). The controls at the top allow for contrast adjustment of the individual channels.

2.2.1 Select features of interest and landmarks

The purpose of the next step in the Matrix MAPS CLEM workflow is to generate coordinate lists that can be transferred to the EM control software.

- The first type of markers (yellow) you place are the landmarks used for aligning the LM and EM coordinate frames. Typical landmarks would be the markings on a finder-grid, grid squares where the support film has ruptured, or similar features that are easily detected in both imaging modalities. When clicking the first button, it will activate, turning red, and you can now select the landmark positions in the *Stage Overview*.

- You can use the slider on the far right of the *Stage Overview* to navigate through the different z slices in your acquisition. You can adjust the displayed contrast of each acquired channel using the tools at the top of the *Stage Overview* (Fig. 4).
- Second, you mark the fluorescent signals of interest. You position these second type markers, the green markers, the same way as before.
- You can also use the third type of marker (blue) to mark the positions of another set of features.
- Click *Export Data Files* to save the selected coordinates and images ready to be imported into the EM control software. In the advanced options, the exact format of the generated output files can be configured.

At the moment Matrix MAPS supports the navigator file format used in SerialEM. We ensure compatibility with versions up to 3.5—*old navigator format*, as well as with the most recent releases starting with version 3.6.

We have previously described a detailed protocol on how to import these coordinate files into SerialEM and continue with the EM data acquisition for the cryo-CLEM case (Schorb et al., 2017). This approach is applicable for many CLEM use cases at room temperature as well. We have successfully employed it to set up the regions for batch acquisition of electron tomograms at multiple positions.

3. DISCUSSION/OUTLOOK

While we have designed the CLEM module of Matrix MAPS with the specific CLEM experiments performed at the EMBL EM core facility in mind, the template-based structuring of experimental settings offers flexibility to incorporate a variety of different CLEM approaches.

When implementing the stage templates to represent the different geometries of CLEM specimen supports, we tried to cover the most commonly used types and shapes. If your experiment is based on a specimen geometry that is not yet present in the module, just contact the developer and we will create a suitable template for you.

To acquire TEM data in a CLEM experiment, we exclusively use SerialEM. As this software is freely available and runs on a variety of different microscope configurations, we prefer it over manufacturer-specific software solutions. Our main focus thus far was to generate coordinate files and images in a format compatible with SerialEM. However, adding another file format as an export option is not a significant modification to the Matrix MAPS CLEM module. Please contact the developer for suggesting alternative formats.

With the module as we present it here, the user defines the positions of landmarks and features of interest manually. The LASX platform, which Matrix MAPS is built on, offers a number of image analysis procedures, such as image filters or stack projection operations. We envision that in the near future, the selection of features of interest for subsequent in-depth EM imaging will be based on automated image processing routines that analyze the LM data. Matrix MAPS already has the built-in capability to communicate with external software solutions that provide powerful image analysis tools via the so-called CAM interface (Tischer et al., 2014). Open-source software compatible with this approach includes CellProfiler (Carpenter et al., 2006; Kamentsky et al., 2011), KNIME (Dietz & Berthold, 2016), and the Fiji (Schindelin et al., 2012) environment.

As the CLEM module and Matrix MAPS are part of LAS X, the software suite that is employed to control all types of light microscopes in Leica's portfolio, there is basically no limitation to a specific microscopy method. While in the experiments described here, we exclusively employ wide-field microscopy using a camera to detect the signal, any other acquisition technique from confocal to STED is fully supported by the software. We will continue to test the application of these advanced microscopy methods for CLEM, and if required generate additional specific workflows for Matrix MAPS to simplify using them.

Matrix MAPS CLEM with its template-based experimental workflow enables acquiring the LM data and obtaining the target coordinates for subsequent EM imaging with minimal user intervention. Combined with powerful image analysis and standardized specimen transfer solutions, we will move yet another step further toward a fully automated CLEM experiment.

ACKNOWLEDGMENTS

We like to thank the Electron Microscopy Core Facility at EMBL Heidelberg and its head Yannick Schwab for providing an ideal environment for the ongoing developments in CLEM. We like to thank William Wan for critically reading and commenting the manuscript.

REFERENCES

Ader, N. R., & Kukulski, W. (2017). triCLEM: Combining high-precision, room temperature CLEM with cryo-fluorescence microscopy to identify very rare events. In T. Mueller-Reichert, & P. Verkade (Eds.), *Correlative light and electron microscopy III* (Vol. 140, pp. 303−320).

Agronskaia, A. V., Valentijn, J. A., van Driel, L. F., Schneijdenberg, C. T. W. M., Humbel, B. M., van Bergen en Henegouwen, P. M. P., … Gerritsen, H. C. (2008). Integrated fluorescence and transmission electron microscopy. *Journal of Structural Biology, 164*, 183−189. http://dx.doi.org/10.1016/j.jsb.2008.07.003.

de Boer, P., Hoogenboom, J. P., & Giepmans, B. N. G. (2015). Correlated light and electron microscopy: ultrastructure lights up! *Nature Methods, 12*, 503−513. http://dx.doi.org/10.1038/nmeth.3400.

Briegel, A., Chen, S., Koster, A. J., Plitzko, J. M., Schwartz, C. L., & Jensen, G. J. (2010). Correlated light and electron cryo-microscopy. *Methods in Enzymology, 481*, 317–341. http://dx.doi.org/10.1016/S0076-6879(10)81013-4.

Bykov, Y. S., Cortese, M., Briggs, J. A. G., & Bartenschlager, R. (2016). Correlative light and electron microscopy methods for the study of virus-cell interactions. *FEBS Letters*. http://dx.doi.org/10.1002/1873-3468.12153.

Carpenter, A. E., Jones, T. R., Lamprecht, M. R., Clarke, C., Kang, I. H., Friman, O., … Sabatini, D. M. (2006). CellProfiler: Image analysis software for identifying and quantifying cell phenotypes. *Genome Biology, 7*, R100. http://dx.doi.org/10.1186/gb-2006-7-10-r100.

Conrad, C., Wünsche, A., Tan, T. H., Bulkescher, J., Sieckmann, F., Verissimo, F., … Ellenberg, J. (2011). Micropilot: Automation of fluorescence microscopy-based imaging for systems biology. *Nature Methods, 8*, 246–249. http://dx.doi.org/10.1038/nmeth.1558.

Dietz, C., & Berthold, M. R. (2016). KNIME for open-source bioimage analysis: A tutorial. In W. H. D. Vos, S. Munck, & J.-P. Timmermans (Eds.), *Focus on bio-image informatics, advances in anatomy, embryology and cell biology* (pp. 179–197). Springer International Publishing. http://dx.doi.org/10.1007/978-3-319-28549-8_7.

Fukuda, Y., Schrod, N., Schaffer, M., Feng, L. R., Baumeister, W., & Lucic, V. (2014). Coordinate transformation based cryo-correlative methods for electron tomography and focused ion beam milling. *Ultramicroscopy, SI: Correlative Microscopy, 143*, 15–23. http://dx.doi.org/10.1016/j.ultramic.2013.11.008.

Hampoelz, B., Mackmull, M.-T., Machado, P., Ronchi, P., Bui, K. H., Schieber, N., … Beck, M. (2016). Pre-assembled nuclear pores insert into the nuclear envelope during early development. *Cell, 166*, 664–678. http://dx.doi.org/10.1016/j.cell.2016.06.015.

Heiligenstein, X., Paul-Gilloteaux, P., Raposo, G., & Salamero, J. (2017). eC-CLEM: A multidimension, multimodel software to correlate intermodal images with a focus on light and electron microscopy. In T. Mueller-Reichert, & P. Verkade (Eds.), *Correlative light and electron microscopy III* (Vol. 140, pp. 335–352).

Iijima, H., Fukuda, Y., Arai, Y., Terakawa, S., Yamamoto, N., & Nagayama, K. (2014). Hybrid fluorescence and electron cryo-microscopy for simultaneous electron and photon imaging. *Journal of Structural Biology, 185*, 107–115. http://dx.doi.org/10.1016/j.jsb.2013.10.018.

Kamentsky, L., Jones, T. R., Fraser, A., Bray, M.-A., Logan, D. J., Madden, K. L., … Carpenter, A. E. (2011). Improved structure, function and compatibility for CellProfiler: Modular high-throughput image analysis software. *Bioinformatics (Oxford, England), 27*, 1179–1180. http://dx.doi.org/10.1093/bioinformatics/btr095.

Keene, D. R., Tufa, S. F., Wong, M. H., Smith, N. R., Sakai, L. Y., & Horton, W. A. (2014). Correlation of the same fields imaged in the TEM, confocal, LM, and microCT by image registration: from specimen preparation to displaying a final composite image. *Methods in Cell Biology, 124*, 391–417. http://dx.doi.org/10.1016/B978-0-12-801075-4.00018-5.

Kolovou, A., Schorb, M., Tarafder, A., Sachse, C., Schwab, Y., & Santarella-Mellwig, R. (2017). A new method for cryo-sectioning cell monolayers using a correlative workflow. In T. Mueller-Reichert, & P. Verkade (Eds.), *Correlative light and electron microscopy III* (Vol. 140, pp. 85–104).

Kukulski, W., Schorb, M., Welsch, S., Picco, A., Kaksonen, M., & Briggs, J. A. G. (2011). Correlated fluorescence and 3D electron microscopy with high sensitivity and spatial precision. *Journal of Cell Biology, 192*, 111–119. http://dx.doi.org/10.1083/jcb.201009037.

Kukulski, W., Schorb, M., Welsch, S., Picco, A., Kaksonen, M., & Briggs, J. A. G. (2012). Precise, correlated fluorescence microscopy and electron tomography of lowicryl sections using fluorescent fiducial markers. *Methods in Cell Biology, 111*, 235–257. http://dx.doi.org/10.1016/B978-0-12-416026-2.00013-3.

Loussert Fonta, C., & Humbel, B. M. (2015). Correlative microscopy. *Archives of Biochemistry and Biophysics, Electron Microscopy in Structural Biology, 581*, 98–110. http://dx.doi.org/10.1016/j.abb.2015.05.017.

Mastronarde, D. N. (2005). Automated electron microscope tomography using robust prediction of specimen movements. *Journal of Structural Biology, 152*, 36–51. http://dx.doi.org/10.1016/j.jsb.2005.07.007.

Neumann, B., Walter, T., Hériché, J.-K., Bulkescher, J., Erfle, H., Conrad, C., … Ellenberg, J. (2010). Phenotypic profiling of the human genome by time-lapse microscopy reveals cell division genes. *Nature, 464*, 721–727. http://dx.doi.org/10.1038/nature08869.

Paul-Gilloteaux, P., Heiligenstein, X., Belle, M., Domart, M.-C., Larijani, B., Collinson, L., … Salamero, J. (2017). eC-CLEM: Flexible multidimensional registration software for correlative microscopies. *Nature Methods, 14*, 102–103. http://dx.doi.org/10.1038/nmeth.4170.

Schellenberger, P., Kaufmann, R., Siebert, C. A., Hagen, C., Wodrich, H., & Grünewald, K. (2014). High-precision correlative fluorescence and electron cryo microscopy using two independent alignment markers. *Ultramicroscopy, 143*, 41–51. http://dx.doi.org/10.1016/j.ultramic.2013.10.011.

Schindelin, J., Arganda-Carreras, I., Frise, E., Kaynig, V., Longair, M., Pietzsch, T., … Cardona, A. (2012). Fiji: An open-source platform for biological-image analysis. *Nature Methods, 9*, 676–682. http://dx.doi.org/10.1038/nmeth.2019.

Schorb, M., & Briggs, J. A. G. (2014). Correlated cryo-fluorescence and cryo-electron microscopy with high spatial precision and improved sensitivity. *Ultramicroscopy, 143*, 24–32. http://dx.doi.org/10.1016/j.ultramic.2013.10.015.

Schorb, M., Gaechter, L., Avinoam, O., Sieckmann, F., Clarke, M., Bebeacua, C., … Briggs, J. A. G. (2017). New hardware and workflows for semi-automated correlative cryo-fluorescence and cryo-electron microscopy/tomography. *Journal of Structural Biology, 197*, 83–93. http://dx.doi.org/10.1016/j.jsb.2016.06.020.

Thévenaz, P., Ruttimann, U. E., & Unser, M. (1998). A pyramid approach to subpixel registration based on intensity. *IEEE Transactions on Image Processing: A Publication of the IEEE Signal Processing Society, 7*, 27–41. http://dx.doi.org/10.1109/83.650848.

Tischer, C., Hilsenstein, V., Hanson, K., & Pepperkok, R. (2014). Adaptive fluorescence microscopy by online feedback image analysis. *Methods in Cell Biology, 123*, 489–503. http://dx.doi.org/10.1016/B978-0-12-420138-5.00026-4.

Wacker, I., Chockley, P., Bartels, C., Spomer, W., Hofmann, A., Gengenbach, U., … Schröder, R. R. (2015). Array tomography: Characterizing FAC-sorted populations of zebrafish immune cells by their 3D ultrastructure. *Journal of Microscopy, 259*, 105–113. http://dx.doi.org/10.1111/jmi.12223.

CHAPTER

16

eC-CLEM: a multidimension, multimodel software to correlate intermodal images with a focus on light and electron microscopy

Xavier Heiligenstein[*,1,a], **Perrine Paul-Gilloteaux**[||,1,a], **Graça Raposo**[*], **Jean Salamero**[*]

**Institut Curie, PSL Research University, CNRS UMR 144 & Cell and Tissue Imaging Facility, Paris, France*

[||]Structure Fédérative de Recherche François Bonamy, INSERM, CNRS, Université de Nantes, Nantes, France

[1]Corresponding authors: E-mail: xavier.heiligenstein@curie.fr; perrine.paul-gilloteaux@univ-nantes.fr

CHAPTER OUTLINE

[a]authors contributed equally to this work.

Methods in Cell Biology, Volume 140, ISSN 0091-679X, http://dx.doi.org/10.1016/bs.mcb.2017.03.014

Abstract

Correlative light and electron microscopy (CLEM) is a scientific method covered by a broad range of techniques. The path taken to explore a scientific question is often driven both by the question and the technology available. Yet, one common step to all CLEM workflows is the registration of the multimodal images to assign a fluorescent signal to an ultrastructure. The manual relocation and registration of light microscopy and electron microscopy images can be challenging and time-consuming (Muller-Reichert & Verkade, 2014). eC-CLEM is a free open-source software to address this step.

eC-CLEM has been designed with an intuitive procedure and the manual registration has been extensively described in step-by-step protocols on the eC-CLEM webpage as well as video tutorials.

In this book chapter, we focus our description on the "automatic registration" procedure, which requires some fine tuning. We recommend the user to first get familiar with eC-CLEM through the aforementioned tutorials. If large volume data sets or automatic tracking and controlling of microscopes are pursued by the user, going through the fine-tuning steps described in this chapter is worth the effort.

INTRODUCTION

The manual relocation and registration of light microscopy (LM) and electron microscopy (EM) images can be challenging and time- consuming (Muller-Reichert & Verkade, 2014). eC-CLEM is a free open-source software to address this step.

In this method chapter, we describe the procedure for automatic registration on eC-CLEM software with natural fiducial markers: melanosomes. The same protocol can be applied with nuclei or artificial beads added during the sample preparation.

To ease the presentation of the automatic method, we use image registration of fluorescence on sections (as described in Nixon et al., 2009 and Kukulski et al., 2011) that we will treat as **2D registration**. The presented software eC-CLEM is also able to perform 3D registrations that will be briefly discussed at the end of this chapter, and it has to be underlined that the exact same protocol can be applied in 3D. The only difference will be to give the 3D images as input, but the tools used will remain the same. When suitable and to keep the message clear, we will refer to some step-by-step protocols or video tutorials already available on the eC-CLEM webpage.

1. MATERIAL

In this chapter, we will use a melanocytic cell line endogenously expressing syntaxin 13 GFP proteins (MNT1-Stx13-GFP).

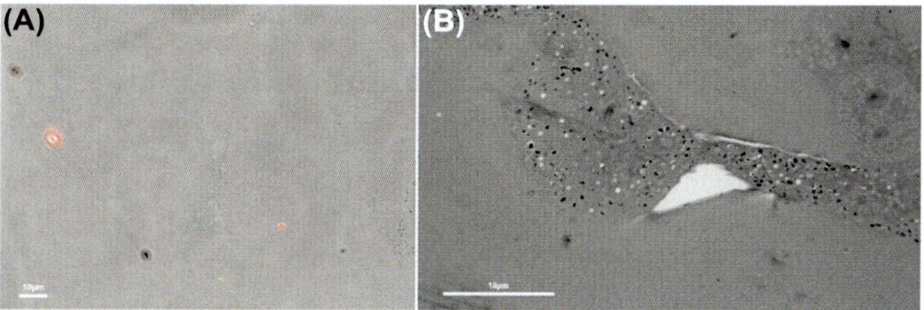

FIGURE 1

Side-by-side fluorescent (A) and electron microscopy (B), raw (not registered, not preprocessed).

The cells were cultured on a CryoCapsule (CryoCapCell, France) for 3 days in Dulbecco's Modified Eagle Medium (Heiligenstein et al., 2014). Three hours before high pressure freezing, the cells were incubated with Transferrin Alexa 546 as a marker of the endosomal system. High pressure freezing was conducted on a HPM100 and the samples were quick freeze substituted with 0.05% uranyl acetate, 0.01% glutaraldehyde, and 1% water in dry acetone. The temperature rose from −90 to −40°C in 22 min on a rocker [quick freeze substitution adaptation to Lowicryl embedding (McDonald & Webb, 2011)]. At −40°C, the fixative was rinsed with dry acetone three times and then embedded in a mix of acetone and Lowicryl HM20 in increasing concentration for 2 h (25%, 50%, 75%, 100%, 100% overnight). After polymerization under UV light (24 h −40°C, temperature rises to +20°C at 5°C per hours, 48 h at +20°C), the cells were sliced in sections of 70 nm on an Ultracut S (Reichert), collected on slot grids coated with Pioloform, immersed in 60% glycerol in distilled water and imaged for fluorescence on section (Kukulski et al., 2012; Nixon et al., 2009). Z-stack fluorescence imaging was done on a Nikon Te2000, by epifluorescence. A z-stack of 21 slices 0.5 μm was acquired, focusing on the cell in the middle of the image. After collection of the fluorescent signal and transmitted light, the sections were contrasted for 10 min with Reynolds lead citrate solution and rinsed in distilled water three times. The grids were loaded in a Tecnai spirit and imaged at 80 kV at various magnifications (from 1200× to 9000×).

Raw data used in this chapter (Fig. 1) are available from the OpenImadis data management system (https://strandls.github.io/openimadis/), hosted at https://cid.curie.fr, with the login *clemreader* and the password *Clem!123*.

Note: whether the transmitted light contains a signal of interest or not, we strongly recommend acquiring a z-stack also for this channel as it often contains peripheral information that will later be convenient for the registration of the electron micrograph (cell border or section defects diffracting light).

2. IMAGE PREPROCESSING AND AUTOMATIC DETECTION OF THE MELANOSOMES IN BOTH IMAGING MODALITIES

2.1 REDUCE DIMENSIONS TO OPTIMIZE THE COMPUTING EFFICIENCY

In fluorescence on section imaging, the data set often comes in more than two dimensions (X,Y). Depth and channels often add information to the sample but also complexity for the latter registration. Fluorescence live cell imaging also add the time dimensionality (Fig. 2).

To save computation time, we recommend conducting the registration:

- on a single plane image,
- on the channels containing landmarks also visible in the EM image (multimodal probes or natural landmarks such as melanosomes), and
- transmitted light is convenient if no specific fluorescent channel is available, as we will see later in this chapter (defects in the section, cell periphery).

2.2 PREPROCESS THE STACK

In this chapter, we use a fluorescent z-stack acquired from a single EM section, 70 nm thick as described in material. The z-stack spans 21 sections with 0.5 μm steps. The very large depth of acquisition reflects the fact that sections collected on a slot grid with a Pioloform support film are not flat (Fig. 3).

To reduce the dimensionality of this data set and ease the registration, eC-CLEM proposes a preprocessing tool ("I want to preprocess my data") to denoise and flatten the image. The denoising uses the "edge preserving denoising and smoothing" Icy plugin (Beck & Teboulle, 2009) and the flattening can be done using various options: maximum intensity projection or the "Extended Depth of Field" ImageJ plugin (automatically called in Icy) for fluorescence data sets and minimum intensity projection for transmitted light.

For detailed instructions on the flattening tools in eC-CLEM, please refer to the online tutorials:

http://icy.bioimageanalysis.org/plugin/ec-CLEM#Preprocess
https://drive.google.com/file/d/0B_nZ8lTqtNtYT3l6M2VZd092dTA/view

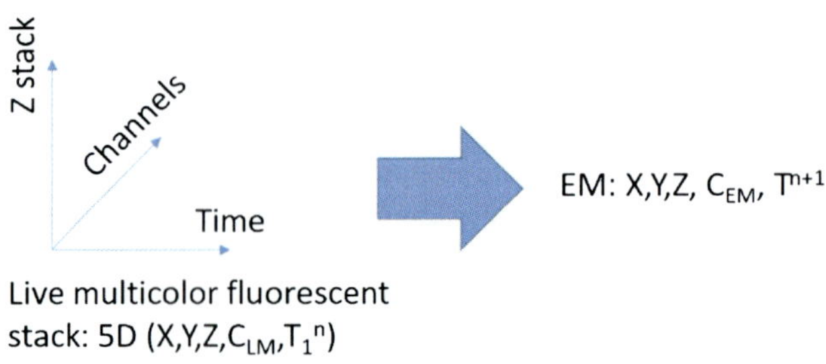

FIGURE 2

Multiple dimensions in correlative light and electron microscopy imaging.

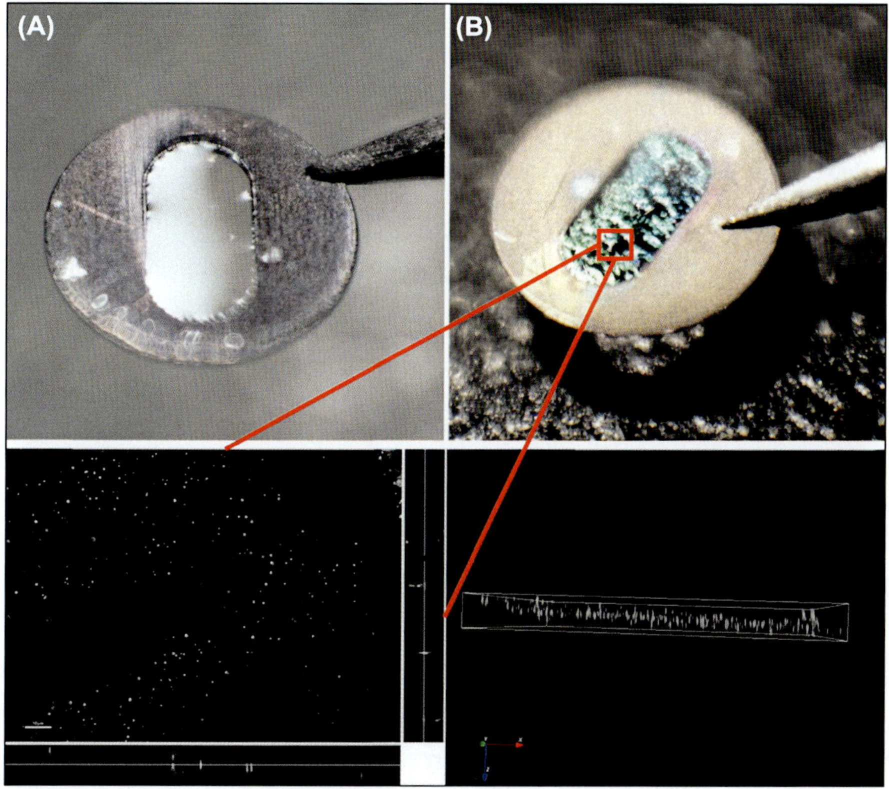

FIGURE 3

Electron microscopy (EM) sections on a support film are not flat. (A) a single hole grid covered with a Pioloform film. The light reflects non perfect flatness of the film. (B) the same slot grid is supporting several EM sections. The region of interest marked as a red square, is imaged by light microscopy and displays a large heterogeneity in the focal plane as illustrated in the two lower images. The lower left image shows top and side views of the fluorescent stack. The lower right image shows a 3D view of the fluorescent signal from the 70 nm thin section spanning accross several micrometers.

For a more technical and mathematical description of the flattening tool, please refer to (Paul-Gilloteaux et al., 2017); Supplementary Note 3: Image preprocessing.

To download the data, login to https://cid.curie.fr (as indicated in the introduction), click on Actions → Download, then on Results → Links. Untar and unzip the data before using it (Fig. 4).

In this protocol, we will extract the information the most in focus of each plane:

- Open Icy (it can be downloaded from http://icy.bioimageanalysis.org/download). This chapter was written using version 1.8.6 of Icy, preferably download this one to follow the different step. Once the reader get familiar with this version, enhancement that may change the user interface are available for more recent version of Icy.

- Open the file called Bloc1_Slot2_S2_cell2_1_w3Trans. STK, which contains the transmission image.
- Launch eC-CLEM by typing eC-CLEM in the search bar of Icy.
- Click on "I want to preprocess my data."
- Select Bloc1_Slot2_S2_cell2_1_w3Trans as the source image to be processed. In step 3 of the plugin, indicate that the image need to be flattened, and select the option Create an optimized in Focus slice (EDF EPFL Plugin), as described in Forster, Van De Ville, Berent, Sage, and Unser (2004).
- Press Play. Depending on the power of your computer, the computation can take up to 10 min.
- Once finished, the stack is replaced by a new image suffixed with ("*_focused"). This resulting image (Fig. 5) is also provided in the available data sets. Check the metadata (0.1 μm per pixels).

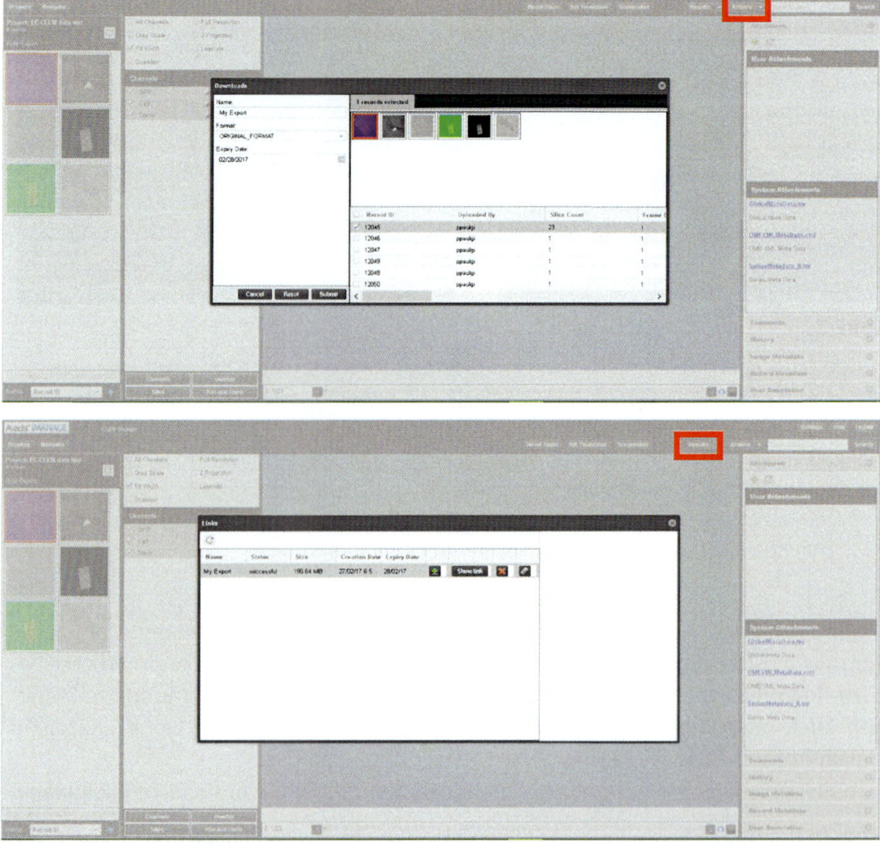

FIGURE 4

Getting the data from the database: login to https://cid.curie.fr (as indicated in the introduction), click on Actions (*red rectangle top figure*) → Download, then on Results (*red rectangle bottom figure*) → Links. Untar and unzip the data before using it.

2.3 AUTOMATIC FEATURE EXTRACTION

2.3.1 Light microscopy image: the melanosomes in bright field

Select the focused image we have just created. From the Icy search bar, launch the "spot-detector" plugin (Fig. 6).

The preprocessing option is not required as processing was already applied during the section projection step (preprocess the stack).

In this protocol, we are using the melanosomes as bimodal fiducial landmarks for the registration. We will therefore set the parameters to detect the dark spots in the bright field image. The size of spot detected should be approximately the size of the melanosomes (about 3 pixels in LM), but do not hesitate to test several combinations to obtain the best results with **your own particles**.

Two parameters can be adjusted (Fig. 7): the scale, corresponding to the size of the tracked particles, and the sensitivity, an arbitrary cut off threshold. To start up, keep the sensitivity to 100 and just test the various scales. Fine-tune the level of detection with the sensitivity (more details about the parameters are available on

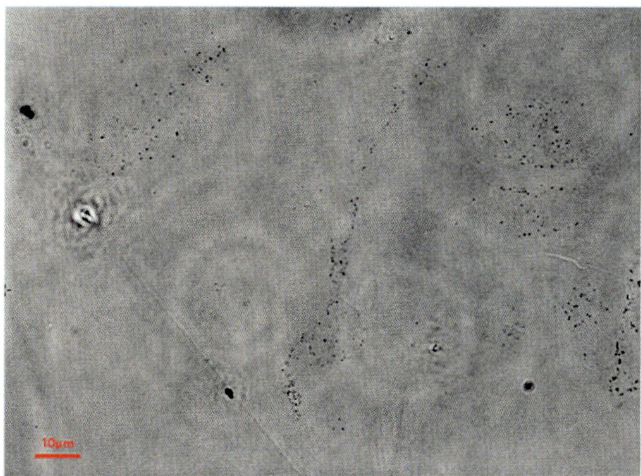

FIGURE 5

Flat image of the melanosomes in bright field. By comparison with Fig. 1, we observe that melanosomes are projected on the entire surface (fine black dots). In Fig. 1, only some melanosomes were visible on the image.

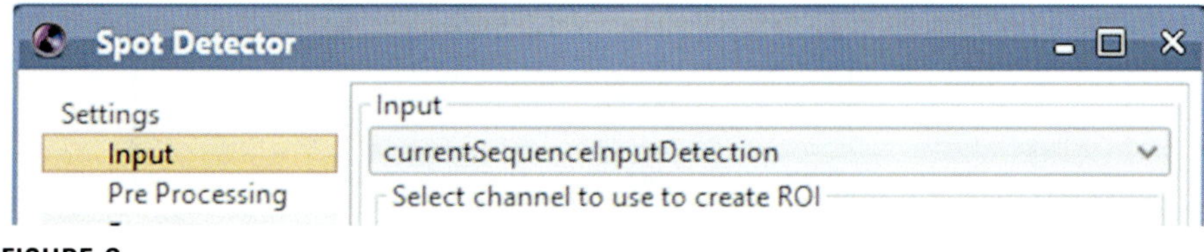

FIGURE 6

Spot detector interface.

the spot detector webpage). **In this example, we select the Scale 2 (about 3 pixels) and keep the default sensitivity**.

In the output window, make sure that "export to ROI" is checked and start detection (Fig. 8).

Setting the parameters to "scale 3, sensitivity 100" detects accurately the melanosomes on our bright field image and all detected spots are exported in the region of interest (ROI) window (Fig. 9).

2.3.2 Electron microscopy image: the melanosomes transmitted EM

Open now the raw data (compressed to 8 bits) for EM Bloc1_G2_S4_20-8bits.tif. On the EM image, the procedure is similar; however, the size of the melanosomes is necessarily different (acquisition scale significantly different between both modalities). To detect accurately the melanosomes, we take "scale 5 (25 pixels), sensitivity 100". Click on "add scale" to have scale bigger than 7 pixels to be added. Beside a large hole in the section (sectioning artifact), the distribution of the particles is accurate. The mathematical strategy of automatic registration makes the plugin robust to some sample artifacts assuming that the majority of detected spot is correct.

Run the Spot-Detector plugin against the EM with "export to ROI" as in the previous example. Melanosomes are now detected in EM image as well (Fig. 10).

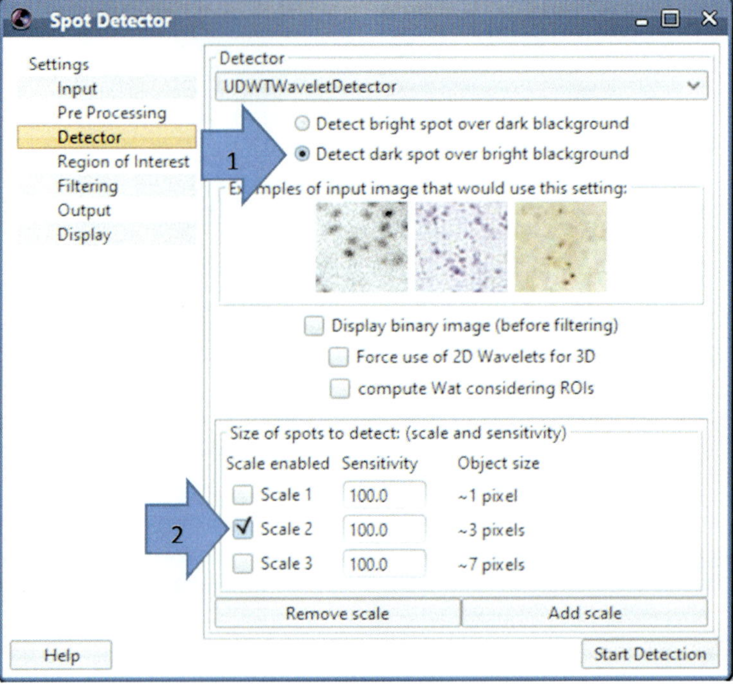

FIGURE 7

Setting parameters of spot detector.

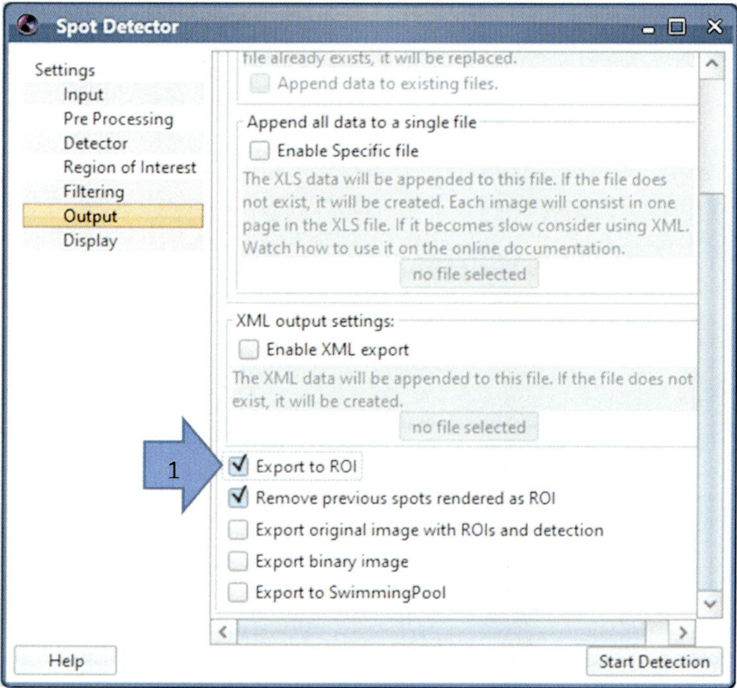

FIGURE 8

Set the output to be exported as region of interest for Autofinder to find the registration spots.

3. IDENTIFY THE TRANSFORMATION PROJECT

3.1 WHICH IS THE SOURCE, WHICH IS THE TARGET IMAGE?

As the registration process takes one image and transforms it to fit the other image, it is important to determine which image will be transformed (source) to match the other (target that remains unchanged).

Because of physics limitations, the fluorescence image is necessarily of lower resolution than the EM image. Most fluorescent images reach 300 nm in lateral resolution for 500 nm in depth, whereas 2D EM commonly has a lateral resolution power of 10 nm or higher and a depth around 80 nm.

As a consequence, the field of view in each modality is also often significantly different, as illustrated in Fig. 11.

3.2 WHAT ARE THE CONSEQUENCES?

Consequently, transforming the source image requires to adapt to the pixel size of the target image. From an EM image to be transformed to a LM image, the pixels will be rendered as the mean value of the pixel cluster beneath the LM pixel. The

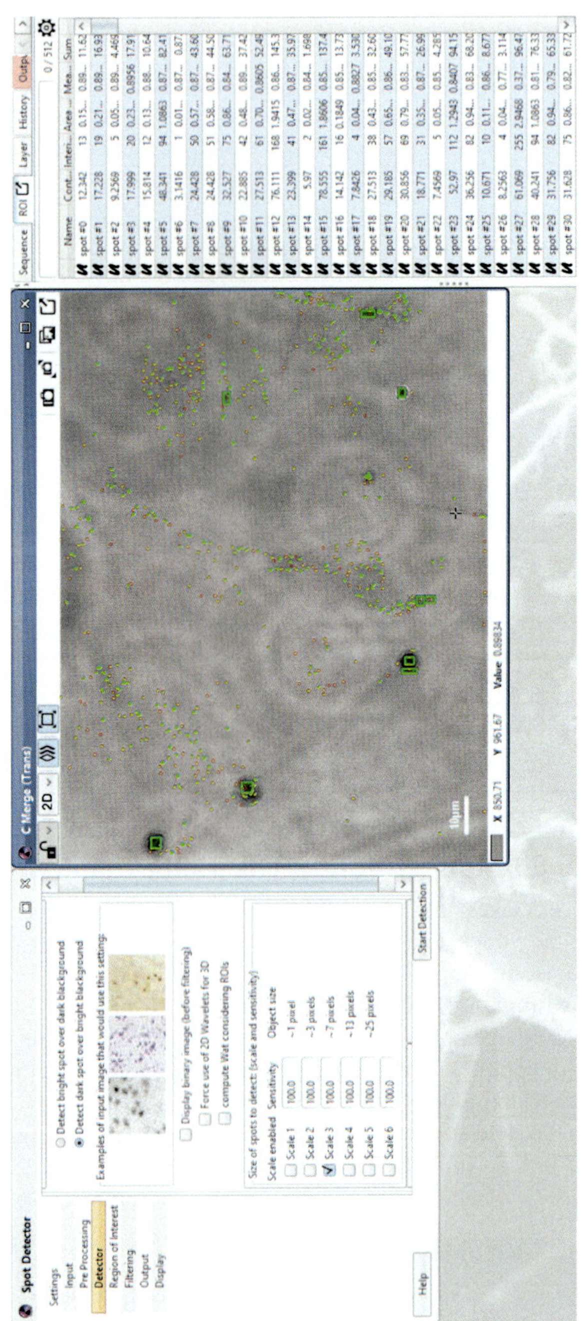

FIGURE 9

Spot detector workflow: once parametrized correctly, the spot detector runs against the selected color channel and automatically exports the detected region of interrest(ROI) to the ROI window of Icy.

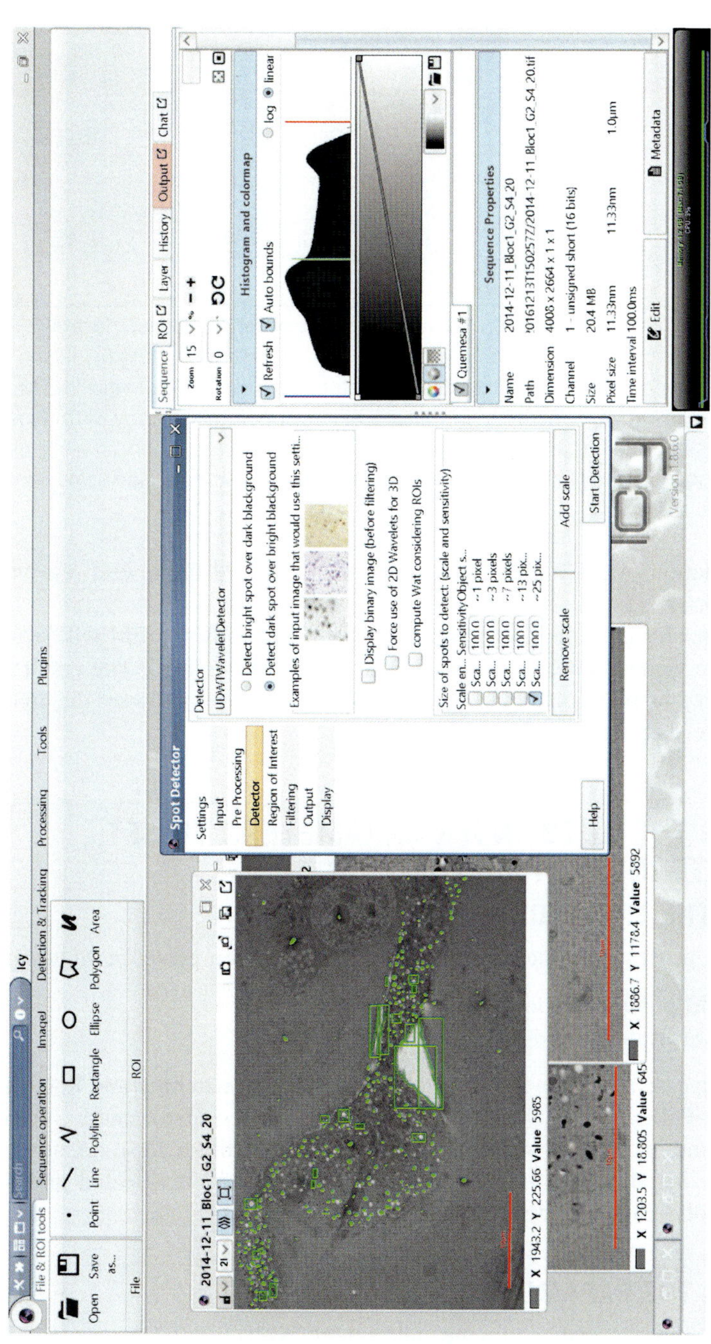

FIGURE 10

Spot Detector also works on melanosomes on electron microscopy pictures: the pixel size is adapted to the resolution of the image. Then the spot detector exports the target point to the ROI window.

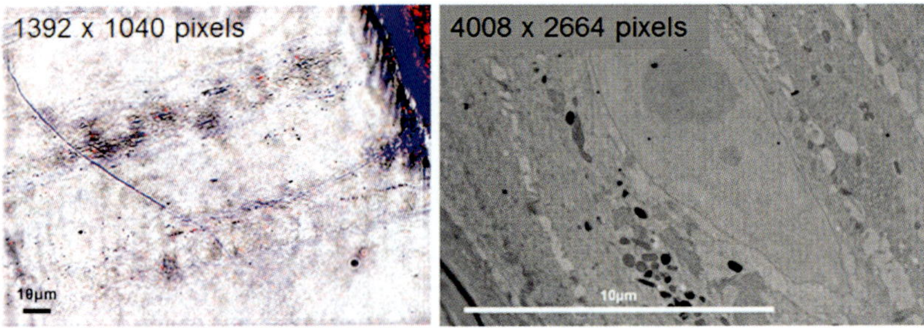

FIGURE 11

Field of view and resolution comparison between light microscopy(LM) and electron microscopy(EM): Owing to the intrinsic nature of photons and electrons, the resolution of images is significantly different. This impacts the resolution and the field of view. In LM, the field of view is significantly broader than in transmission electron microscopy, with a lower resolving power (see scale bar and number of pixels). In EM, the resolution power is much higher. As consequence, the field of view is lower beside larger cameras (see scale bar and number of pixels).

result is a pixelation of the highest resolution image toward the lowest resolution image (Fig. 12).

Inversely, transforming the LM to match the fine details of the EM will require interpolating the values between the pixels that have been created. The result is a general smoothing by interpolation of the lowest resolution image toward the highest resolution image (Fig. 13).

4. SETTING THE AUTOFINDER PARAMETERS TO GET ACCURATE INITIAL REGISTRATION

4.1 DEFINE THE TRANSFORMATION

To optimize the processing time, several scenarios have been preset in Autofinder:

- Locate an EM picture in a larger LM frame.
- Project an LM frame on the EM picture.

The successive density scan followed by principal axis matching and RANSAC procedure (Paul-Gilloteaux et al., 2017) is optimized for the first scenario. Once a transform is found, eC-CLEM will automatically propose the reverse transform if the checkbox "Also show the target transformed on source" is selected and will also save it to solve the second scenario.

4.2 SETTING THE PARAMETERS

- In eC-Clem, go to "Advanced Usage" and select "AUTOFINDER." A graphical user interface as shown on Fig. 14 is displayed.

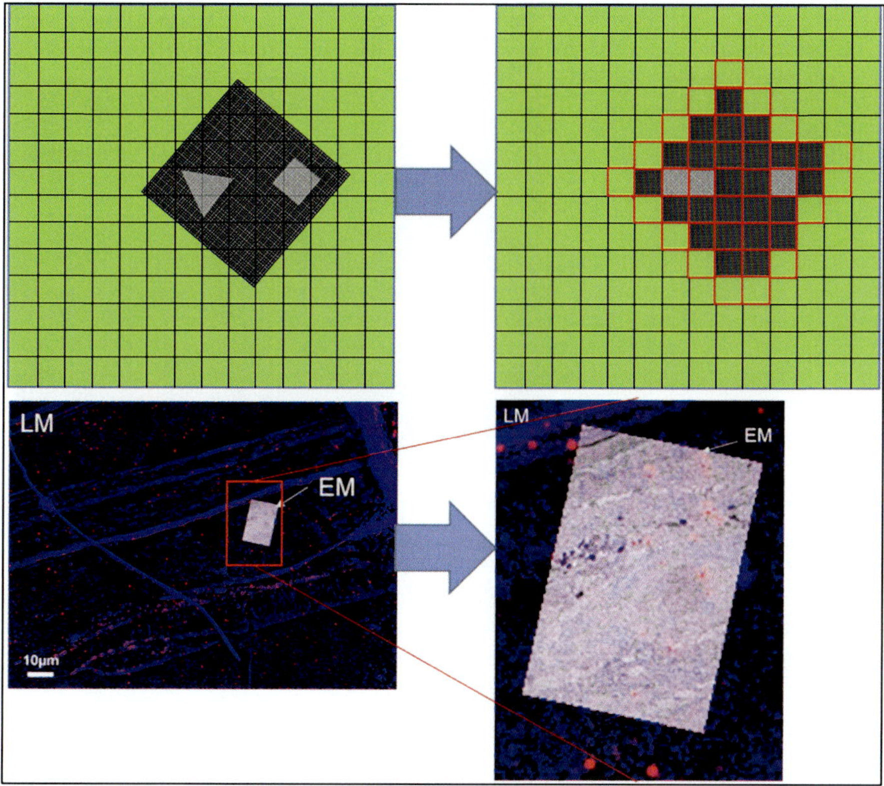

FIGURE 12

The transformation impacts the resolution of the source image: As registration consists of transforming one image to match the second one, the source image must adopt the pixel size of the target image.

- Select the EM image as the Source Image.
- Select the LM image (focused) as the Target Image.
- Assert that the metadata are correctly set (read from the picture headers), and adjust the values if required. Note that a wrong estimation will prevent any good registration. In this example, EM pixel size should be 11.33 nm, and LM pixel size should be 0.1 µm. Slice spacing is not important in 2D–2D registration.
- In the "Transform Mode," select "Find small part in Bigger field of view."
- If desired, check "show the transformed target on source" (it will display both the EM on LM and the LM on EM)
- In our example data set, keep the parameters by default (i.e., "max of error for testing in microns" = 1, and "Percentage of target point to keep for test" = 70).
- Run the software.

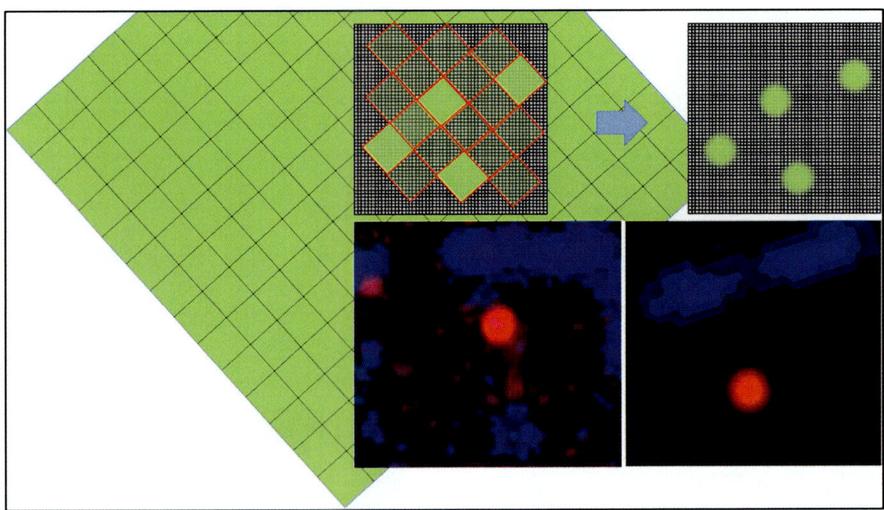

FIGURE 13

The transformation impacts the resolution of the source image: in this example, the interpolation of the light microscopy to match the fine resolution of the electon microscopy image imposes to increase the number of pixels from the source image. The final image is smoothed by this procedure as pixel values have been interpolated to add new pixels.

The results shown on Fig. 15 indicate that 190 points have been correctly matched, with an average error of 0.33 μm, i.e., 3 pixels in LM.

The orange circles (Fig. 15B) indicate the candidate areas that are tested. In this example, only one candidate area had a distribution of points similar to the one in EM. These parameters might require extensive testing. The two important parameters are the "percentage of target point to keep for test" and the "max error allowed for testing in microns." The default values (70% and 10 times the bigger pixel size) should give good results in most of the cases. If a lot of candidates were found but a bad candidate was finally elected, setting the value of percentage of point to keep to 90%, and reducing the maximum error allowed will make the selection more restrictive (more points need to be nearer from each other after registration). This would be the case where the spot detections lead to similar detection in both images. If several candidates were identified (blue circle) but none become orange, trying to be less restrictive can solve the problem. In that case, setting the percentage of target points to be tested to 30% or 50% can help. If no circle was found (i.e., no area has a similar organization of density of points), checking the correctness of metadata can help, otherwise the detection steps (spot detector plugin, see above) must be performed again with different parameters for one of the two images.

Different cases are presented in the online tutorials available on http://icy.bioimageanalysis.org/plugin/Ec-Clem_AUTOFINDER.

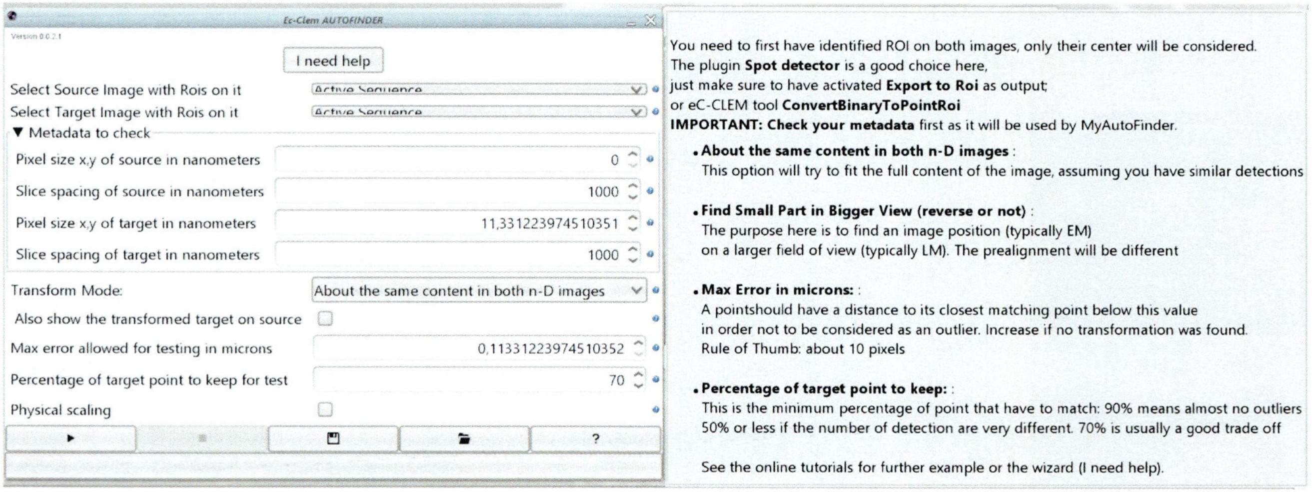

Ec-Clem AUTOFINDER

Version 0.0.2.1

I need help

Select Source Image with Rois on it — Active Sequence

Select Target Image with Rois on it — Active Sequence

▼ Metadata to check

Pixel size x,y of source in nanometers — 0

Slice spacing of source in nanometers — 1000

Pixel size x,y of target in nanometers — 11,331223974510351

Slice spacing of target in nanometers — 1000

Transform Mode: — About the same content in both n-D images

Also show the transformed target on source

Max error allowed for testing in microns — 0,11331223974510352

Percentage of target point to keep for test — 70

Physical scaling

?

You need to first have identified ROI on both images, only their center will be considered.
The plugin **Spot detector** is a good choice here,
just make sure to have activated **Export to Roi** as output;
or eC-CLEM tool **ConvertBinaryToPointRoi**
IMPORTANT: Check your metadata first as it will be used by MyAutoFinder.

- **About the same content in both n-D images** :
  This option will try to fit the full content of the image, assuming you have similar detections
- **Find Small Part in Bigger View (reverse or not)** :
  The purpose here is to find an image position (typically EM)
  on a larger field of view (typically LM). The prealignment will be different
- **Max Error in microns:** :
  A pointshould have a distance to its closest matching point below this value
  in order not to be considered as an outlier. Increase if no transformation was found.
  Rule of Thumb: about 10 pixels
- **Percentage of target point to keep:** :
  This is the minimum percentage of point that have to match: 90% means almost no outliers
  50% or less if the number of detection are very different. 70% is usually a good trade off

See the online tutorials for further example or the wizard (I need help).

FIGURE 14

Graphical User Interface of AutoFinder. The plugin automatically detects and displays the metadata of interest for the automated registration. The scaling factors are essential to allow the algorithm to function: detecting density pattern is strongly dependent on the scale and therefore the pixel size. When starting the plugin, a guiding window pops-up from the side to guide the user.

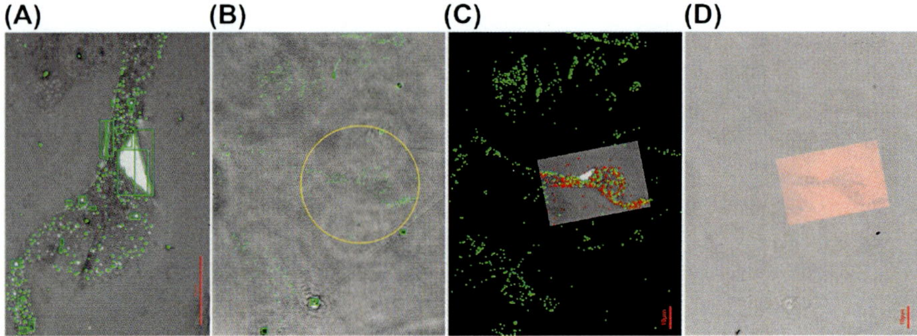

FIGURE 15

Autofinder application: (A) the melanosomes have been detected in the EM image (low magnification with a large field of view). (B) The melanosomes are detected in the transmitted LM image and the Autofinder is initiated. Only one region contains the appropriate density in this example and therefore is used to continue the Autofinder procedure. (C) After a density mapping, the main axis distribution is detected and the points are paired by a RANSAC procedure. (D) The best match detected is used to create the overlay.

Autofinder is primarily designed to locate a structure of interest in a larger field of view. Although accurate localization should be obtained, a more accurate registration doing a manual point by point pairing, as described in the online tutorials, might be required.

Furthermore, the computation from AutoFinder does not "compute the whole predicted error map" as the manual pairing offers (Paul-Gilloteaux et al., 2017). To compute an error map, you can switch to eC-CLEM by checking the box "Export results for further analysis in eC-CLEM", that will automatically load matched fiducials on source and target image files. The nonrigid option (elastic transformation) can then also be called if needed, as explained in (Paul-Gilloteaux et al., 2017), Supplementary Note 5. An example of this advanced usage is demonstrated in the online tutorials.

Note that all transforms are saved in files called "*_TRANSFOAUTO.xml" both for target to source and source to target registrations and can be reapplied.

5. MANUAL REGISTRATION VERSUS AUTOFINDER

The Autofinder becomes a critical technology when addressing cryo-EM samples. The sensitivity to the beam prevents an exhaustive search. Spending time to fine-tune the Autofinder option becomes more critical as it will open premapping of ROIs and allow better efficiency in cryo-CLEM approaches (Schorb & Briggs, 2014).

The Autofinder option requires some trials and errors and is not always recommended. For plastic sections, where electron beam damage is not critical, a manual registration may be faster and more accurate.

In the case of manual registration, please refer to online tutorials on the Icy webpage. http://icy.bioimageanalysis.org/plugin/ec-CLEM.

6. 3D TRANSFORMATIONS

Finding the 3D transformation of a 2D EM section into a 3D fluorescence section can sometimes be very challenging, as well as the registration of two 3D volumes. A blind identification of the multimodal fiducial in each data set prior to the Autofinder procedure allows to preregister the data sets efficiently. Fine tuning the parameters to an optimal registration can be worth the investment, especially if several data sets have been collected.

The landmarks selection procedure should be realized using the spot-detector plugin in both imaging modalities or operated manually according to the type of landmarks.

Once accurate registration with Autofinder is obtained, it can be transposed to several data sets of the same sample type to ease the registration process.

REFERENCES

Beck, A., & Teboulle, M. (2009). Fast gradient-based algorithms for constrained total variation image denoising and deblurring problems. *IEEE Transactions on Image Processing, 18*(11), 2419–2434.

Forster, B., Van De Ville, D., Berent, J., Sage, D., & Unser, M. (2004). Complex wavelets for extended depth-of-field: A new method for the fusion of multichannel microscopy images. *Microscopy Research and Technique, 65*(1–2), 33–42. http://dx.doi.org/10.1002/jemt.20092.

Heiligenstein, X., Heiligenstein, J., Delevoye, C., Hurbain, I., Bardin, S., Paul-Gilloteaux, P., … Raposo, G. (2014). The CryoCapsule: Simplifying correlative light to electron microscopy. *Traffic, 15*(6), 700–716. http://dx.doi.org/10.1111/tra.12164.

Kukulski, W., Schorb, M., Welsch, S., Picco, A., Kaksonen, M., & Briggs, J. A. (2011). Correlated fluorescence and 3D electron microscopy with high sensitivity and spatial precision. *Journal of Cell Biology, 192*(1), 111–119. http://dx.doi.org/10.1083/jcb.201009037.

Kukulski, W., Schorb, M., Welsch, S., Picco, A., Kaksonen, M., & Briggs, J. A. (2012). Precise, correlated fluorescence microscopy and electron tomography of lowicryl sections using fluorescent fiducial markers. *Methods in Cell Biology, 111*, 235–257. http://dx.doi.org/10.1016/B978-0-12-416026-2.00013-3.

McDonald, K. L., & Webb, R. I. (2011). Freeze substitution in 3 hours or less. *Journal of Microscopy, 243*(3), 227–233. http://dx.doi.org/10.1111/j.1365-2818.2011.03526.x.

Muller-Reichert, T. F., & Verkade, P. (2014). Preface. Correlative light and electron microscopy II. *Methods in Cell Biology*. ISSN: 0091-679X (Print).

Nixon, S. J., Webb, R. I., Floetenmeyer, M., Schieber, N., Lo, H. P., & Parton, R. G. (2009). A single method for cryofixation and correlative light, electron microscopy and tomography of zebrafish embryos. *Traffic, 10*(2), 131–136. http://dx.doi.org/10.1111/j.1600-0854.2008.00859.x.

Paul-Gilloteaux, P., Heiligenstein, X., Belle, M., Domart, M.-C., Larijani, B., Collinson, L., … Salamero, J. (2017). eC-CLEM: Flexible multidimensional registration software for correlative microscopies. *Nature Methods, 14*(2), 102–103. http://dx.doi.org/10.1038/nmeth.4170.

Schorb, M., & Briggs, J. A. (2014). Correlated cryo-fluorescence and cryo-electron microscopy with high spatial precision and improved sensitivity. *Ultramicroscopy, 143*, 24–32. http://dx.doi.org/10.1016/j.ultramic.2013.10.015.

CPI Antony Rowe
Eastbourne, UK
May 31, 2017